AF333347

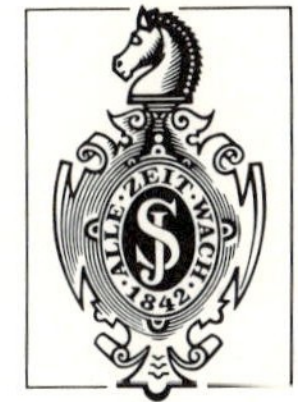

Oncogenes and Growth Control

Edited by
Patricia Kahn and Thomas Graf

With 35 Figures

Springer-Verlag
Berlin Heidelberg New York
London Paris Tokyo

Patricia Kahn
Thomas Graf

European Molecular
Biology Laboratory
Postfach 10 22 09
6900 Heidelberg, FRG

ISBN 3-540-16839-7 Springer-Verlag Berlin Heidelberg New York
ISBN 0-387-16839-7 Springer-Verlag New York Berlin Heidelberg

Library of Congress Cataloging in Publication Data. Oncogenes and growth control. In-cludes index. 1. Oncogenes. 2. Cancer cells—Growth—Regulation. 3. Cells—Growth—Regulation. I. Kahn, Patricia. II. Graf, T. (Thomas) RC268.4.0525 1986 616.99′407 86-17730

Typesetting: K + V Fotosatz GmbH, Beerfelden.
Offsetprinting and Bookbinding: Konrad Triltsch, Graphischer Betrieb, Würzburg.
2131/3130-543210

Preface

Work in the oncogene and growth control fields is proceeding at such a rapid rate that it has become increasingly difficult to keep abreast of the newest developments. For this reason we chose to produce a collection of mini-reviews which present an overview of the current concepts without a lengthy description of primary data.

In a volume with the relatively wide scope of this one, the references alone could equal, if not exceed, the length of the articles. We therefore pressed the contributors not only to keep their articles below a certain length but, even more to their dismay, to drastically limit the number of references. The result is, of course, a compromise in which the brevity is paid for by the inevitable subjectivity of such an approach. We apologize to all authors who feel that their work should have been cited; the missing references might well have been included in the original version but fell to the axe of the editors.

The choice of topics for a collection of 47 papers is likely to be somewhat arbitrary, and this volume is no exception. Rather than attempting to be as comprehensive as possible, we decided to limit the scope of the book to the better-described experimental systems to which molecular approaches have been applied. The book therefore concentrates on fibroblast and hematopoietic cell systems and largely ignores epithelial cells. We also neglected certain oncogenes in favor of those which are being studied most intensively. For example, *src*, *myc* and *ras* are discussed in different contexts and in various chapters of the book. Likewise, we have emphasized the well-studied epidermal growth factor (EGF), the EGF receptor and its transforming counterpart, the *erbB* oncogene.

In putting this volume together we have been aided and encouraged by several colleagues. We are especially grateful to Birgit Blanasch, Claire Brady, Ged Brady, Achim Leutz, and Scott Ness for their generous help.

Heidelberg
June, 1986

Patricia Kahn and Thomas Graf

Contents

V Malignant Transformation as a Multistep Process

Contents

VI Oncogenesis in Transgenic Mice

Contributors

BALMAIN, ALLAN, Beatson Institute for Cancer Research, Garscube Estate, Switchback Road, Bearsdon, Glasgow G61 1BD, Scotland

BALTIMORE, DAVID, Whitehead Institute for Biomedical Research, Nine Cambridge Center, Cambridge MA 02142, USA

BEATO, MIGUEL, Interdisziplinäres Zentrum für Molekularbiologie und Tumorforschung, Emil-Mannkopff-Str. 1, D-3550 Marburg, FRG

BEN-NERIAH, YINON, Whitehead Institute for Biomedical Research, Nine Cambridge Center, Cambridge MA 02142, USA

BERRIDGE, MICHAEL J., A.F.R.C. Unit of Insect Neurophysiology and Pharmacology, Department of Zoology, University of Cambridge, Downing Street, Cambridge CB2 3EJ, UK

BEUG, HARTMUT, Differentiation Programme, European Molecular Biology Laboratory, Postfach 10.2209, D-6900 Heidelberg, FRG

BLAIR, DONALD G., Laboratory of Molecular Oncology, NIH, National Cancer Institute, Frederick Cancer Research Facility, Frederick, MD, USA

BOURNE, HENRY R., Departments of Pharmacology and Medicine and the Cardiovascular Research Institute, University of California, San Francisco, CA 94143, USA

BRAVO, RODRIGO, Differentiation Programme, European Molecular Biology Laboratory, Postfach 10.2209, D-6900 Heidelberg, FRG

CARPENTER, GRAHAM, Department of Biochemistry and Division of Dermatology, Vanderbilt University School of Medicine, Nashville, TN 37232, USA

CELANDER, DANIEL, Dana-Farber Cancer Institute, Department of Pathology, Harvard Medical School, Boston, MA 02115, USA

CHARNAY, PATRICK, Differentiation Programme, European Molecular Biology Laboratory, Postfach 10.2209, D-6900 Heidelberg, FRG

CHENG, SENG H., Protein Engineering Group, Integrated Genetics Inc., 31 New York Avenue, Framingham, MA 01701, USA

CUZIN, FRANÇOIS, Unité Inserm 273, Centre de Biochimie, Université de Nice, F-06034 Nice, France

DAYTON, ANDREW, Dana-Farber Cancer Institute, Department of Pathology, Harvard Medical School, Boston, MA 02115, USA

DERYNCK, RIK, Department of Molecular Biology, Genentech, Inc., 460 Point San Bruno Boulevard, South San Francisco, CA 94080, USA

DEXTER, T. MICHAEL, Department of Experimental Haematology, Paterson Laboratories, Christie Hospital and Holt Radium Institute, Withington, Manchester M20 9B4, UK

DOERFLER, WALTER, Institut für Genetik, Universität Köln, Weyertal 121, D-5000 Köln 41, FRG

FAHRLANDER, PAUL D., Department of Biochemistry, State University of New York at Stony Brook, NY 11794, USA

FASANO, OTTAVIO, Differentiation Programme, European Molecular Biology Laboratory, Postfach 10.2209, D-6900 Heidelberg, FRG

FOULKES, J. GORDON, Laboratory of Eukaryotic Molecular Genetics, National Institute for Medical Research, Mill Hill, London NW71AA, UK

GEBHARDT, ANGELIKA, Laboratory of Eukaryotic Molecular Genetics, National Institute for Medical Research, Mill Hill, London NW71AA, UK

GOH, WEI CHUN, Institute for Molecular and Cell Biology, Osaka University, 1–3, Yamadaoka, Suita-shi, Osaka 565, Japan

GOODMAN, LINDA, Department of Biochemistry and Division of Dermatology, Vanderbilt University School of Medicine, Nashville, TN 37232, USA

GOUGH, NICHOLAS M., Melbourne Tumour Biology Branch, Ludwig Institute for Cancer Research, P.O. Royal Melbourne Hospital, Victoria 3050, Australia

GRAF, THOMAS, Differentiation Programme, European Molecular Biology Laboratory, Postfach 10.2209, D-6900 Heidelberg, FRG

GREEN, A. RICHARD, Imperial Cancer Research Fund Laboratories, St. Bartholomew's Hospital, Dominion House, Bartholomew Close, London EC1A7BE, UK

GRUSS, PETER, Zentrum für Molekulare Biologie, Im Neuenheimer Feld 282, D-6900 Heidelberg, FRG

HALL, DAVID, Department of Microbiology and Molecular Genetics, Harvard Medical School and the Dana-Farber-Cancer Institute, 44 Binney Street, Boston, MA 02115, USA

HANAFUSA, HIDESABURO, The Rockefeller University, 1230 York Avenue, New York, NY 10021, USA

HANAHAN, DOUGLAS, Cold Spring Harbor Laboratory, Cold Spring Harbor, NY 11794, USA

HASELTINE, WILLIAM A., Dana-Farber-Cancer Institute, Department of Pathology, Harvard Medical School, Boston, MA 02115, USA

HATAKEYAMA, MASANORI, Institute for Molecular and Cell Biology, Osaka University, 1 − 3, Yamadaoka, Suita-shi, Osaka 565, Japan

HAYMAN, MICHAEL J., Department of Microbiology, State University of New York at Stony Brook, Stony Brook, NY 11794, USA

HELDIN, CARL-HENRIK, Uppsala Branch of the Ludwig Cancer Institute for Cancer Research, Biomedical Center, S-751 23 Uppsala, Sweden

HEYWORTH, CLARE M., Department of Experimental Haematology, Paterson Laboratories, Christie Hospital and Holt Radium Institute, Withington, Manchester M20 9B4, UK

HUNTER, TONY, Molecular Biology and Virology Laboratory, Salk Institute, P.O. Box 85800, San Diego, CA 92138, USA

JAKOBOVITS, AYA, Department of Anatomy, School of Medicine, University of California, San Francisco, CA 94143, USA

JENUWEIN, THOMAS, Differentiation Programme, European Molecular Biology Laboratory, Postfach 10.2209, D-6900 Heidelberg, FRG

KAHN, PATRICIA, Biocomputing Programme, European Molecular Biology Laboratory, Postfach 10.2209, D-6900 Heidelberg, FRG

KLEIN, GEORGE, Department of Tumor Biology, Karolinska Institute, S-104 01 Stockholm, Sweden

LAND, HARTMUT, Imperial Cancer Research Fund, Lincoln's Inn Fields, London WC2A 3PX, UK

LEOF, EDWARD B., Department of Cell Biology, School of Medicine, Vanderbilt University, Nashville, TN 37232, USA

LEUTZ, ACHIM, Differentiation Programme, European Molecular Biology Laboratory, Postfach 10.2209, D-6900 Heidelberg, FRG

MARCU, KENNETH B., Department of Biochemistry, State University of New York at Stony Brook, Stony Brook, NY 11794, USA

MARKLAND, WILLIAM, Protein Engineering Group, Integrated Genetics Inc., 31 New York Avenue, Framingham, MA 01701, USA

MARSHALL, CHRISTOPHER J., Institute of Cancer Research, Royal Cancer Hospital, Chester Beatty Research Institute, Fulham Road, London SW3 6JB, UK

MASTERS, SUSAN B., Departments of Pharmacology and Medicine and the Cardiovascular Research Institute, University of California, San Francisco, CA 94143, USA

MINAMOTO, SEIJIRO, Institute for Molecular and Cell Biology, Osaka University, 1−3, Yamadaoka, Suita-shi, Osaka 565, Japan

MOELLING, KARIN, Max-Planck-Institut für Molekulare Genetik, Ihnestraße 73, D-1000 Berlin 33, FRG

MOOLENAAR, WOUTER H., Hubrecht Laboratory, Uppsalalaan 8, NL-3584 CT Utrecht, The Netherlands

MORI, HISASHI, Institute for Molecular and Cell Biology, Osaka University, 1−3, Yamadaoka, Suita-shi, Osaka 565, Japan

MOSES, HAROLD L., Department of Cell Biology, School of Medicine, Vanderbilt University, Nashville, TN 37232, USA

MOUGNEAU, EVELYNE, Unité Inserm 273, Centre de Biochimie, Université de Nice, F-06034 Nice, France

MÜLLER, ROLF, Differentiation Programme, European Molecular Biology Laboratory, Postfach 10.2209, D-6900 Heidelberg, FRG

OREN, MOSHE, Department of Chemical Immunology, The Weizmann Institute of Science, Rehovot 76100, Israel

PARKER, PETER J., Imperial Cancer Research Fund, Lincoln's Inn Fields, London WC2A 3PX, UK

PFEFFER, SUZANNE, Department of Biochemistry, Stanford University School of Medicine, Stanford, CA 94305, USA

PHILIPSON, LENNART, European Molecular Biology Laboratory, Postfach 10.2209, D-6900 Heidelberg, FRG

ROHRSCHNEIDER, LARRY R., Division of Basic Sciences, Fred Hutchinson Cancer Research Center, 1124 Columbia Street, Seattle, WA 98104, USA

ROLLINS, BARRETT, Department of Medicine, Harvard Medical School and the Dana-Farber Cancer Institute, 44 Binney Street, Boston, MA 02115, USA

ROSEN, CRAIG, Dana-Farber-Cancer Institute, Department of Pathology, Harvard Medical School, Boston, MA 02115, USA

SCHLESSINGER, JOSEPH, Biotechnology Research Center, Meloy Laboratories, 4 Research Court, Rockville, MD 20850, USA

SCHLOKAT, UWE, Zentrum für Molekulare Biologie, Im Neuenheimer Feld 282, D-6900 Heidelberg, FRG

SCHWAB, MANFRED, Institut für Pathologie, Deutsches Krebsforschungszentrum, Im Neuenheimer Feld 280, D-6900 Heidelberg, FRG

SHAVER, LYNN, Department of Biochemistry and Division of Dermatology, Vanderbilt University School of Medicine, Nashville, TN 37232, USA

SHERR, CHARLES J., Department of Tumor Cell Biology, St. Jude Children's Research Hospital, Memphis, TN 38105, USA

SMITH, ALAN E., Protein Engineering Group, Integrated Genetics Inc., 31 New York Avenue, Framingham, MA 01701, USA

SODROSKI, JOSEPH, Dana-Farber Cancer Institute, Department of Pathology, Harvard Medical School, Boston, MA 02115, USA

STANLEY, E. RICHARD, Department of Microbiology and Immunology, and Department of Cell Biology, Albert Einstein College of Medicine, Bronx, NY 10461, USA

STILES, CHARLES D., Department of Microbiology and Molecular Genetics, Harvard Medical School and the Dana-Farber-Cancer Institute, 44 Binney Street, Boston, MA 02115, USA

TANIGUCHI, TADATSUGA, Institute for Molecular and Cell Biology, Osaka University, 1 – 3, Yamadaoka, Suita-shi, Osaka 565, Japan

THOMAS, GEORGE, Friedrich Miescher-Institut, P.O. Box 2543, CH-4002 Basel, Switzerland

ULLRICH, AXEL, Department of Developmental Biology, Genentech., Inc., 460 Point San Bruno Boulevard, South San Francisco, CA 94080, USA

VENNSTRÖM, BJÖRN, Differentiation Programme, European Molecular Biology Laboratory, Postfach 10.2209, D-6900 Heidelberg, FRG

WAGNER, ERWIN F., Differentiation Programme, European Molecular Biology Laboratory, Postfach 10.2209, D-6900 Heidelberg, FRG

WESTERMARK, BENGT, Department of Pathology, University Hospital, S-751 85 Uppsala, Sweden

WHETTON, ANTHONY D., Department of Biochemistry and Applied Molecular Biology, University of Manchester Institute of Science and Technology, Manchester, USA

WYKE, JOHN A., Imperial Cancer Research Fund Laboratories, St. Bartholomew's Hospital, Dominion House, Bartholomew Close, London EC1A 7BE, UK

ZULLO, JOHN, Department of Microbiology and Molecular Genetics, Harvard Medical School and the Dana-Farber Cancer Institute, 44 Binney Street, Boston, MA 02115, USA

Abbreviations

aa	amino acids
A-MuLV	Abelson murine leukemia virus
AC	adenylate cyclase
Ad	adenovirus
AEV	avian erythroblastosis virus
AIDS	aquired immune deficiency syndrome
ALL	acute lymphoblastic leukemia
ATL	adult T-cell leukemia
BL	Burkitt's lymphoma
bp	base pairs
BPV	bovine papilloma virus
cAMP	cyclic AMP
CEF	chicken embryo fibroblasts
C_H	constant region of immunoglobulin heavy chain
CLL	chronic lymphocytic leukemia
CML	chronic myelocytic leukemia
cMGF	chicken myelomonocytic growth factor
c-*onc*	cellular or proto-oncogene
CSF	colony-stimulating factor
CSF-1	see M-CSF
CT	cholera toxin
DH	DNase I hypersensitive sites
DMBA	dimethylbenzanthracene
EBV	Epstein-Barr virus
EC	embryonal carcinoma
ECDGF	embryonal carcinoma derived growth factor
EF-Tu	elongation factor Tu
EGF	epidermal growth factor
EIA	early region A of adenoviruses
EPO	erythropoietin
ER	endoplasmic reticulum

FeLV	feline leukemia virus
FeSV	feline sarcoma virus
FGF	fibroblast growth factor
gag	group-specific antigen of retroviruses
G-CSF	granulocyte colony-stimulating factor
GM-CSF	granulocyte-macrophage colony-stimulating factor
G_s, G_p, G_i, G_o	G proteins
GR	glucocorticoid receptor
GRE	glucocorticoid regulatory element
HER	human EGF receptor
HRE	hormone regulatory element
HTLV(-I, II, III)	human T-cell leukemia virus
Ig	immunoglobulin
IGF	insulin-like growth factor
IL-2	interleukin-2
IL-2R	interleukin-2 receptor
IL-3	interleukin-3 or Multi-CSF
Ins 1,4,5P	inositol 1,4,5-triphosphate, also IP_3
IP_3	inositol triphosphate
kb	kilobases
K_d	dissociation constant
kDa	kilodaltons
LTR	long terminal repeat
MW	molecular weight
M-CSF	macrophage colony-stimulating factor, CSF-1
Multi-CSF	see IL-3
MEF	mouse embryo fibroblasts
MHC	major histocompatibility complex
MMTV	mouse mammary tumor virus
MNNG	N-methyl-N-nitro-N-nitrosoguanidine
MPC	mouse plasmacytoma
MSV	murine sarcoma virus
MuLV (MLV)	murine leukemia virus
neo	neomycin
NMU	N-nitroso-N-methylurea
ODGF	osteosarcoma-derived growth factor

PDE	phosphodiesterase
PDGF	platelet-derived growth factor
pH_i	cytoplasmic pH
PI	phosphatidylinositol, also PtdIns
PKC	protein kinase C
PLC	phospholipase C
PS	phosphatidylserine
PT	pertussis toxin
plt	polyoma virus large T protein
pmt	polyoma virus middle T protein
pst	polyoma virus small T protein
PtdIns	phosphatiylinositol (also PI)
pol	polymerase gene of retroviruses
PR	progesterone receptor
PRE	progesterone regulatory element
REF	rat embryo fibroblasts
RFLP	restriction fragment length polymorphism
RSV	Rous sarcoma virus
SCGF	stem cell growth factor
SSV	simian sarcoma virus
SV40	simian virus 40
SDS-PAGE	sodium dodecylsulfate polyacrylamide gelelectrophoresis
TGF	transforming growth factor
T1, T2	transducin 1,2
TPA	12-o-tetradecanoyl-phorbol-13 acetate, a tumor promoter
v-*onc*	viral oncogene
$\alpha_s\ \alpha_i\ \alpha_0\ \alpha_{t1}$	α subunits of $G_s\ G_i\ G_0$ and T1 respectively
$\beta\gamma_s,\ \beta\gamma_i,\ \beta\gamma_0,\ \beta\gamma_{t1}$	$\beta\gamma$ subunits of $G_s,\ G_i,\ G_0$ and T1, respectively

Oncogenes Discussed in this Book

See also tables in articles by WAGNER and MÜLLER and by LAND.

Oncogene	Subcellular localization of product	Properties
1. Cellular and retroviral		
abl	plasma membrane	protein tyrosine kinase
erbA	cytoplasm; nucleus?	related to steroid receptors
erbB	plasma membrane; endoplasmic reticulum	protein tyrosine kinase; truncated form of EGF receptor
ets	?	?
fms	plasma membrane; endoplasmic reticulum	protein tyrosine kinase; mutated form of CSF-1 receptor
fos	nucleus	binds DNA (?)
fps/fes	plasma membrane	protein tyrosine kinase
mil/raf	cytoplasm	protein serine-threonine kinase
mos	cytoplasm	protein serine-threonine kinase
myb	nucleus	binds DNA
myc	nucleus	binds DNA
neu	plasma membrane	protein tyrosine kinase; mutated form of unidentified receptor
p53	nucleus	binds DNA
ras	plasma membrane	binds GTP
ros	plasma membrane	protein tyrosine kinase; truncated receptor (?)
sea	plasma membrane	protein tyrosine kinase
sis	secreted	corresponds to B chain of PDGF
src	plasma membrane	protein tyrosine kinase
yes	plasma membrane	protein tyrosine kinase
2. DNA tumor viral		
E1A	nucleus; cytoplasm	regulates transcription
plt	nucleus	initiates DNA synthesis and regulates transcription
pmt	plasma membrane	binds and stimulates $pp60^{c\text{-}src}$
SV40 large T	nucleus; plasma membrane	initiates DNA synthesis; regulates transcription; binds and stabilizes p53

Introduction

In recent years there has been a virtual explosion in our understanding of the mechanisms which regulate vertebrate cell proliferation. This applies both to normal cells, in which growth is tightly controlled, as well as to cancer cells, which divide in an uncontrolled fashion. Molecular and biochemical studies have led to the identification of a number of genes whose products are involved in regulating normal cell growth. In addition, many genes which are capable of inducing a transformed phenotype have been identified. Perhaps most important in fueling the remarkable progress of the past few years was the demonstration of something which was believed by many, but for a long time remained speculative: that these two groups of genes are in fact largely one and the same. This realization has been tremendously catalytic for both areas of research, and is the central concept around which this book is organized.

Although unequivocal evidence for this notion was obtained only recently, the concept itself emerged gradually from analyses of the genes which are responsible for malignant transformation by certain tumor viruses. Several strains of retroviruses, the family of viruses which carry an RNA genome but replicate via a DNA intermediate, rapidly induce fatal tumors following infection of experimental animals of the appropriate host species. The same strains also induce the transformation of cells in culture. Analysis of these retroviral genomes has led to the identification of some two dozen different transforming genes, known as viral oncogenes (v-*onc*), which are distinct from the genes required for viral replication. In 1976, Stéhelin, Varmus, Bishop, and Vogt discovered that oncogenes are not unique to retroviruses but that nearly identical sequences are present in the genome of all vertebrate cells (Stéhelin et al. 1976). These so-called cellular oncogenes (c-*onc* or proto-oncogenes, as they are also designated) show a remarkable degree of evolutionary conservation, suggesting that they serve essential functions. The fact that retroviruses which have captured cellular oncogenes are able to induce grossly abnormal cell proliferation raised the speculation that proto-oncogenes participate in regulating the growth of normal cells.

Evidence linking proto-oncogenes to the induction of neoplasias by nonviral mechanisms was obtained using the technique of DNA transfection. The groups of R. Weinberg and G. Cooper found that DNA from a human bladder tumor cell line contained a gene which induced a tumorigenic phenotype following transfer into nonmalignant cultured cells (Murray et al. 1981;

Oncogenes and Growth Control
Edited by P. Kahn and T. Graf
© Springer-Verlag Berlin Heidelberg 1986

Krontiris and Cooper 1981). Molecular cloning and analysis of this gene revealed it to be the cellular homolog of a familiar viral oncogene, the *ras* oncogene present in the Harvey strain of murine sarcoma virus. Furthermore, the critical difference between the normal *ras* proto-oncogene and its "activated" counterpart from the tumor cells resided in a single nucleotide change which led to one amino acid substitution in the *ras* gene product (Tabin et al. 1982; Reddy et al. 1982; Taparowsky et al. 1982). Since these initial experiments, analysis of many other neoplasms has demonstrated that some contain mutated, amplified and/or translocated proto-oncogenes, supporting the notion that alterations in proto-oncogenes play a role in malignant transformation.

Parallel with these studies, progress was also being made in elucidating the biochemical function of certain viral oncogenes and, by implication, of the homologous proto-oncogenes. In 1978 the groups of R. Erikson and M. Bishop demonstrated that the oncogene present in the Rous sarcoma virus (RSV), known as the v-*src* oncogene, encodes a product which catalyzes the transfer of a phosphate group from a molecule of ATP onto various protein substrates (Collett and Erikson 1978; Levinson et al. 1978). Protein phosphorylation had long been known to occur in normal cells and was thought to play a role in regulating protein function. However, Hunter and Sefton (1980) found that the v-*src* protein kinase differs from all other protein kinases previously described: the viral enzyme invariably transfers the phosphate group onto a tyrosine residue and not onto threonine or serine. This novel kinase activity, which was also exhibited by the c-*src* protein, was subsequently found to be shared by the products of several other retroviral oncogenes (for review, see Hunter and Cooper 1985).

The link between an oncogene and a cellular gene with a known role in growth control was suggested shortly thereafter. In 1962 S. Cohen had described the isolation of a molecule present in mouse submaxillary glands, (Cohen 1962) which he designated as the epidermal growth factor (EGF). EGF later proved to have potent mitogenic activity on fibroblasts in culture and to deliver its mitogenic signal via specific transmembrane receptors present in responsive cells. Analysis of the EGF receptor led Cohen's group to discover the relationship with oncogenes: the receptor possesses protein kinase activity specific for tyrosine residues. Furthermore, they found that the receptor kinase is stimulated several fold by the binding of EGF (Ushiro and Cohen 1980). These exciting findings raised the possibility that tyrosine kinases could be responsible for transmitting growth factor-induced mitogenic signals to the interior of the cell, and that oncogenes might stimulate uncontrolled cell proliferation via some perturbation of this pathway.

Since the discovery of EGF, more than a dozen other polypeptide growth factors which stimulate the proliferation of fibroblasts or of other cell types in culture have been described; indeed, specific growth factors are probably required for the growth and differentiation (and in some cases for the very survival) of most vertebrate cells. Like EGF, these factors transmit a mitogenic signal to their target cells via specific cell surface receptors, several of which have been shown to possess tyrosine kinase activity. While the physiological roles of growth factors have been more difficult to establish, it is likely that they participate in regulating proliferation and/or differentiation during embryogenesis, differentiation, growth, and wound healing.

Additional evidence relating growth factors to oncogenesis came from an entirely different line of investigation. It had long been observed that transformed cells in culture are generally able to grow in much lower concentrations of serum, the usual source of growth factors for cultured cells, than are nontransformed cells. In an attempt to determine whether transformed cells themselves produce growth factor-like activities, DeLarco and Todaro (1978) discovered that fibroblasts transformed in culture with certain oncogene-containing retroviruses secrete factors which transiently induce normal cells to express a transformed phenotype. The discovery of these so-called transforming growth factors led Sporn and Todaro (1980) to suggest that the ability of a cell to produce a factor(s) which can support its own growth might be a general mechanism contributing to the unlimited growth capacity of tumor cells.

Strong support for this prediction was obtained several years later by a finding which led the gradually converging paths of growth factors and oncogenes to finally intersect. The simian sarcoma virus (SSV) carries an oncogene known as v-*sis*, which encodes the capacity of the virus to induce a variety of neoplasms in monkeys and to transform fibroblasts in culture. DNA sequence analysis of the v-*sis* oncogene revealed that it is virtually identical to the B chain of platelet-derived growth factor (PDGF), a potent mitogen for fibroblasts (Doolittle et al. 1983; Waterfield et al. 1983). SSV-induced transformation thus appears to be due to an autocrine mechanism; that is, to the capacity of the cells to synthesize their own growth factor. This finding established unambiguously that a growth factor gene can function as an oncogene.

Soon after this discovery, DNA sequence analysis revealed the origin of another oncogene. V-*erbB*, a tyrosine kinase-encoding oncogene present in an avian retrovirus which induces erythroleukemias and sarcomas, was found to be a mutated version of the gene encoding the EGF receptor (Downward et al. 1984). The v-*erbB* protein appears to function as a growth factor receptor gone amok: by virtue of its unregulated kinase activity, it is believed to deliver a constitutive mitogenic signal rather than one which is regulated by the growth factor. Shortly thereafter, the v-*fms* oncogene was shown to be a modified version of the receptor for the macrophage colony-stimulating factor (CSF-1 or M-CSF) (Sherr et al. 1985).

The precise functions of the other known viral oncogenes are less clear. It is likely that most of them will not turn out to be mutated growth factor receptor genes, since the products of the remaining oncogenes which encode tyrosine kinase activities generally reside at the cytoplasmic face of the plasma membrane, while growth factor receptors are transmembrane proteins (for review, see Bishop 1985). Even less is known about the functions of the oncogenes that do not encode tyrosine kinases, which include some whose products are localized in the cytoplasm and others which encode nuclear proteins. Our relative ignorance about these oncogenes is paralleled by the paucity of information regarding how growth factor-induced mitogenic signals are transmitted beyond the growth factor receptors through the interior of the cell and ultimately to the nucleus, which must somehow be stimulated to undergo a round of DNA synthesis. These questions form the basis for a great deal of current research.

One last finding deserves mention in discussing how research on oncogenes and growth factors has merged. Studies designed to probe the nature of the mitogenic response have revealed that stimulation of cultured cells with certain mitogens and growth factors induces the expression in the nucleus of two proto-oncogenes, *myc* and *fos* (Kelly et al. 1983; Greenberg and Ziff 1984; Cochran et al. 1984; Kruijer et al. 1984; Müller et al. 1984). Although the precise functions of these proto-oncogene products are not known, their early activation following growth factor stimulation suggests that they participate in the induction of cell division, lending further support to the notion that oncogenes and growth factors utilize related signaling pathways.

Clearly, many loose ends have come together. We now recognize that mechanisms of oncogenesis by retroviruses and by nonviral carcinogens are likely to have the same molecular basis, and that in principle any gene encoding a growth factor, a receptor, an intracellular signaling molecule, or a protein which regulates transcription may be considered as a proto-oncogene. Perhaps the most important consequence is that disciplines such as virology, oncology, growth control, differentiation, and the regulation of gene expression have found a new common denominator: oncogenes.

References

Bishop JM (1985) Viral oncogenes. Cell 42:23 – 38

Cochran BH, Zullo J, Verma IM, Stiles CD (1984) Expression of the c-*fos* oncogene and a newly discovered c-*fox* is stimulated by platelet-derived growth factor. Science 226:1080 – 1082

Cohen S (1962) Isolation of a submaxillary gland protein accelerating incisor eruption and eyelid opening in the newborn animal. J Biol Chem 237:1555 – 1562

Collett MS, Erikson RL (1978) Protein kinase activity associated with the avian sarcoma virus *src* gene products. Proc Natl Acad Sci USA 75:2021 – 2024

DeLarco JE, Todaro GJ (1978) Growth factors from murine sarcoma virus-transformed cells. Proc Natl Acad Sci USA 75:4001 – 4005

Doolittle RF, Hunkapiller MW, Hood LE, DeVare SG, Robbins KC, Aaronson SA, Antoniades HN (1983) Simian sarcoma virus *onc* gene, v-*sis*, is derived from the gene (or genes) encoding a platelet-derived growth factor. Science 221:275 – 276

Downward J, Yarden Y, Mayes E, Scrace G, Totty N, Stockwell P, Ullrich A, Schlessinger J, Waterfield MD (1984) Close similarity of epidermal growth factor receptor and v-*erbB* oncogene protein sequences. Nature 307:521 – 527

Greenberg ME, Ziff EB (1984) Stimulation of 3T3 cells induces transcription of the c-*fos* proto-oncogene. Nature 311:433 – 438

Hunter T, Cooper JA (1985) Protein-tyrosine kinases. Annu Rev Biochem 54:897 – 930

Hunter T, Sefton BM (1980) Transforming gene product of Rous sarcoma virus phosphorylates tyrosine. Proc Natl Acad Sci USA 77:1311 – 1315

Kelly K, Cochran BH, Stiles CD, Leder P (1983) Cell-specific regulation of the c-*myc* gene by lymphocyte mitogens and platelet-derived growth factor. Cell 35:603 – 610

Krontiris TG, Cooper GM (1981) Transforming activity of human tumor DNAs. Proc Natl Acad Sci USA 78:1181 – 1184

Kruijer W, Cooper JW, Hunter T, Verma IM (1984) Platelet-derived growth factor induces rapid but transient expression of the c-*fos* gene and protein. Nature 312:711 – 716

Levinson AD, Oppermann H, Levintow L, Varmus HE, Bishop JM (1978) Evidence that the transforming gene of avian sarcoma virus encodes a protein kinase associated with a phosphoprotein. Cell 15:561 – 572

Müller R, Bravo R, Burckhardt J, Curran T (1984) Induction of c-*fos* gene on protein by growth factor precedes activation of c-*myc*. Nature 312:716 – 720

Murray MJ, Shilo B-Z, Shih C, Cowing D, Hsu HW, Weinberg RA (1981) Three different human tumor cell lines contain different oncogenes. Cell 25:355 – 361

Reddy EP, Reynolds R, Santos E, Barbacid M (1982) A point mutation is responsible for the acquisition of transforming properties by the T24 human bladder carcinoma oncogene. Nature 300:149 – 152

Sherr CJ, Rettenmier CW, Sacca R, Roussel MF, Look AT, Stanley ER (1985) The c-*fms* proto-oncogene product is related to the receptor for the mononuclear phagocyte growth factor, CSF-1. Cell 41:665 – 676

Sporn MB, Todaro GJ (1980) Autocrine secretion and malignant transformation of cells. N Engl J Med 303:878 – 880

Stéhelin D, Varmus HE, Bishop JM, Vogt PK (1976) DNA related to the transforming gene(s) of avian sarcoma viruses is present in normal avian DNA. Nature 260:170 – 173

Tabin CJ, Bradley SM, Bargmann CI, Weinberg RA, Papageorge AG, Scolnick EM, Dhar R, Lowy DR, Chang EH (1982) Mechanism of activation of a human oncogene. Nature 300:143 – 149

Taparowsky E, Suard Y, Fasano O, Shimizu K, Goldfarb M, Wigler M (1982) Activation of the T24 bladder carcinoma transforming gene is linked to a single amino acid change. Nature 300:762 – 765

Ushiro H, Cohen S (1980) Identification of phosphotyrosine as a product of epidermal growth factor-activated protein kinase in A431 cell membranes. J Biol Chem 255:8363 – 8365

Waterfield MD, Scarce GJ, Whittle N, Strooband P, Johnson A, Wasteson A, Westermark B, Heldin C-H, Huang JS, Deuel TF (1983) Platelet-derived growth factor is structurally related to the putative transforming protein p28^{sis} of simian sarcoma virus. Nature 304:35 – 39

I Growth Factors and Proto-Oncogenes in Development and Differentiation

Before embarking on a discussion about the functions of individual proto-oncogenes and growth factors at the molecular level, this section considers their possible physiological roles. Embryogenesis requires extensive proliferation and differentiation, and it is therefore probable that growth factors play an important part in this process. The first article reviews the potential involvement of some well-defined growth factors during the early stages of embryonic development. One example is the transforming growth factor-α, which is expressed in mid-gestation embryos but not in normal adult tissues, and whose expression is re-induced in a variety of tissues upon malignant transformation. Recently, several novel growth factor activities have been detected in cultures of murine embryonic carcinoma cells, the neoplastic counterparts of early embryo cells.

The subsequent articles evaluate the possible roles of proto-oncogenes (also called cellular oncogenes or c-*onc* genes) in differentiation. The strongest evidence for the existence of a link between proto-oncogenes and differentiation comes from the demonstration that certain c-*onc* genes are highly expressed in specific types of terminally differentiated, post-mitotic cells. For example, c-*fms* and c-*fos* are highly expressed in macrophages, while neuronal cells and platelets express elevated levels of the c-*src* protein kinase activity. The meaning of this expression is an open question. One possibility is that some c-*onc* genes play a role in inducing the differentiated phenotype. The finding that the v-*src* oncogene is a potent inducer of differentiation in a neuronal cell line seems to support this notion, although the fact that v-*ras* has a similar effect but c-*ras* proto-oncogene expression is not increased in nerve cells casts doubts about the specificity of these viral oncogene effects. Another possibility is that some c-*onc* genes participate in maintaining the differentiated phenotype, that is, they might serve a specialized function in a particular cell type. Expression in macrophages of the c-*fms* gene, which encodes the receptor for a macrophage-specific growth factor (see Sect. II), is one such example. In other instances, proto-oncogene expression could be irrelevant to the differentiated phenotype. Definitive answers await experiments in which the action of proto-oncogene products can be blocked in the appropriate cell types and the effects on the differentiated phenotype examined.

The Expression of Growth Factors and Growth Factor Receptors During Mouse Embryogenesis

Aya Jakobovits

Elucidation of molecular mechanisms underlying embryonic growth control is a key step to understanding embryonic development, as well as the regulation of cellular proliferation and its impairment in malignancy. Since extensive proliferation and differentiation take place during development, it seems likely that growth factors have a major role in embryogenesis. Identification of specific growth factors involved in embryonic development and their characterization is a difficult task because the events that occur at each developmental step are complex and the quantities of embryonic material available are limited. In the mouse, for example, the first 3 days after fertilization are devoted primarily to continuous multiplication. During the next 2 days, the first two differentiation steps take place to form the layers which will give rise to the fetus and to extra-embryonic structures (see Fig. 1). After implantation, at about day 7 of gestation, gastrulation begins, followed by major morphogenetic and organogenetic processes during the next 5 days. The remainder of the 20-day gestational period is devoted primarily to terminal differentiation and enlargement of the fetus. Therefore, elucidation of the key developmental processes requires investigation of the embryo during the peri-implantation period-stages in which small numbers of cells undergo extensive differentiation.

An alternative to studying the embryo in vivo is to use in vitro model systems such as teratocarcinoma stem cells (embryonal carcinoma or EC cells) which share many biochemical, morphological, and immunological properties with normal early embryonic cells (for review, see Martin 1980). Some EC cell lines are pluripotent and upon differentiation, either in vivo or in vitro, will give rise to a variety of cell types that derive from the three primitive germ layers. Therefore, EC cells provide a system accessible to the analysis of growth control mechanisms analogous to those in early mouse embryogenesis (3.5 to 7 days of gestation) and in particular to the search for unique embryonic growth factors.

The role of growth factors in the process of mouse embryogenesis has been studied by two approaches described in this review: (1) Elucidation of the role that growth factors derived from adult tissues play in embryogenesis. (2) Identification of growth factors produced by EC cells that might be unique to embryonic processes.

Oncogenes and Growth Control
Edited by P. Kahn and T. Graf
© Springer-Verlag Berlin Heidelberg 1986

Day 3
Day 4
Day 5
Day 6
Day 7
morula
ICM
primitive ectoderm
primitive endoderm
ectoderm
mesoderm
endoderm
embryo proper
extraembryonic mesoderm
(yolk sac mesoderm, chorionic mesoderm, allantois, amnion)
parietal endoderm
visceral embryonic ectoderm
visceral extraembryonic endoderm
(yolk sac endoderm)
trophectoderm
mural trophectoderm
primary giant cells
polar trophectoderm
secondary giant cells
ectoplacental cone
extraembryonic ectoderm
(chorionic ectoderm)
ICM
TE
ECpr
ENpr
TE
ENve
ENp
ECe
ENv
GC
EC
AF
RM
YSC
PS

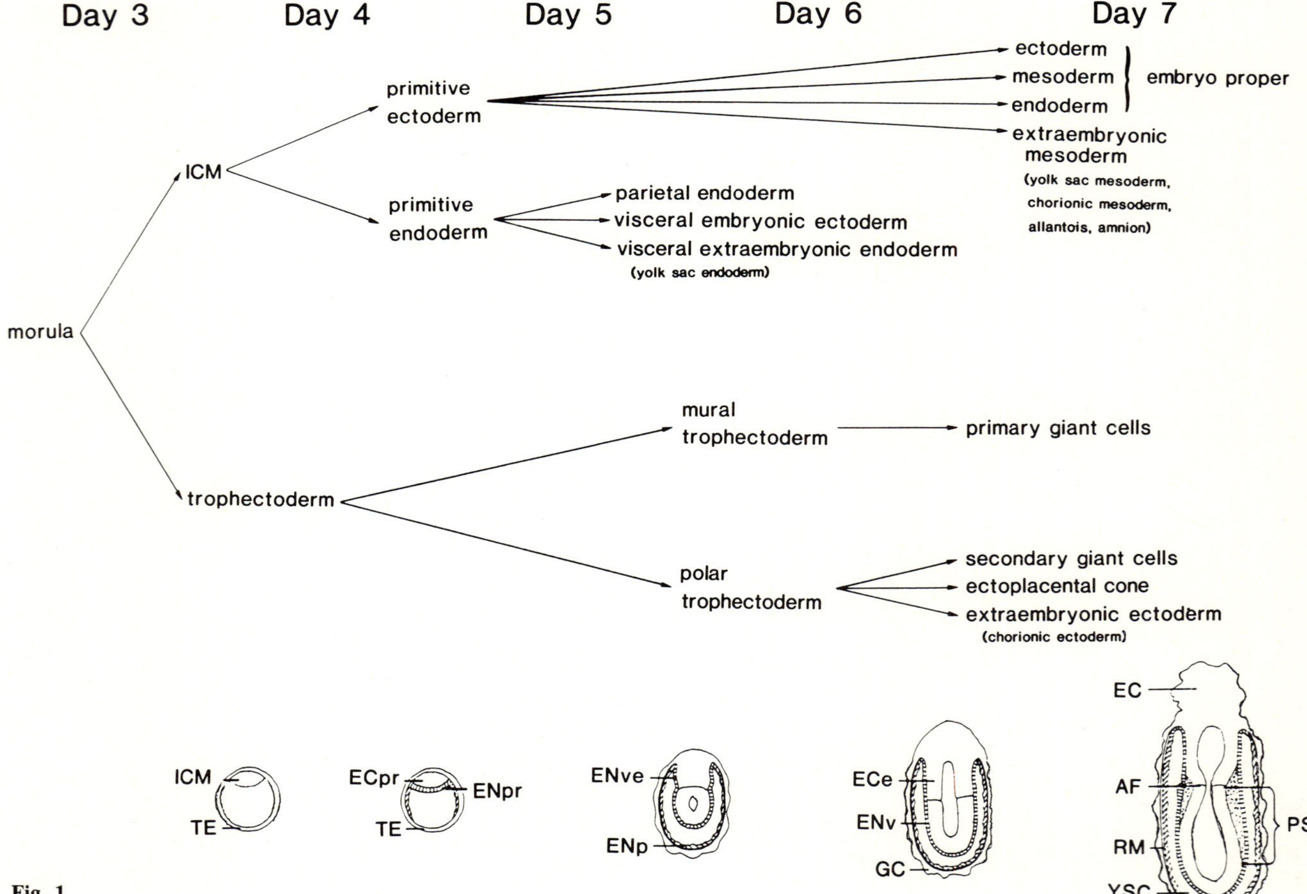

Fig. 1

Expression of Defined Growth Factors in Mouse Embryogenesis

Sufficient data have been accumulated on four growth factors to suggest their involvement in embryonic development.

A. Insulin-like growth factors (IGFs), IGF-I and IGF-II are polypeptides that share structural similarities with insulin and that can be distinguished by specific radioimmunoassays. Their biological activities in vitro include stimulation of proliferation of fibroblasts, weak insulin-like activity in adipose tissue, and stimulation of sulfate uptake by cartilage (for review, see Nissley and Rechler 1984).

Several studies suggest that the IGFs play a role in embryogenesis. Rat IGF-II was detected at high levels in fetal serum (Moses et al. 1980), and in supernatants of fetal liver explants (Rechler et al. 1979) and embryo fibroblasts (Adams et al. 1983). Similarly, explants of multiple organs from mouse embryos were found to secrete IGF-I immunoreactivity (D'Ercole et al. 1980). Of the two IGFs, IFG-II is considered to be the predominant fetal growth factor. IGF-II levels have been shown to be 20 – 100-fold higher in fetal serum than in maternal serum and decline a few days after birth (Moses et al. 1980), whereas IGF-I exhibits the reciprocal developmental pattern: the levels are low in the neonate and increase to adult levels within 4 weeks of birth (Sara et al. 1980; Daughaday et al. 1982).

Support of the possible role of IGF-II as an embryonic growth factor was obtained from studies with EC cells. Undifferentiated EC cells (F9, PC13) do not produce IGFs. When induced to differentiate by retinoic acid, the differentiated cells, F9 – Dif5, an endodermal cell line derived from F9, and PC13-END, a mesodermal cell line derived from PC13, secreted IGF-II-like molecules (Nagarajan et al. 1985; Heath and Shi 1986). IGF-II-like molecules were found to be secreted by the extra embryonic yolk sac mesoderm and the amnion of 9.5 day mouse conceptuses, but not by the parietal or visceral endoderm, suggesting that IGF-II is expressed in a mesodermal lineage-specific manner. On the other hand, Nagarajan et al. (1985) have reported the secretion of IGF-II molecules by established cell lines related to visceral endoderm and parietal endoderm. However, secretion of IGF-II molecules by the endodermal cell lines could be a property acquired by these cells during in vitro establishment. Expression of IGF-II by extra-embryonic membranes may account for the presence of relatively high levels of IGF and IGF binding proteins in fetal amniotic fluid and supports the hypothesis that these tissues play a role in supporting embryonic growth in utero (Shi and Heath 1984).

Fig. 1. Schematic representation of the peri-implantation stages of mouse embryogenesis, and the cell lineages in the mouse embryo. *ICM* inner cell mass; *TE* trophectoderm; *ECpr* primitive ectoderm; *ENpr* primitive endoderm; *ENve* visceral extraembryonic endoderm; *ENp* parietal endoderm; *ECe* extraembryonic ectoderm; *ENv* visceral endoderm; *GC* giant cells; *EC* ectoplacental cone; *AF* amniotic fold; *RM* Reichert's membrane; *YSC* yolk sac cavity; *PS* primitive streak. (Lock and Martin 1986)

In addition, both type I and type II IGF receptors (for review, see Nissley and Rechler 1984) are expressed on F9 and PC13 cells (Nagarajan et al. 1982; Heath and Shi 1986). Upon differentiation, their number decreases dramatically. The apparent loss of receptors might be explained by their being occupied by IGF-like growth factors secreted by the differentiated cells, possibly by an autocrine mechanism. However, the finite lifespan of PC13-END cells and the inability of exogenously added IGF-II to promote the stimulation of growth-arrested F9-Dif5 cells (Heath and Rees 1985; Nagarajan et al. 1985, respectively), suggests that other factors are required to induce cell growth. It is possible that IGF produced by differentiated cells acts as a mitogen on nearby stem cells (Heath and Rees 1985).

B. Epidermal growth factor (EGF) is a potent polypeptide mitogen for epidermal cells in vivo and for a variety of cultured cells in vitro, whose mitogenic effect is mediated through a specific receptor (Carpenter et al., and Schlessinger, this Vol. and references therein). The initial identification of EGF by its ability to induce premature eyelid opening and incisor eruption in newborn mice, as well as the observations that it is involved in development of embryonic lung in rabbits and growth of human secondary palate (Catterton et al. 1979; Hassell 1975), suggested EGF as an embryonic growth factor. The first indication that EGF-like material is synthesized by the mouse embryo was obtained by Nexø et al. (1980), who found that extracts from embryos, 11 days and older, could compete with human EGF for EGF binding sites. However, the lack of detectable immunoreactive EGF in embryos before 14 days of gestation suggested that the EGF-like substance noted in mid-gestation embryos was different from authentic adult EGF. In addition, EGF-specific mRNA could not be detected in fetuses, fetal membranes, or placenta, from day 9 of gestation through early postnatal periods (Papliker et al. 1986). These findings imply that while EGF is not synthesized in mouse embryos, there is an embryonic EGF-like growth factor, now believed to correspond to the transforming growth factor-α. The detection of EGF in older embryos (15 and 17 days old) (Nexø et al. 1980; Twardzik et al. 1982; Proper et al. 1982), could be explained by transport of maternal EGF to the fetus (Papliker et al. 1986).

Unlike EGF, receptors for EGF are readily detectable at various stages during mouse embryogenesis. Binding assays using radiolabelled EGF revealed the presence of EGF receptors in membranes prepared from 11.5-day and older embryos, as well as in various tissues from mid-gestation embryos (Adamson et al. 1981; Adamson and Meek 1984). In 13-day-old embryos, the amnion showed the greatest binding activity, followed by lungs, trophoblast, limbs, and visceral yolk sac. Heart, brain, and liver showed weak binding activity. The embryonic tissues, when cultured in vitro, were shown to respond to EGF with increased thymidine incorporation (Adamson et al. 1981). Immunoprecipitable EGF receptor kinase activity was also found in embryos beginning with the 9th day of gestation (Hortsch et al. 1983). The embryonic liver acquires detectable EGF receptors only at about 18–19 days of gestation

(Hortsch et al. 1983; Adamson and Meek 1984) suggesting that this change either reflects the transition at about 18 days of gestation from a hematopoeitic organ to a site for glycolysis or changes in the metabolic effects of EGF on this tissue.

The presence and the possible function of EGF receptors in early embryogenesis has also been studied in the teratocarcinoma model system. As judged by EGF binding and mitogenic response experiments, none of the EC cell lines examined (PC13, F9, PSA-1, OC15) nor primary endoderm cells obtained from differentiated PSA-1 embryoid bodies appear to express EGF receptors (Rees et al. 1979; A Jakobovits and GR Martin, unpublished data). However, EGF receptors are detected on both PSA-5E cells, which are related to visceral endoderm and on PC13-END cells, which resemble extra-embryonic mesoderm of the postimplantation embryo (Rees et al. 1979).

In conclusion, the potential ligand for the detected EGF receptors in the mouse embryo could be either maternally-derived EGF or an EGF-like growth factor such as TGF-α. As EGF was not detected in embryos younger than 14 days, it is proposed that TGF-α is the physiological ligand during mid-gestation, and perhaps other stages as well.

C. Transforming growth factors (TGFs) are polypeptides capable of inducing non-transformed cells to express attributes of the transformed phenotype, such as anchorage-independent growth in soft agar. Two classes of TGFs (α and β) have been identified. TGF-α competes with EGF for the EGF receptor and induces soft agar-growth of cells such as normal rat kidney (NRK) cells in synergism with TGF-β. TGF-β does not bind to the EGF-receptor, and requires either EGF or TFG-α to stimulate soft agar-growth (Derynck, this Vol., and Moses and Leof, this Vol. and references therein).

The detection of TGF-α in transformed cells, but not in normal adult cells, suggests that TGF-α is an embryonic growth factor which is expressed inappropriately during tumor formation. Indeed, peptides with properties characteristic of TGF-α (binding to EGF receptors and stimulation of anchorage-independent growth) have recently been isolated from normal mouse embryos (Twardzik et al. 1982; Twardzik 1985). In these studies, TGF-α was found to be differentially expressed during embryogenesis: levels were highest at day 7 followed by a smaller peak at day 13, an observation supported by mRNA studies in rat embryos (Lee et al. 1984). These findings indicate that TGF-α plays an important role in the development of the mid-gestational embryo. Since the ability to induce soft agar cell growth was shown to be the result of a synergism between TGF-α and TGF-β, the transforming activity detected in the embryo extract suggested the simultaneous presence of TGF-β or related growth factors. Indeed, additional transforming growth activities were detected in mouse embryos and in EC cells (Proper et al. 1982; Rizzino 1983; Jakobovits et al. 1985). Their properties as unique embryonic growth factors are discussed below.

D. Platelet-derived growth factor (PDGF), is a potent polypeptide mitogen for fibroblasts and other connective tissue cells in vitro that is proposed to play a role in stimulating cell proliferation in vivo at sites of vascular damage. The cellular response to PDGF is mediated by specific high affinity receptors (for review, see Antoniades and Owen 1984). The possible involvement of PDGF or its embryonic homologs in mouse embryogenesis has so far been examined mainly in the EC cell system. Gudas et al. (1983) have found that PSA-1-G cells, a line selected from PSA-1 cells by their ability to grow without a feeder layer and to differentiate spontaneously into fibroblast-like cells, secretes a PDGF-like growth factor. This activity can compete with radiolabeled human PDGF for binding on BALB/C 3T3 cells and can stimulate the growth of these cells. On the other hand, medium conditioned by the parental PSA-1 cells was found to contain no PDGF-like molecules (Jakobovits et al. 1985), a finding that might reflect the different natures of the EC cell lines examined. F9 cells have also been found to produce a PDGF-like growth factor (Gudas et al. 1983; Rizzino and Pope 1985), while contradictory results have been reported for PC13 cells (Gudas et al. 1983; Heath and Isacke 1984).

EC cells and their differentiated derivatives were also studied for the expression of PDGF receptors. Of all EC cell lines examined (F9, PC13, PSA-1, PSA-1-G), none bound significant amounts of PDGF (Rizzino and Bowen-Pope 1985; Jakobovits et al. 1985; Gudas et al. 1983). Rizzino and Bowen-Pope (1985) reported that differentiation of F9 cells resulted in a 30-fold increase in the binding of PDGF, while Gudas et al. (1983) detected only a two-fold increase. However, the latter authors observed a 20- to 30-fold increase in PDGF binding by PSA-1-G-derived fibroblast-like cells. These data suggest that the expression of PDGF-like molecules and PDGF receptors is not characteristic of pluripotent stem cells but might be induced upon differentiation.

Embryonal Carcinoma-Derived Growth Factors

Since identification and isolation of growth factors directly from the embryo is very difficult, investigators have turned to EC cells as an alternative source.

A. Heath and Isacke (1984) have isolated an embryonal carcinoma-derived growth factor (ECDGF) from PC13-derived conditioned medium. ECDGF is a single-chain, 17 kDa protein that stimulates the proliferation of several fibroblastic cell lines as well as PC13-END cells. On the basis of its chemical properties and specificity, ECDGF differs from EGF, PDGF and IGF-II. Its ability to promote the growth of PC13-END cells raises the possibility that normal stem cells produce growth factors that can support the survival, proliferation and/or differentiation of their progeny.

B. PSA-1-derived conditioned medium was found to contain several growth-promoting and differentiation-inhibiting activities which are attributable to three separate growth factors, termed stem cell growth factors, or SCGFs (Jakobovits et al. 1985): (1) SCGF-1 stimulates the proliferation of the pluripotent cells that produce it (autostimulatory activity); (2) SCGF-2 stimulates the proliferation of fibroblasts and induces them to express properties of the transformed phenotype, including growth in soft agar; (3) SCGF-3 stimulates the proliferation of Friend erythroleukemia cells and also inhibits their induced differentiation. As PSA-1 cells are more similar to normal embryonic stem cells than most other EC cells (Martin 1980), their secreted growth factors are likely to resemble those produced by pluripotent normal embryonic cells and to differ from those produced by other EC cell lines.

SCGFs are unrelated to EGF or PDGF as they do not compete with them in receptor-binding assays. Furthermore, EGF and PDGF do not mimic either SCGF-1 or SCGF-3 activities. The lack of similarity to EGF also suggests that SCGF-2 is not TGF-α. The growth-promoting activities produced by PSA-1 cells appear to be developmentally regulated, since they are not detected in medium conditioned by PSA-1- derived endoderm cells (A. Jakobovits and G. R. Martin, unpublished data). In addition, unlike the situation in the PC13 cell system, PSA-1 conditioned medium does not promote the proliferation of its endoderm cell derivatives.

C. Other EC cell lines, such as F9, PC13, and OC15, release factors that promote the growth of NRK cells in soft agar (Rizzino et al. 1983), but are unrelated to the defined TGFs. Proper et al. (1982) have characterized a transforming growth factor from extracts of 17-day-old embryos which is unrelated to TGF-α. The nature of these activities can only be assessed after their purification.

Concluding Remarks

Our understanding of the role that growth factors play in mouse embryogenesis is still in a very preliminary phase. Little is known about which cells secrete these growth factors, their in vivo targets, or their mechanism of action. However, even in this phase, we can make some hypotheses regarding the role of growth factors in murine development and the regulatory mechanisms involved. TFG-α or EC-derived growth factors, whose expression is restricted to short periods of embryogenesis, might represent a class of factors that control the proliferation and/or differentiation of a cell population, unique to that developmental stage. On the other hand, IGF II, expressed at multiple developmental stages, might represent a second class of factors that exert their effects on a variety of cell types which appear sequentially during gestation, or on distinct subsets of cells whose response to growth factors vary as development progresses.

The autostimulation of growth by an activity secreted by PSA-1 cells and the stimulation of PC13-END cells by a factor from PC13 cells implicates autocrine and paracrine growth mechanisms as possible regulatory processes in early embryogenesis. Autocrine mechanisms might be essential in early embryogenesis, where extensive multiplication of pluripotent embryonic cells is required. Paracrine growth control, in which regulatory molecules are locally diffused to nearby target cells, might be essential at a stage where different cell types appear but no circulatory system is yet established. Finally, later in development endocrine processes become a major growth control mechanism. The role which regulated expression of soluble growth factors plays in different stages of embryogenesis is yet to be established.

Acknowledgements. I would like to thank Drs. G. R. Martin, E. B. Jakobovits, and M. A. Frohman for critical reading of the manuscript and valuable suggestions and Mr. Ed Cleary for the preparation of the manuscript. The author's own work was supported by postdoctoral fellowships from Dr. Chaim Weizmann Fund and from the California Division of the American Cancer Society.

References

Adams SO, Nissley SP, Greenstein LA, Yang YW-H, Rechler MM (1983) Synthesis of multiplication-stimulating activity (rat insulin-like growth factor II) by rat embryo fibroblasts. Endocrinology 112:979 – 987

Adamson ED, Meek J (1984) The ontogeny of epidermal growth factor receptors during mouse development. Dev Biol 103:62 – 70

Adamson ED, Deller MJ, Warshaw JB (1981) Functional EGF receptors are present on mouse embryo tissues. Nature 291:656 – 657

Antoniades HN, Owen AJ (1984) Human platelet-derived growth factor. In: Li CH (ed) Hormonal proteins and peptides, vol XII. Academic Press, London New York, pp 232 – 277

Catterton WZ, Escobedo MB, Sexson WR, Gray ME, Sundel HW, Stahlman MT (1979) Effect of epidermal growth factor on lung maturation in fetal rabbits. Pediatr Res 13:104 – 108

Daughaday WH, Parker KA, Borowsky S, Trivedi B, Kapadia M (1982) Measurement of somatomedin-related peptides in fetal, neonatal, and maternal rat serum by insulin-like growth factor (IGF-I) radioimmunoassay, IGF-II radioreceptor assay (RRA) and multiplication-stimulating activity RRA after acid ethanol extraction. Endocrinology 110:575 – 581

D'Ercole AJ, Applewhite GT, Underwood LE (1980) Evidence that somatomedin is synthesized by multiple tissues in the fetus. Dev Biol 75:315 – 328

Gudas LF, Singh JP, Stiles CD (1983) Secretion of growth regulatory molecules by teratocarcinoma stem cells. In: Silver LM, Martin GR, Strickland S (eds) Teratocarcinoma stem cells, Cold Spring Harbor Conf Cell Proliferation 10. Cold Spring Harbor Lab, Cold Spring Harbor, NY, pp 229 – 236

Hassell JR (1975) The development of rat palate shelves in vitro. Dev Biol 45:90 – 102

Heath JK, Isacke CM (1984) PC13 embryonal carcinoma-derived growth factor. EMBO J 3:1957 – 1962

Heath JK, Rees AR (1985) Growth factors in mammalian embryogenesis. In: Evered D (ed) Growth factors in biology and medicine. Ciba Symp 116. Longmas, London, pp 1 – 32

Heath JK, Shi W-K (1986) Developmentally regulated expression of insulin-like growth factors by differentiated murine teratocarcinomas and extra-embryonic mesoderm. J Embryol Exp Morphol, in press

Hortsch M, Schlessinger J, Gootwine E, Webb CG (1983) Appearance of functional EGF receptor kinase during rodent embryogenesis. EMBO J 2:1937 – 1941

Jakobovits A, Banda MJ, Martin GR (1985) Embryonal carcinoma-derived growth factors: specific growth-promoting and differentiation-inhibiting activities. In: Feramisco J, Ozanne B, Stiles C (eds) Growth factors and transformation/cancer cells 3. Cold Spring Harbor Lab, Cold Spring Harbor, NY, pp 393 – 399

Lee DC, Rockford R, Todaro GJ, Villarreal LP (1984) Developmental expression of rat transforming growth factor-α mRNA. Mol Cell Biol 5:3644 – 3646

Lock F, Martin GR (1986) (1986) In: Risley MS (ed) Chromosome structure and function. Nostrand Reinhold, New York, p 193

Martin GR (1980) Teratocarcinomas and mammalian embryogenesis. Science 709:768 – 776

Moses AC, Nissley SP, Short PA, Rechler MM, White RM, Knight AB, Higa OZ (1980) Elevated levels of multiplication-stimulating activity, an insulin-like growth factor, in fetal rat serum. Proc Natl Acad Sci USA 77:3649 – 3653

Nagarajan L, Nissley SP, Rechler MM, Anderson WB (1982) Multiplication-stimulating activity stimulates the multiplication of F9 embryonal carcinoma cells. Endocrinology 110:1231 – 1237

Nagarajan L, Anderson WB, Nissley SP, Rechler MM, Jetten AM (1985) Production of insulin-like growth factor-II (MSA) by endodermal-like cells derived from embryonal carcinoma cells: possible mediator of embryonic cell growth. J Cell Physiol 124:199 – 206

Nexø E, Hollenberg MO, Figueroa A, Pratt RM (1980) Detection of epidermal growth factor – urogastrone and its receptors during fetal mouse development. Proc Natl Acad Sci USA 77:2782 – 2785

Nissley SP, Rechler MM (1984) Insulin-like growth factors: biosynthesis, receptors, and carrier proteins. In: L CH (ed) Hormonal proteins and peptides, vol XII. Academic Press, London New York, pp 128 – 203

Papliker N, Shatz A, Aviv A, Ullrich A, Schessinger J, Webb CG (1986) Lack of endogenous synthesis of epidermal growth factor in mouse embryos. Dev Biol, in press

Proper JA, Bjornson CL, Moses HL (1982) Mouse embryos contain polypeptide growth factor(s) capable of inducing a reversible neoplastic phenotype in nontransformed cells in culture. J Cell Physiol 110:169 – 174

Rechler MM, Eisen HJ, Higa OZ, Nissley SP, Moses AC, Schilling EE, Fennoy I, Bruni CB, Phillips LS, Baird KL (1979) Characterization of a somatomedin (insulin-like growth factor) synthesized by fetal rat liver organ cultures. J Biol Chem 254:7942 – 7950

Rees AR, Adamson ED, Graham CF (1979) Epidermal growth factor receptors increase during the differentiation of embryonal carcinoma cells. Nature 281:309 – 311

Rizzino A (1983) Model systems for studying the differentiation of embryonal carcinoma cells. Cell Biol Int Rpts 7:559 – 566

Rizzino A, Bowen-Pope DF (1985) Production of PDGF-like growth factors by embryonal carcinoma cells and binding of PDGF to their endoderm-like differentiated cells. Dev Biol 110:15 – 22

Rizzino A, Orme LS, DeLarco JE (1983) Embryonal carcinoma cell growth and differentiation: production of and response to molecules with transforming growth factor activity. Exp Cell Res 143:143 – 152

Sara VR, Hall K, Lins PE, Fryklund L (1980) Serum levels of immunoreactive somatomedin A in the rat: some developmental aspects. Endocrinology 107:622 – 625

Shi W-K, Heath JK (1984) Apolipoprotein expression by murine visceral yolk sac mesoderm. J Embryol Exp Morphol 81:143 – 152

Twardzik DR (1985) Differential expression of transforming growth factor-α during prenatal development of the mouse. Cancer Res 45:5413 – 5416

Twardzik DR, Rachlais JE, Todaro GJ (1982) Mouse embryonic transforming growth factors related to those isolated from tumor cells. Cancer Res 42:590 – 593

A Role for Proto-Oncogenes in Differentiation?

ERWIN F. WAGNER and ROLF MÜLLER

Ever since the discovery of proto-oncogene products in the normal cell, their biological role and molecular function have been of major interest in molecular and cellular biology. To date more than 30 different c-*onc* genes are known which encode proteins localized in the nucleus or cytoplasm, are associated with the plasma membrane or even secreted (see Table 1). It has been presumed from the outset that proto-oncogenes play a role in growth control, mainly because of their potential to induce uncontrolled cell proliferation. This notion is now strongly supported by evidence that the products of several proto-oncogenes are either growth factors or growth-factor receptors (see Hunter, Sherr and Stanley, Heldin, Beug et al., all this Vol.).

In addition to their role in growth control, proto-oncogenes may also be involved in cellular differentiation. The strongest evidence in support of this hypothesis stems from their remarkable tissue-, cell-type- and stage-specific expression both in embryos and in tissues of adult organisms from different species (see Table 1 and Varmus 1984). Furthermore, proliferation and differentiation appear to be related processes, in that cell division often seems to be a prerequisite for differentiation, and "growth factors" are frequently also "differentiation factors" (Sachs 1986). Therefore, it is conceivable that the same protein may function in cellular differentiation and proliferation. In our view, differentiation is defined as a multistage process which turns a precursor cell into a terminally differentiated cell. Proto-oncogenes may act as positive or negative elements at different branch points in such a pathway, where the consequences of previous steps become the causes of subsequent steps. In this chapter we review some of the data which suggest that proto-oncogene products participate in the process of cell differentiation.

c-*src* Expression and Terminal Differentiation

The specific expression of the c-*src* product, $pp60^{c\text{-}src}$, in neuronal cells provides one of the best examples of a proto-oncogene which appears to be involved in cellular differentiation. In both chick neural retina and cerebellum, $pp60^{c\text{-}src}$ expression is first detectable in developing neurons at the onset of differentiation when proliferation ceases and persists in terminally differentiated neurons. In agreement with these findings, primary cultures of either neurons or astrocytes derived from rat brain contain elevated c-*src* protein and kinase

Oncogenes and Growth Control
Edited by P. Kahn and T. Graf
© Springer-Verlag Berlin Heidelberg 1986

Table 1. Expression of proto-oncogenes[a]

Proto-oncogene	Gene product		Special properties	Tissue with highest mRNA expression (species)[c]
	Molecular mass (kDa)	Localization[b]		
src-Family of protein kinases				
c-*src*	60	PM	Tyrosine kinase	Neuronal tissues (ck, m, D), smooth muscle cells of gut (D)
c-*ros*		PM	Tyrosine kinase	Kidney (ck)
c-*fps*/c-*fes*	98/92	PM	Tyrosine kinase	Myeloid cells (ck, m, h)
c-*yes*	60		Tyrosine kinase	Kidney (ck)
c-*erb-B*	170	PM	Tyrosine kinase, transmembrane glyco-protein, homologous to EGF-receptor	Embryo (ck, m)
c-*fms*	140	PM	Tyrosine kinase, transmembrane glyco-protein, homologous to CSF-1 receptor	Extra-embryonic tissues, macrophages (ck, h)
c-*abl*	150	PM	Tyrosine kinase, tissue-specific splicing of transcript	Thymus, testes (m)
c-*neu*	185	PM	Tyrosine kinase, transmembrane glyco-protein, partial homology with c-*erbB*	
c-*raf*/c-*mil*	75	C	Threonine kinase	
c-*mos*		C	Extremely low level of expression	Gonades (m)
Other membrane-associated proteins				
H-*ras*	21	PM	GTP/GDP binding, GTPase activity, involved in regulation of adenylate cyclase in yeast	Ubiquitous
K-*ras*	21	PM		Ubiquitous
N-*ras*	21	PM		Ubiquitous

Table 1 (continued)

Proto-oncogene	Gene product		Special properties	Tissue with highest mRNA expression (species)[c]
	Molecular mass (kDa)	Localization[b]		
Other cytoplasmic proteins				
c-*erbA*		C	Partial homology with steroid hormone receptors	Embryo (ck)
c-*ets*	56	C		Lymphoid organs (ck)
Growth factor-like protein				
c-*sis*	28	secreted and membrane associated	Homology with PDGF-β-chain	Embryo (m), activated monocytes
Nuclear proteins				
c-*myc*	60	N (SnRNPs?)	DNA-binding	Ubiquitous in proliferating cells
N-*myc*, L-*myc*	50	N		Brain, kidney
c-*myb*	75	N (chromatin?)	DNA-binding	Immature, proliferating or quiescent, T-lymphocytes (m, h, ck) cell cycle dependent expression in B-cells, fibro-blasts and other cells
c-*fos*	55 – 65	N (chromatin?)	Complexed with p39, extensively post-translationally modified	Fetal membranes, bone marrow, neutrophils (m, h)
c-*ski*		N		Cartilage, muscle, skin (ck)

[a] This table does not include all proto-oncogenes identified to date, but rather gives examples of c-*onc* genes whose expression has been analyzed in some detail or which encode proteins exhibiting interesting biochemical properties.

[b] PM, plasma membrane; c, cytoplasm; N, nucleus.

[c] ck, chicken; m, mouse; h, human; D, *Drosophila*.

activity compared to fibroblasts (Sorge et al. 1984; Brugge et al. 1985; see article by Rohrschneider for additional references). These findings suggest a function for c-*src* during neuronal differentiation rather than a role in cellular proliferation. Whether this function is associated with the induction of differentiation or whether pp60$^{c\text{-}src}$ provides a function required by differentiating (and/or by differentiated) neurons remains unclear, although the latter seems more likely.

A gene homologous to v-*src* has recently been isolated from *Drosophila*. This gene is expressed in all cells of the early embryo (Simon et al. 1985). Interestingly, after the first 8 h of development its expression seems to be restricted to neural tissues and smooth muscle cells of the gut, correlating again with cellular differentiation rather than with proliferation. The analysis of mutants in the *Drosophila* c-*src* gene could provide an excellent opportunity to study the function of pp60^{src} in eukaryotes.

A correlation of pp60$^{c\text{-}src}$ expression with the state of differentiation has also been observed in an in vitro differentiation system involving non-neuronal cells. Induction of the human myeloid leukemia cell line HL60 to either macrophage differentiation by TPA or to neutrophil-like differentiation by dimethyl sulfoxide (DMSO) is paralleled by an increase in the steady-state level of the c-*src* protein and its associated tyrosine kinase activity (Barnekow and Gessler 1986). The following observations suggest that these changes are related to cellular differentiation and not simply to a nonspecific response of the cells to the inducing agents: (1) An increase in pp60^{src} protein levels and kinase activity is observed following treatment with a variety of inducing agents. (2) pp60$^{c\text{-}src}$ expression and kinase activity correlate with the appearance of differentiated cells. (3) Activation of the c-*src* kinase is also found when HL60 cells are committed to macrophage differentiation by a short exposure to TPA followed by 2 days culture in normal growth medium. Taken together, these findings link c-*src* with the differentiation of several cell types. In certain other cells, however, c-*src* has also been implicated in growth control. This is based on the finding that growth stimulating compounds like PDGF or TPA induce post-translational modification(s) in pp60$^{c\text{-}src}$ in fibroblasts (Gould et al. 1985).

Correlation Between c-*fos* Expression and Hematopoietic Differentiation

Another proto-oncogene implicated in cellular differentiation is the c-*fos* gene, which shows a highly tissue-, cell type- and stage-specific pattern of expression. The highest levels of c-*fos* expression occur in the late-gestation extra-embryonal membranes surrounding the fetus (amnion and yolk sac), in the mid-gestation fetal liver and, after birth, in whole bone marrow (reviewed by Müller 1986). Among the various hematopoietic cells analyzed, high levels of c-*fos* mRNA have been detected in mast cell lines, differentiated macrophages, and blood neutrophils.

In an attempt to correlate c-*fos* expression with differentiation, three myeloid cell lines that can be induced to monocyte/macrophage differentiation in vitro have been studied: The WEHI-3B(D$^+$) murine leukemia cell line, which is inducible by the granulocyte-colony stimulating factor (G-CSF) (Gonda and Metcalf 1984) and the human cell lines U-937 and HL60, which are inducible by TPA (Mitchell et al. 1985; Müller 1986). In all three cases, the induction of differentiation is paralleled by the rapid induction of c-*fos* expression, suggesting that c-*fos* plays a role in the commitment to macrophage differentiation. However, several lines of evidence do not support this conclusion. For instance, TPA appears to be a more or less universal inducer of c-*fos* expression, since it is equally effective in various cell types. Moreover, no rapid c-*fos* induction is observed when HL60 cells are treated with another inducer of macrophage differentiation, 1,25-dihydroxy vitamin D$_3$. Finally, HL60 cells induced to neutrophil-like differentiation do not express c-*fos*, while blood neutrophils in vivo do so. (For further discussion and references see Müller 1986).

When using chemical inducers of differentiation it is important to distinguish the pathway transducing the induced signal from the differentiation pathway affected by this signal. It appears that in the HL60 and U937 models c-*fos* is associated with the first mechanism, the transduction of the TPA-mediated signal, since c-*fos* is not induced by other differentiation inducers. However, since it has been shown that c-*fos* expression is dependent on external signals in a variety of cell types, including fibroblasts, macrophages, and amnion cells (see Bravo and Müller, this Vol., Müller 1986) analysis of c-*fos* expression in factor-independent cell lines such as HL60 and U937 may be misleading. Therefore, the WEHI-3B/G-CSF system seems to provide the strongest evidence for a correlation of c-*fos* expression with macrophage differentiation. In these cells, c-*fos* expression increases during late stages of differentiation, suggesting that c-*fos* induction may be a consequence of terminal differentiation rather than a causative event. It is therefore possible that the c-*fos* protein is required specifically by the differentiated cell for processes other than growth control or the induction of differentiation. In the case of myelomonocytic cells, the funtion of c-*fos* may be associated with the activation of macrophage-specific functions, since modulators of macrophage activity (e.g., lipopolysaccharides, chemotactic factors) also modulate c-*fos* expression (unpublished observations). These findings suggest multi-functional properties for the c-*fos* gene product which has also been implicated in growth control mechanisms, as suggested by its transient induction by various polypeptide growth factors in fibroblasts (reviewed by Müller 1986).

c-*fms* Expression in Differentiated Macrophages and Extraembryonal Cells

The c-*fms* gene represents another proto-oncogene whose expression changes during the macrophage-like differentiation of HL60 cells in vitro. Expression of c-*fms* is very low in undifferentiated HL60 cells, but is readily detectable at terminal stages of TPA- or vitamin D_3-induced HL60 differentiation (Sariban et al. 1985), and is high in differentiated monocytes in vivo. This finding is consistent with the recent discovery that c-*fms* probably encodes the receptor for the macrophage growth factor CSF-1 (see Sherr and Stanley, this Vol. for references), which is expressed at much higher levels in mature macrophages than on precursor cells. This increase in c-*fms* expression may thus represent a consequence of differentiation rather than an inductive signal.

The observation that c-*fms* expression is also high in placenta, yolk sac, and amnion (Müller 1986) suggests that CSF-1 receptors are not restricted to myelomonocytic cells, but also function in cells of the placenta and fetal membranes. In addition, the finding that the highest levels of c-*fms* mRNA are detected close to term supports the notion that the c-*fms* gene product may play a role in differentiating (or differentiated) cells.

Inverse Correlation Between Proto-Oncogene Expression and Differentiation

The HL60 in vitro differentiation system has also been used to study the expression of c-*myc* and c-*myb* proto-oncogenes following the induction of macrophage-like or neutrophil-like differentiation. Any induction of HL60 differentiation leads to decreased or undetectable levels of c-*myc* and c-*myb* mRNA (Westin et al. 1982; Mitchell et al. 1985; Müller et al. 1986). This suggests a role for their encoded gene products in switching on proliferation or inhibiting differentiation.

Furthermore, an inverse correlation of c-*myc* and N-*myc* expression has been observed during the in vitro differentiation of neuroblastoma cells (Thiele et al. 1985) and during retinoic acid-induced differentiation of F9 teratocarcinoma cells (Campisi et al. 1984), respectively. In vivo, the c-*myc* gene is expressed in a large variety of growing cells, while its cognates N-*myc* and L-*myc* are selectively expressed in mouse brain and kidney at early stages of postnatal life, suggesting a role in mouse development (Jakobovits et al. 1985; Zimmerman et al. 1986).

The downregulation of c-*myc*, N-*myc* and c-*myb* mRNA levels observed in different systems could be responsible for, or a consequence of, the loss of proliferative activity which accompanies the induction of differentiation, since these genes have been implicated in growth control mechanisms (see Bravo and Müller, this Vol.). However, it is also possible that switching off c-*myc* and c-*myb* expression is required for, or may even trigger differentiation in certain lineages. The issue is even more complex in that c-*myc* expression in certain in vitro differentiation model systems is transiently increased following the induction of differentiation. This has been observed in erythroid differentiation (DMSO-induced Friend cells; Lachman and Skoultchi 1984) as well as in neuronal differentiation (NGF-induced PC12 pheochromocytoma cells; Curran and Morgan 1985). It is therefore possible that the consequences

of c-*myc* expression depend on the cell lineage and even on the differentiation
state in this lineage, so that c-*myc* may be able to contribute to different (but
perhaps related) processes such as proliferation and differentiation. Likewise,
the c-*myb* gene product may play a role in the differentiation of thymic lymphocytes (where it is constitutively expressed at high levels), while in other
cells its expression is modulated during the cell cycle and may thus be involved
in growth control (Thompson et al. 1986).

Modulation of the Differentiation State by Expression of Exogenous Oncogenes

The implication of c-*fos* in hematopoietic differentiation, together with its
high expression in extraembryonic tissues and the possible role of c-*src* in neuronal differentiation, prompted investigators to analyze the biological effect
of exogenous oncogene products in various in vitro differentiation systems
such as the embryonal carcinoma (EC) cells and PC12 pheochromocytoma
cells.

The F9 EC cell line has been used to obtain direct evidence for an involvement of c-*fos* in cellular differentiation. F9 stem cells do not differentiate
spontaneously in culture, but treatment with retinoic acid and dibutyryl cyclic
AMP can lead to endoderm-like differentiation. When normal c-*fos* genes or
metallothionein promoter c-*fos* constructs were introduced by DNA transfection into F9 stem cells, colonies of differentiated cells expressing certain differentiation-associated proteins were obtained (Müller and Wagner 1984; Rüther et al. 1985). Although these observations demonstrate a certain differentiation-promoting potential of c-*fos*, their relevance for the normal function
of the gene is hard to assess. F9 cells and perhaps even their normal in vivo
counterparts may not represent a cell type where a high expression of c-*fos*
normally plays a role and it is therefore possible that an aberrant type of differentiation is triggered in these cells. A similar differentiated cell type was obtained when other transforming genes such as E1a or a c-Ha-*ras* oncogene was
transfected into F9 cells (U. Rüther, personal communication). These data
suggest that the observed differentiation response is an intrinsic property of
this cell line.

The PC12 cells on the other hand exhibit a chromaffin-like phenotype and
respond to nerve-growth factor (NGF) by differentiating into neuron-like
cells. It was intriguing to find that the differentiation-inducing effect of NGF
could be mimicked, at least in part, by the v-*src* gene product introduced into
the cells by infection with Rous sarcoma virus (Alema et al. 1985). Although
these results may support the idea that $pp60^{c-src}$ plays a role in neuronal differentiation, it is possible that the effects of v-*src* and c-*src* on PC12 differentiation are fundamentally different. Moreover, even though expression of an
exogenous oncogene may have certain effects on cellular processes, these effects may be distinct from the normal function of the same gene product,

which is subjected to normal regulatory mechanisms and expressed in a particular cell type in a defined state of differentiation (see Rohrschneider, this Vol.).

Neuronal differentiation of PC12 cells was also observed after microinjection of the *ras* oncogene protein (Bar-Sagi and Feramisco 1985) or after infection of these cells with *ras* containing retroviruses (Noda et al. 1985), suggesting a function for c-*ras* proteins in PC12 differentiation. In agreement with this hypothesis is the observation that microinjection of *ras*-specific antibodies inhibits the NGF-induced differentiation of PC12 cells (Hagag et al. 1986). It is possible that the mechanism(s) by which exogenous v-*src* and *ras* products work in PC12 is to eliminate the requirement of these cells for NGF. If this assumption is correct, it can be predicted that other oncogene products will turn out to be able to "induce" PC12 cell differentiation.

Future Perspectives

It will be important for the future to establish in vitro and in vivo systems which would allow the selective inactivation of proto-oncogene products through targeted integration, anti-sense mRNA expression or injection of antibodies. Of great importance in the search for proto-oncogene function will be the use of transgenic animals to direct inducible or constitutive c-*onc* expression to certain cell types or tissues at a particular stage of development, as has been shown for other genes (for a review see Wagner and Stewart 1986). Recently, transgenic mice carrying the c-*myc* oncogene linked to either the MMTV promoter (Stewart et al. 1984) or to the immunoglobulin enhancer were produced (Adams et al. 1985). In the former case, lactating females developed adenocarcinomas and the latter ones developed lymphomas within a few months from birth. In addition, various transgenic mouse lines were established with c-*fos* genes and the consequence(s) of exogenous c-*fos* expression in these animals is presently under investigation (Rüther and Wagner, unpublished observations). The approaches in mammalian systems have to be complemented with studies in genetically more versatile systems such as *Drosophila* and yeast to unravel the role(s) proto-oncogenes may play in the fundamental biological processes of development, growth control and cellular differentiation.

Acknowledgments. The authors are very grateful to Drs. Norman Iscove, Colin Stewart and Ulrich Rüther for critically reading the manuscript and to Ines Benner for typing.

References

Adams JM, Harris AW, Pinkert CA, Corcoran LM, Alexander WS, Cory S, Palmiter RD, Brinster RL (1985) The c-*myc* oncogene driven by immunoglobulin enhancers induces lymphoid malignancy in transgenic mice. Nature 318:533–538

Alema S, Casalbore P, Agostini E, Tato F (1985) Differentiation of PC12 phaeochromocytoma cells induced by v-*src* oncogene. Nature 316:557−559

Barnekow A, Gessler M (1986) Activation of the pp60^{c-src} kinase during differentiation of monomyelocytic cells in vitro. EMBO J 5:701−705

Bar-Sagi D, Feramisco J (1985) Microinjection of the *ras* oncogene protein into PC12 cells induces morphological differentiation. Cell 42:841−848

Brugge JS, Cotton PC, Queral AE, Barrett JN, Nonner D, Keane RW (1985) Neurones express high levels of a structurally modified, activated form of pp60^{c-src}. Nature 316:554−557

Campisi J, Gray HE, Pardee AB, Dean M, Sonenshein GE (1984) Cell-cycle control of c-*myc* but not c-*ras* expression is lost following chemical transformation. Cell 36:241−247

Curran T, Morgan JI (1985) Superinduction of c-*fos* by nerve growth factor in the presence of peripherally active benzodiazepines. Science 229:1265−1268

Gonda TJ, Metcalf D (1984) Expression of *myb*, *myc* and *fos* proto-oncogenes during the differentiation of a murine myeloid leukaemia. Nature 310:249−251

Gould KL, Woodgett JR, Cooper JA, Buss JE, Shalloway D, Hunter T (1985) Protein kinase C phosphorylates pp60^{src} at a novel site. Cell 42:849−857

Hagag N, Halegoua S, Viola M (1986) Inhibition of growth factor-induced differentiation of PC12 cells by microinjection of antibody to *ras* p21. Nature 319:680−682

Jakobovits A, Schwab M, Bishop MJ, Martin GR (1985) Expression of N-*myc* in teratocarcinoma stem cells and mouse embryos. Nature 318:188−191

Lachman HM, Skoultchi AI (1984) Expression of c-*myc* changes during differentiation of mouse erythroleukaemia cells. Nature 310:592−594

Mitchell RL, Zokas L, Schreiber RD, Verma IM (1985) Rapid induction of the expression of proto-oncogene *fos* during human monocytic differentiation. Cell 40:209−217

Müller R (1986) Cellular and viral *fos* genes: structure, regulation of expression and biological properties of their encoded products. Biochem Biophys Acta Reviews on Cancer 823:207−225

Müller R, Wagner EF (1984) Differentiation of F9 teratocarcinoma stem cells after transfer of c-*fos* proto-oncogenes. Nature 311:438−442

Noda M, Ko M, Ogura A, Liu D, Amano T, Takano T, Ikawa Y (1985) Sarcoma viruses carrying *ras* oncogenes induce differentiation-associated properties in a neuronal cell line. Nature 318:73−75

Rüther U, Wagner EF, Müller R (1985) Analysis of the differentiation-promoting potential of inducible c-*fos* genes introduced into embryonal carcinoma cells. EMBO J 4:1775−1781

Sachs L (1986) Growth, differentiation and the reversal of malignancy. Sci Am 254:30−37

Sariban E, Mitchell T, Kufe D (1985) Expression of the c-*fms* proto-oncogene during human monocytic differentiation. Nature 316:64−66

Simon MA, Drees B, Kornberg T, Bishop JM (1985) The nucleotide sequence and the tissue-specific expression of *Drosophila* c-*src*. Cell 42:831−840

Sorge LK, Levy BT, Maness PF (1984) pp60^{c-src} is developmentally regulated in the neural retina. Cell 36:249−257

Stewart TA, Pattengale PK, Leder P (1984) Spontaneous mammary adenocarcinomas in transgenic mice that carry and express MTV/*myc* fusion genes. Cell 38:627−637

Thiele CJ, Reynolds CP, Israel MA (1985) Decreased expression of N-*myc* precedes retinoic acid-induced morphological differentiation of human neuroblastoma. Nature 313:404−406

Thompson CB, Challoner PB, Neiman PB, Groudine M (1986) Expression of the c-*myb* proto-oncogene during cellular proliferation. Nature 319:374−380

Varmus HE (1984) The molecular genetics of cellular oncogenes. Annu Rev Genet 18:553−612

Wagner EF, Stewart CL (1986) Integration and expression of genes introduced into mouse embryos. In: Rossant J, Peterson R (eds) Experimental approaches to mammalian embryonic development. Cambridge Univ Press (in press)

Westin EH, Wong-Staal F, Gelmann EP, Dalla Favera R, Papas TS, Lautenberger JA, Eva A, Reddy EP, Tronick SR, Aarsonson SA, Gallo RC (1982) Expression of cellular homologues of retroviral *onc* genes in human hematopoietic cells. Proc Natl Acad Sci USA 79:2490−2494

Zimmerman KA, Yancopoulos GD, Collum RG, Smith RK, Kohl NE, Denis KA, Nau MM, Witte ON, Toran-Allerand D, Gee CE, Minna JD, Alt FW (1986) Differential expression of *myc* family genes during murine development. Nature 319:780−783

Tissue-Specific Expression and Possible Functions of pp60$^{c\text{-}src}$

LARRY R. ROHRSCHNEIDER

Although some remarkable progress has been made in identifying the function of a few proto-oncogene products such as c-*erbB*, c-*fms*, c-*sis* (see Wagner and Müller; Beug et al.; Sherr and Stanley; Heldin and Westermark, all this Vol.), the function of the c-*src* protein pp60$^{c\text{-}src}$ has remained an enigma despite the fact that it was the very first proto-oncogene product identified (Collett et al. 1978). The difficulty in studying pp60$^{c\text{-}src}$ is that it occurs at a very low abundance and that it is not easily accessible to the investigator since it is neither secreted nor located at the cell surface (as are the proto-oncogene products listed above). Nevertheless, several analyses of pp60$^{c\text{-}src}$ expression in various tissues and stages of embryogenesis have recently been carried out, and the results promise to shed new light on the regulation of cell growth and differentiation. In the present review, I discuss the significance of these findings for a functional role of the c-*src* gene product.

The viral *src* protein (pp60$^{v\text{-}src}$) releases a variety of cell types from normal growth restraints and blocks their differentiation. This fact led to the assumption that the c-*src* protein exerts a similar function. It was therefore surprising to discover that pp60$^{c\text{-}src}$ is expressed in certain differentiated cells which had ceased dividing. Furthermore, although species from sponges to man express the c-*src* gene in a variety of tissues (Collett et al. 1978; Oppermann et al. 1979; Rohrschneider et al. 1979; Schartl and Barnekow 1982; Lev et al. 1984), the highest levels are found in brain and related neural tissues that do not proliferate (Barnekow et al. 1982).

In embryonic tissues one might expect that pp60$^{c\text{-}src}$ protein would be expressed in the most actively dividing and differentiating cell masses. Again, this does not appear to be the case, since this protein shows highest expression in neural tissues of the chick embryo including brain, neural tube, neural crest, neural retina, and neural ectoderm (Levy et al. 1984; Cotton and Brugge 1983; Fults et al. 1985; Sorge et al. 1984; Jacobs and Ruebsamen 1983). The expression of pp60$^{c\text{-}src}$ occurs in two phases during chick embryogenesis. Upon neural closure (around stage 12) pp60$^{c\text{-}src}$ expression declines in the neural tube but remains high in the neural crest region. Re-expression occurs in terminally differentiated neurons at about stage 21 in the neural retina and stage 17 in the developing cerebellum. Very similar patterns of c-*src* mRNA expression are seen during *Drosophila* development (Simon et al. 1985).

The specific cells expressing pp60$^{c\text{-}src}$ in the embryonic brain have been identified as postmitotic central nervous system neurons and astrocytes (Brug-

Oncogenes and Growth Control
Edited by P. Kahn and T. Graf
© Springer-Verlag Berlin Heidelberg 1986

ge et al. 1985). Both cell types contained 15 – 20-fold more c-*src* protein than normal fibroblasts but the neuronal pp60^{c-src} exhibited a tyrosine kinase specific activity which was up to tenfold higher than that found in astrocytes. Recent evidence suggests that this activation of the neuronal pp60^{c-src} kinase activity may involve phosphorylations within the N-terminal end of the protein (Bolen et al. 1985; Brugge et al. 1985; Ralston and Bishop 1985).

What could be the function of pp60^{c-src} expressed in nondividing cells such as neural ectoderm? One possibility is that it serves as a switch which directs the differentiation program in these cells. An alternative hypothesis can be derived by analogy with the actions of the transforming v-*src* protein in fibroblasts and other cell types. Thus, cell shape changes within the neural ectoderm, presumably driven by alterations in the cellular cytoskeleton, initiate the formation of folds that develop into the neural tube. The v-*src* protein is known to alter cell shape and the cytoskeleton in fibroblasts, and perhaps this reflects the attempt to re-enact neurulation in these cells.

Further examples which demonstrate that c-*src* expression can be associated with differentiation rather than with cell proliferation come from both chick and *Drosophila* embryonic systems (Sorge et al. 1984; Simon et al. 1985). Within the developing retina of the chick, the first cells that exhibit detectable pp60^{c-src} levels are postmitotic ganglion and amacrine neurons, which are at the onset of differentiation. At stages 21 – 23, these pp60^{c-src}-expressing cells are adjacent to the pigmented epithelial layer, but soon migrate toward the inner surface of the retina and develop into pp60^{c-src}-positive axons and dendrites. The c-*src* protein is present within the body of these cells but not in their processes. Early development of *Drosophila* involves a process known as germ-band extension, in which ventral mesodermal and ectodermal cells undergo cell division and migration. This process is complete by 6 h, and at this stage all cells of the embryo express c-*src* transcripts. During subsequent germ-band retraction (at approximately 9.5 h), cells migrate back along the ventral midline of the embryo, and localized regions of c-*src* transcripts appear in cells that had completed their final divisions by 8 h and will form the smooth muscle of the gut. Thus, in both cases, the expression of c-*src* occurs in postmitotic cells ready to undergo terminal differentiation.

In the above two examples, it is striking that expression of elevated c-*src* levels coincides with cellular migration. Furthermore, cells within the neural crest region of the developing chick, which migrate extensively and give rise to almost all outlying components of the nervous system, also express pp60^{c-src} (Maness et al. 1986). Perhaps in these particular instances, pp60^{c-src} directs cellular migration either independently of, or as part of the differentiation program. Certainly, transformation of fibroblasts and other cell types by the viral *src* gene leads to a "reprogramming" of cell migration.

An alternative hypothesis concerning the function of pp60^{c-src} within certain developmentally active tissues suggests an involvement in electrical or neurochemical signaling. In early development, pp60^{c-src} is expressed in the neuroectoderm, neural folds, and crest region, as well as in the ventral floor of the neural tube overlying the notochord. These regions may be active sites of neurotransmitter uptake (Maness and Fults 1985; Harris 1981) that could act as inductive signals for differentiation. The detection of pp60$^{c\text{-}src}$ in the neuronal processes within the molecular layer of the cerebellum (Fults et al. 1985) and in plexiform layers of the retina (Sorge et al. 1984) again suggests an involvement in electrical or neurochemical signaling between cells. Perhaps such signals are important not only for the fully differentiated function of neurons, but also for induction of differentiation along this lineage. That differentiation signals can actually be induced by the v-*src* gene was demonstrated using PC12 rat pheochromocytoma cells (Alemà et al. 1985). These cells respond to nerve-growth factor (NGF) by shifting from a chromaffin cell-like phenotype to a sympathetic neuron-like phenotype with neurite outgrowth. Expression of pp60$^{v\text{-}src}$ in PC12 cells induces neurite outgrowth in the absence of NGF, suggesting that the v-*src* gene product is able to induce these cells to progress further along their differentiation pathway.

In contrast to the expression of pp60$^{c\text{-}src}$ in embryos, the expression of pp60$^{c\text{-}src}$ in fully differentiated adult tissue suggests that it also has functions which are necessary for the maintenance of these tissues. Two specific examples are of considerable interest: chromaffin cells and blood platelets. Chromaffin cells of the adrenal medulla are developmental derivatives of the neural crest and have evolved for the very specific function of releasing transmitters by exocytosis (Livett 1984); PC12 cells are related to these cells; in addition, neuroblastomas, which also express high levels of pp60$^{c\text{-}src}$, may arise from this tissue. It was recently shown that pp60$^{c\text{-}src}$ kinase activity is present within adrenal medullary chromaffin cells and that it localizes to the membrane of the secretory vesicles (Parsons and Creutz 1986). Furthermore, soluble proteins called chromobindins (Creutz et al. 1983) bind to the chromaffin granule membrane in the presence of Ca^{2+}. One such protein has been shown to be phosphorylated on tyrosine and antigenically related to the 36 K substrate of pp60$^{v\text{-}src}$ (S. Parsons, personal communication). Clearly, further studies in the chromaffin cell system are needed to establish the molecular and ionic bases of neurotransmitter secretion and the possible role of c-*src* in this process.

Another site of high pp60$^{c\text{-}src}$ expression within adult tissue occurs in the plasma membrane of blood platelets (Golden et al. 1986). This activity in the nonproliferating platelets may be either a remnant of their genesis from megakaryocytes (which also possess elevated pp60^{src} kinase activity), a reflection of a secretory process as in the chromaffin cells, or part of a mechanism that transduces or responds to extracellular signals across the membrane. The tyrosine kinase domain of pp60$^{c\text{-}src}$ on the cytoplasmic face of the plasma membrane could represent a more mobile form of related domains found in several other growth factor receptors.

In summary, a number of potential functions for pp60$^{c\text{-}src}$ can be proposed based on the pattern of expression of this protein during embryogenesis and in adult tissues. Although no clear function can yet be defined, pp60$^{c\text{-}src}$ may have a role in cell shape changes, cell migrations within the embryo, inductive signaling of differentiation, neurotransmitter release, potential regulation of ion channel activity, and participation in growth factor receptor function.

The tissue-specific expression of pp60$^{c\text{-}src}$ suggests that it induces different cellular responses in different cell types (e.g., neurons vs. neuroectoderm cells). To reconcile these various responses, a single function for pp60$^{c\text{-}src}$ could be proposed which induces different responses depending upon the cell type involved. An alternative possibility is that different forms of c-*src* gene products exist which carry out different functions. Three different c-*src* mRNAs have indeed been detected in *Drosophila* (Simon et al. 1983). Yet another possibility is that pp60$^{c\text{-}src}$ acts together with different proto-oncogene products to induce different phenotypic effects. These possibilities need not be mutually exclusive. Considerably more research will be necessary to refine our present concepts.

References

Alemà S, Casalbore P, Agostini E, Tato F (1985) Differentiation of PC12 phaeochromocytoma cells induced by v-*src* oncogene. Nature 316:557–559

Barnekow A, Schartl M, Anders F, Bauer H (1982) Identification of a fish protein associated with a kinase activity and related to the Rous sarcoma virus transforming protein. Cancer Res 42:2429–2433

Bolen JB, Rosen N, Israel MA (1985) Increased pp60$^{c\text{-}src}$ tyrosyl kinase activity in human neuroblastomas is associated with amino-terminal tyrosine phosphorylation of the *src* gene product. Proc Natl Acad Sci USA 82:7275–7279

Brugge JS, Cotton PC, Queral AE, Barrett JN, Nonner D, Keane RW (1985) Neurones express high levels of a structurally modified, activated form of pp60$^{c\text{-}src}$. Nature 316:554–557

Collett MS, Brugge JS, Erikson RL (1978) Characterization of a normal avian cell protein related to the avian sarcoma virus-transforming gene product. Cell 15:1363–1369

Cotton PC, Brugge JS (1983) Neural tissues express high levels of the cellular *src* gene product pp60$^{c\text{-}src}$. Mol Cell Biol 3:1157–1162

Creutz CE, Dowling LG, Sando JJ, Villar-Palasi C, Whipple JH, Zaks WJ (1983) Characterization of the chromobindins. J Biol Chem 258:14664–14674

Fults DW, Towle AC, Lauder JM, Maness PF (1985) pp60$^{c\text{-}src}$ in the developing cerebellum. Mol Cell Biol 5:27–32

Golden A, Nemeth SP, Brugge JS (1986) Blood platelets express high levels of the pp60$^{c\text{-}src}$-specific tyrosine kinase activity. Proc Natl Acad Sci USA 83:852–856

Harris WA (1981) Neural activity and development. Annu Rev Physiol 43:689–691

Jacobs C, Ruebsamen H (1983) Expression of pp60$^{c\text{-}src}$ protein kinase in adult and fetal human tissue: high activities in some sarcomas and mammary carcinomas. Cancer Res 43:1696–1702

Lev Z, Leibowitz N, Segev O, Shilo BZ (1984) Expression of the *src* and *abl* oncogenes during development of *Drosophila melanogaster*. Mol Cell Biol 4:982–984

Levy BT, Sorge LK, Meymandi A, Maness PF (1984) pp60$^{c\text{-}src}$ kinase is in chick and human embryonic tissues. Dev Biol 104:9–17

Livett BG (1984) Adrenal medullary chromaffin cells in vitro. Physiol Rev 64:1103–1161

Maness PF, Fults DW (1985) Immunocytochemical mapping of pp60$^{c\text{-}src}$ in the developing nervous system. Cancer Cells 3:425–432

Maness PF, Sorge LK, Fults DW (1986) An early developmental phase of pp60$^{c\text{-}src}$ expression in the neural ectoderm. Dev Biol (in press)

Oppermann H, Levinson A, Varmus H, Levintow L, Bishop JM (1979) Uninfected vertebrate cells contain a protein that is closely related to the product of the avian sarcoma virus transforming gene (*src*). Proc Natl Acad Sci USA 76:1804–1808

Parsons SJ, Creutz CE (1986) p60$^{c\text{-}src}$ activity detected in the chromaffin granule membrane. BBRC (in press)

Ralston R, Bishop JM (1985) The product of the proto-oncogene c-*src* is modified during the cellular response to platelet-derived growth factor. Proc Natl Acad Sci USA 82:7845 – 7849

Rohrschneider LR, Eisenman RN, Leitch CR (1979) Identification of a Rous sarcoma virus transformation-related protein in normal avian and mammalian cells. Proc Natl Acad Sci USA 76:4479 – 4483

Schartl M, Barnekow A (1982) The expression in eukaryotes of a tyrosine kinase which is reactive with pp60$^{v\text{-}src}$ antibodies. Differentiation 23:109 – 114

Simon MA, Kornberg TB, Bishop JM (1983) Three loci related to the *src* oncogene and tyrosine-specific kinase activity in *Drosophila*. Nature 302:837 – 839

Simon MA, Drees B, Kornberg T, Bishop JM (1985) The nucleotide sequence and the tissue-specific expression of *Drosophila* c-*src*. Cell 42:831 – 840

Sorge LK, Levy BT, Maness PF (1984) pp60$^{c\text{-}src}$ is developmentally regulated in the neural retina. Cell 36:249 – 257

II Growth Factors, Receptors, and Related Oncogenes

This section begins with several articles which describe the molecular structure and biological function of some of the better-characterized growth factors and receptors, as well as their involvement in the transformed phenotype, particularly in the autocrine stimulation of tumor cell growth. The articles cover the following systems: the murine and human hematopoietic colony-stimulating factors (CSFs); the T-cell growth factor, or interleukin-2 (IL-2); the platelet-derived growth factor (PDGF) and its oncogenic counterpart, the v-*sis* oncogene; the transforming growth factors (TGFs) α and β; and the epidermal growth factor (EGF).

Sequence comparisons between various growth factor and receptor genes have yielded important insights into their potential function and have raised intriguing ideas about their evolution. The structural similarity of the predicted EGF precursor molecule to growth factor receptors raises the possibility that growth factors and receptors have evolved from common ancestors which were involved in cell-cell recognition in primitive multicellular organisms by virtue of being anchored in the plasma membrane. Two homeotic-type genes which were recently found to resemble the EGF precursor at the protein level in their predicted overall structure could represent this type of ancestral molecule. Furthermore, the finding that the receptor-like EGF precursor gene may not be processed in certain tissues could mean that it serves a function distinct from that of mature EGF.

A variety of growth factor receptors and oncogenes contain a domain which encodes a tyrosine kinase activity and is believed to play a crucial role in the action of their gene products. The EGF receptor is discussed as a prototype for this class of receptor, with emphasis on how a signal initiated by the binding of growth factor to a transmembrane receptor is transduced across the plasma membrane and then into the interior of the cell. The mechanisms which can activate the transforming potential of receptors and the ways in which activated receptor molecules transform cells are discussed for the EGF and CSF-1 receptors (the products of the c-*erbB* and c-*fms* genes, respectively), as well as for other related proto-oncogenes whose normal roles have not been identified. These examples illustrate the diversity of mechanisms which can generate transforming genes: in the cases of the c-*erbB*, c-*fms*, c-*abl* and c-*src* tyrosine kinase encoding genes mutations and/or deletions lead to critical changes in the coding sequence, while the serine/threonine kinase-encoding c-*mos* gene can be activated by overexpression of the gene product. Another class of receptors, for which the IL-2 receptor serves here as an example, does not encode a kinase domain and therefore must transduce signals by a different mechanism. One possibility is that the IL-2 receptor associates with an as-yet unidentified intracellular protein which mediates signal transduction. The products of the tyrosine kinase-encoding proto-oncogenes which associate with the inner plasma membrane are provocative candidates for accessory proteins that may interact with the IL-2 receptor.

The Granulocyte-Macrophage Colony-Stimulating Factors

Nicholas M. Gough

Haemopoiesis is the process whereby a small population of multipotential stem cells continuously gives rise to a large number of mature blood cells which comprise eight distinct cellular lineages. In normal health, the circulating levels of mature cells are remarkably invariant, suggesting that their production is tightly regulated. However, the haemopoietic system is also flexible, allowing fluctuations in the levels of various cell types to meet emergency situations such as blood loss, infection or reduced oxygen tension. Some of the mechanisms controlling haemopoiesis, particularly those concerned with stem cell populations, appear to involve contact between haemopoietic cells and other cells in the micro-environment at the sites of blood cell formation (e.g. Allen 1981). However, the ability to grow colonies of mature haemopoietic cells from single progenitor cells in semi-solid culture systems has also implicated a number of soluble glycoprotein growth factors. These factors, known as colony-stimulating factors (CSFs), have been shown in vitro to stimulate the proliferation, differentiation and functional activation of cells within different haemopoietic lineages (Metcalf 1984).

Murine Colony-Stimulating Factors

For the murine granulocyte and macrophage population, which is of particular interest since two quite different cell types originate from a common bipotential progenitor cell, four different factors that stimulate colony formation in vitro have been identified (Metcalf 1984) (Table 1). Each of these factors displays a different spectrum of activities. Macrophage-CSF (M-CSF) has the most restricted activity, stimulating almost exclusively the formation of macrophage colonies. At low concentrations granulocyte-CSF (G-CSF) stimulates the formation of exclusively granulocytic colonies, but can at higher concentrations also stimulate the formation of mixed granulocyte-macrophage colonies (the progeny of bipotential progenitors) and pure macrophage colonies. Granulocyte-macrophage-CSF (GM-CSF) stimulates the formation of granulocyte, macrophage and mixed granulocyte-macrophage colonies and, in cultures of foetal liver progenitor cells, eosinophil colonies. Multi-CSF also stimulates the formation of granulocyte, macrophage and mixed granulocyte-macrophage colonies. However, Multi-CSF also stimulates the growth and differentiation of multipotential stem cells and of progenitor cells committed

Oncogenes and Growth Control
Edited by P. Kahn and T. Graf
© Springer-Verlag Berlin Heidelberg 1986

Table 1. Murine granulocyte-macrophage colony-stimulating factors

Name	Synonyms	Mature cells produced in vitro	Cellular sources	Molecular mass (kDa)
Multi-CSF	Interleukin-3 Burst-promoting activity Haemopoietic cell growth factor Mast cell growth factor P cell-stimulating factor Haemopoietin-2 CSF-2α	Granulocytes Macrophages Eosinophils Erythrocytes Megakaryocytes Mast cells	T lymphocytes WEHI-3B myelomonocytic leukaemia	19 to 29
GM-CSF	MGI-1G CSF-2	Granulocytes Macrophages Eosinophils	T lymphocytes Endothelial cells Fibroblasts WEHI-274 monocytic leukaemia Krebs II ascites cells	23
G-CSF	MGI-2 Differentiation factor	Granulocytes (Macrophages)	Monocytes/macrophages RIII-T3 mammary tumor Krebs II ascites cells	25
M-CSF	MGI-1M CSF-1	Macrophages (Granulocytes)	Fibroblasts	70

Multi-CSF, GM-CSF and G-CSF are monomers whereas M-CSF is a disulfide-bonded dimer; all four CSFs are glycoproteins (see Kelso and Gough 1986 for references)

to all of the non-lymphoid lineages, including the erythroid, mast, eosinophil and megakaryocyte lineages.

All of the factors listed in Table 1 have been extensively purified and their biochemical distinction both from each other and from a number of other growth factors is well established. Molecular clones encoding murine GM-CSF and Multi-CSF have been obtained (Fung et al. 1984; Gough et al. 1984; Yokota et al. 1984; Dunn et al. 1985), and clonally pure factor produced (Delamarter et al. 1985; Dunn et al. 1985; Gough et al. 1985; Hapel et al. 1985). All of the biological activities previously ascribed to highly purified preparations of these factors are displayed in the recombinant material, demonstrating that all of these activities are indeed intrinsic to single gene products.

Human Colony-Stimulating Factors

Until recently, detailed information regarding human haemopoietic growth factors was rather limited. Certainly human bone marrow progenitor cells can be stimulated to form colonies of mature haemopoietic cells in agar cultures, but the number of different human CSFs influencing this process and their range of activities is unclear (Metcalf 1984). The inadequate sources of these factors has until recently prevented their purification and the assays for human CSFs have been much less definitive than for their murine counterparts. However, human cDNA clones encoding factors which are analogues of murine GM-CSF, M-CSF and G-CSF have recently been obtained (Wong et al. 1985; Kawasaki et al. 1985; Nagata et al. 1986). Nucleotide sequence analysis of these clones indicates that the human factors bear significant amino acid sequence homology to their murine counterparts (Ben-Avram et al. 1985; Gough et al. 1985; Kawasaki et al. 1985; Wong et al. 1985; Nagata et al. 1986; N. A. Nicola, R. J. Simpson and E. C. Nice, personal communication). However, the identification of a human analogue of murine Multi-CSF, by biological, biochemical or molecular biological techniques has so far proven unsuccessful.

Unique Structures of the CSFs

Since the granulocyte-macrophage CSFs display a range of overlapping activities, and since three of the factors (GM-, Multi- and G-CSF) are similar to one another in that they are all glycoproteins of around 25 kDa, it has been speculated that these factors might represent a multigene family (e.g. Sachs 1982; Staber et al. 1982; Krammar et al. 1983) perhaps analogous to the insulin and glucagon families of pancreatic hormones. However, the molecular cloning of cDNAs and genes encoding these factors has revealed that this is not the case. The complete amino acid sequences of murine and human GM-CSF, murine Multi-CSF and human G-CSF and M-CSF have been deduced from the nucleotide sequences of cloned cDNAs (Fung et al. 1984; Gough et al. 1984,

1985; Yokota et al. 1984; Wong et al. 1985; Nagata et al. 1986; Kawasaki et al. 1985). Despite the observation that these factors display similar biological activities within the granulocyte-macrophage lineages, their primary sequences show no statistically significant similarities, suggesting that the genes encoding these factors are evolutionarily unrelated and therefore do not constitute a multigene family.

Furthermore, despite the considerable biochemical heterogeneity that has been observed for each of these factors (e.g. Nicola et al. 1979), none is a member of independent multigene families, since Southern blot hybridization experiments for murine GM- and Multi-CSF and human GM-, G- and M-CSF indicate that all of these factors are encoded by unique genes and that there are no other closely related genes in the murine or human genomes. Thus, the biochemically distinct forms of these factors produced by different sources probably result from modifications that occur post-translationally.

CSF Receptors

Receptor-binding studies have revealed that each of the four CSFs binds to a unique receptor (Walker et al. 1985). Thus, if the action of the CSF is mediated wholly by the receptor to which it binds, it need not be surprising that CSFs with different structures can elicit very similar biological effects. The biological specificities of each CSF could then be explained on the basis of the distribution of the cognate receptor on CSF-responsive cells.

All four receptors show complete specificity in that they do not bind other growth factors. Moreover, no CSF competes *directly* for binding by another CSF at 0 °C. However, binding experiments performed at 37 °C have revealed a hierarchical ability of CSF-receptor complexes to down-modulate other, unrelated CSF receptors (Walker et al. 1985). The specificity of the down-modulation pattern has led to the suggestion that the spectrum of haemopoietic activities displayed by each of the four CSFs might be determined by their ability either to occupy or to down-modulate lineage-specific receptors (Walker et al. 1985).

Control of CSF Production

The cell types that synthesize the CSFs in vivo and in vitro have generally been difficult to identify, in part because extracts and conditioned media from all organs and tissues contain some colony-stimulating activity. Based largely on studies with cultured cell lines, tumours and hybridomas, a minimal list of CSF-producing cells can be compiled (Table 1). The ability of monocytes and macrophages, fibroblasts, lymphocytes and endothelial cells to synthesize one or more CSF may account for the ubiquity of these factors in freshly excised tissues, but it is clear that definitive identification of the producing cells in vivo must await application of techniques for detecting CSF mRNA and protein in situ. Of the factors listed in Table 1, the regulation of production of GM-CSF and Multi-CSF in T lymphocytes has been most thoroughly studied.

The observation that some monoclonal T-cell populations can be induced to secrete several different lymphokines and haemopoietic growth factors concomitantly suggests that the expression of these factors may be coordinately regulated and that lectins or antigen may induce the synthesis of different lymphokines by a common pathway (for review, see Kelso and Gough 1986). In several gene systems, short DNA segments upstream to the transcriptional start site have been implicated in the induction or tissue-specific expression of the corresponding gene (Davidson et al. 1983). Based on sequence homology, candidate sequences which may play a role in coordinate activation of these genes have been identified within the $5'$-flanking regions of the murine GM-CSF, Multi-CSF and interleukin-2 genes and human interleukin-2 and γ-interferon genes (Stanley et al. 1985). However, as discussed in detail elsewhere (Kelso et al. 1986; Kelso and Gough 1986), the production of GM-CSF and Multi-CSF by T-cell clones is not obligatorily linked, but can in fact be dissociated at several levels. First, the appearance of secreted GM-CSF and cytoplasmic GM-CSF mRNA precedes that of Multi-CSF after concanavalin A stimulation of a T-lymphocyte clone. Second, in certain clones which secrete both GM-CSF and Multi-CSF when cultured with concanavalin A or the relevant antigen, stimulation with interleukin-2 preferentially induces the production of GM-CSF. Third, unrelated clones exhibit marked variation in total CSF production and in the relative amounts of GM-CSF and Multi-CSF produced, whereas sibling clones show much smaller variations in both parameters. Fourth, GM-CSF can be produced by a number of cell types that do not produce Multi-CSF (Table 1). In these cell types, the inductive signals differ from those described for T lymphocytes. For example, whereas T cells can be stimulated to synthesize GM-CSF and Multi-CSF by interleukin-2 or by antigen (or lectins which mimic the action of antigen), synthesis of GM-CSF (but not of Multi-CSF) is induced in Krebs ascites cells by lipopolysaccharide (Metcalf and Nicola 1985) and possibly in fibroblasts by diterpene esters (Koury et al. 1983) and retroviruses (Koury and Pragnell 1982). Thus, although a common element in the DNA adjacent to the GM-CSF and Multi-CSF genes may be involved in mediating lectin- or antigen-mediated induction in a T-lymphocyte, there must be additional factors which regulate the levels to which these genes are expressed, the cell types in which they are expressed and the signals by which they are induced.

Autocrine CSF Production and Leukaemogenesis

The autocrine model is one of the current concepts concerning the origin and nature of cancer and envisages that an affected cell produces an inappropriate level of a growth factor to which it can respond, thus leading to sustained, self-stimulated proliferation (for review see Sporn and Roberts 1985 and Heldin and Westermark; Derynck; Kahn et al., all this Vol.). Proliferation and terminal differentiation are tightly coupled in normal haemopoietic cells (Metcalf 1984) and therefore autonomous production of CSFs by otherwise normal haemopoietic cells could not in itself lead to the emergence of a leukaemic population. However, leukaemogenicity could ensue if autocrine stimulation were a secondary event occurring in a pre-leukaemic clone that had already incurred a lesion causing a differentiation block. How such a differentiation block might be achieved is unclear. However, continuous cell lines (generally mast cells, but occasionally granulocyte or multipotent) can be derived from apparently normal long-term murine bone marrow cultures. Such cell lines are absolutely dependent for growth upon Multi-CSF, or sometimes GM-CSF, and are invariably non-leukaemogenic in syngeneic recipients (Metcalf 1984). However, they can be rendered CSF-independent and leukaemogenic by spontaneous mutation (Schrader and Crapper 1983), by transformation with Abelson murine leukaemia virus (Cook et al. 1985) or by infection with a recombinant retrovirus encoding GM-CSF (Lang et al. 1985). In the latter case, GM-CSF is synthesized and secreted by the cells and it is likely that autocrine stimulation is involved in leukaemogenesis. Whereas the retrovirus-borne GM-CSF gene could be viewed as a second oncogene required to transform cells from an immortalized to a leukaemic state, another interpretation is that the immortalized parental cells are already fully transformed but fail to grow as leukaemias upon transplantation simply because there is insufficient GM-CSF or Multi-CSF at the site of injection. Interestingly, the A-MuLV transformed, factor-independent cells (Cook et al. 1985) synthesize neither GM-CSF nor Multi-CSF, and it has been suggested that the v-*abl* gene product is able to circumvent the requirement for CSF by precipitating those intracellular events normally induced by the CSF-receptor complex.

The correlation between leukaemogenicity and factor independence in haemopoietic cells contrasts with the observation that both murine and human primary myeloid leukaemias are invariably dependent upon exogenous CSF for their proliferation in vitro (Metcalf 1984). However, it is probably significant that establishment of primary myeloid leukaemias in continuous culture eventually results in loss of dependence upon exogenous CSF and usually in endogenous synthesis of CSF. This suggests that most primary myeloid leukaemias are able to behave as malignant cells at a relatively early stage in their progression, and that acquisition of factor-independence and continuous in vitro growth represent later stages in the evolution of a leukaemic cell population.

Acknowledgments. I am grateful to Nick Nicola for access to his unpublished G-CSF amino acid sequence, to Ed Stanley for comments on the manuscript and, as always, to Sue Blackford and Joanne Wright for its preparation.

References

Allen TD (1981) Haemopoietic microenvironments *in vitro*: Ultrastructural aspects. In: Porter R, Whelan J (eds) Microenvironments in haemopoietic and lymphoid differentiation. Pitman Medical, London, pp 38 – 67 (Ciba foundation symposium, vol 84)

Ben-Avram CM, Shively JE, Shadduck RK, Waheed A, Rajavashisth T, Lusis AJ (1985) Amino-terminal amino acid sequence of murine colony-stimulating factor 1. Proc Natl Acad Sci USA 82:4486 – 4489

Cook WD, Metcalf D, Nicola NA, Burgess AW, Walker F (1985) Malignant transformation of a growth factor-dependent myeloid cell line by Abelson virus without evidence of an autocrine mechanism. Cell 41:677 – 683

Davidson EH, Jacobs HT, Britten RJ (1983) Very short repeats and co-ordinate induction of genes. Nature 301:468 – 470

Delamarter JF, Mermod J-J, Liang C-M, Eliason JF, Thatcher D (1985) Recombinant murine GM-CSF from *E. coli* has biological activity and is neutralized by a specific antiserum. EMBO J 4:2575 – 2581

Dunn AR, Metcalf D, Stanley E, Grail D, King J, Nice EC, Burgess AW, Gough NM (1985) Biological characterization of regulators encoded by cloned hemopoietic growth factor gene sequences. In: Feramisco J, Ozanne B, Stiles C (eds) Growth factors and transformation. Cold Spring Harbor, New York, pp 227 – 234 (Cancer cells, vol 3)

Fung MC, Hapel AJ, Ymer S, Cohen DR, Johnson RM, Campbell HD, Young IG (1984) Molecular cloning of cDNA for mouse Interleukin-3. Nature 307:233 – 237

Gough NM, Gough J, Metcalf D, Kelso A, Grail D, Nicola NA, Burgess AW, Dunn AR (1984) Molecular cloning of cDNA encoding a murine haemopoietic growth regulator, granulocyte-macrophage colony stimulating factor. Nature 309:763 – 767

Gough NM, Metcalf D, Gough J, Grail D, Dunn AR (1985) Structure and expression of the mRNA for murine GM-CSF. EMBO J 4:645 – 653

Hapel AJ, Fung MC, Johnson RM, Young IG, Johnson G, Metcalf D (1985) Biologic properties of molecularly cloned and expressed murine Interleukin-3. Blood 65:1453 – 1459

Kawasaki ES, Ladner MB, Wang AM, Van Arsdell J, Warren MK, Coyne MY, Schweikart VL, Lee M-T, Wilson KJ, Boosman A, Stanley ER, Ralph P, Mark DF (1985) Molecular cloning of a complementary DNA encoding human macrophage-specific colony-stimulating factor (CSF-1). Science 230:291 – 296

Kelso A, Gough N (1986) Expression of haemopoietic growth-factor genes in murine T lymphocytes. In: Webb DR, Goeddel D (eds) Molecular cloning and analysis of lymphokines. Academic Press, New York (in press) (The lymphokines, vol 13)

Kelso A, Metcalf D, Gough NM (1986) Independent regulation of granulocyte-macrophage colony-stimulating factor and multi-lineage colony-stimulating factor production in T lymphocyte clones. J Immunol 136:1718 – 1725

Koury MJ, Pragnell IB (1982) Retroviruses induce granulocyte-macrophage colony stimulating activity in fibroblasts. Nature 299:638 – 640

Koury MJ, Balmain A, Pragnell IB (1983) Induction of granulocyte-macrophage colony-stimulating activity in mouse skin by inflammatory agents and tumor promoters. EMBO J 2:1877 – 1882

Krammar PH, Echtenacher B, Gemsa D, Hamann V, Hultner L, Kaltman B, Kees U, Kubelka C, Marcucci F (1983) Immune-interferon (IFN-γ), macrophage-activating factors (MAFs) and colony-stimulating factors (CSFs) secreted by T cell clones in limiting dilution microcultures, long-term cultures, and by T cell hybridomas. Immunol Rev 76:5 – 28

Lang RA, Metcalf D, Gough NM, Dunn AR, Gonda TJ (1985) Expression of a haemopoietic growth factor cDNA in a factor-dependent cell line results in autonomous growth and tumorigenicity. Cell 43:531 – 542

Metcalf D (1984) The haemopoietic colony stimulating factors. Elsevier, Amsterdam

Metcalf D, Nicola NA (1985) Role of the colony stimulating factors in the emergence and suppression of myeloid leukemia populations. In: Wahren B, Holm G, Hammarstrom S, Perlmann P (eds) Molecular biology of tumour cells. Raven, New York, pp 215 – 232 (Progress in cancer research and therapy, vol 32)

Nagata S, Tsuchiya M, Asano S, Kaziro Y, Yamazaki T, Yamamoto O, Hirata Y, Kubota N, Oheda M, Nomursa H, Ono M (1986) Molecular cloning and expression of cDNA for human granulocyte colony stimulating factor (G-CSF). Nature 319:415 – 418

Nicola NA, Burgess AW, Metcalf D (1979) Similar molecular properties of granulocyte-macrophage colony-stimulating factors produced by different mouse organs in vitro and in vivo. J Biol Chem 254:5290 – 5299

Sachs L (1982) Control of growth and normal differentiation in leukemic cells: Regulation of the developmental program and restoration of the normal phenotype in myeloid leukemia. J Cell Physiol Suppl 1:151 – 164

Schrader JW, Crapper RM (1983) Autogenous production of a haemopoietic growth factor "P-cell stimulating factor" as a mechanism for transformation of bone marrow-derived cells. Proc Natl Acad Sci USA 80:6892 – 6896

Sporn MB, Roberts AB (1985) Autocrine growth factors and cancer. Nature 313:745 – 747

Staber FG, Hultner L, Marcucci F, Krammer PH (1982) Production of colony-stimulating factors by murine T cells in limiting dilution and long-term cultures. Nature 298:79 – 82

Stanley E, Metcalf D, Sobieszczuk P, Gough NM, Dunn AR (1985) The structure and expression of the murine gene encoding granulocyte-macrophage colony stimulating factor: Evidence for utilization of alternative promoters. EMBO J 4:2569 – 2573

Walker F, Nicola NA, Metcalf D, Burgess AW (1985) Hierarchical down-modulation of hemopoietic growth factor receptors. Cell 43:269 – 275

Wong GG, Witek JS, Temple PA, Wilkens KM, Leary AC, Luxenberg DP, Jones SS, Brown EL, Kay RM, Orr EC, Shoemaker C, Golde DW, Kaufman RJ, Hewick RM, Wang EA, Clark SC (1985) Human GM-CSF: Molecular cloning of the complementary DNA and purification of the natural and recombinant proteins. Science 228:810 – 815

Yokota T, Lee F, Rennick D, Hall C, Arai N, Mosmann T, Nabel G, Cantor H, Arai K (1984) Isolation and characterization of a full-length cDNA for mast cell growth factor from a mouse T-cell clone: Expression in monkey cells. Proc Natl Acad Sci USA 81:1070 – 1074

Role of PDGF-Like Growth Factors in Autocrine Stimulation of Growth of Normal and Transformed Cells

CARL-HENRIK HELDIN and BENGT WESTERMARK

The normal cellular homologs of several retroviral oncogenes have recently been identified as structural genes for proteins which are proven or thought to play a role in mitogenesis. It is therefore likely that the viral oncogene products stimulate uncontrolled cell proliferation by subverting the mitogenic pathway at key regulatory points. An indication that similar mechanisms may operate in nonviral cell transformation comes from the old observation that cell lines established from malignant tumors grow more or less independently of exogenously added growth factors. One possibility, which is supported by recent experimental data, is that transformed cells may produce growth factors that stimulate their own growth in an autocrine manner. This review focuses on platelet-derived growth factor (PDGF) and related factors and their possible role in autocrine and paracrine mechanisms in both normal and transformed cells.

Platelet-Derived Growth Factor and the Simian Sarcoma Virus (SSV) Transforming Gene

PDGF is the major mitogen in serum for connective tissue-derived cells (for a review on PDGF see Heldin et al. 1985). It has a molecular weight of about 30,000 and is composed of two different disulfide-bonded polypeptide chains, denoted A and B. The observation that the biological activity of PDGF is irreversibly lost after reduction indicates that the dimer structure is functionally important. It is not known, however, whether PDGF is a heterodimer or a mixture of homodimers. PDGF exerts its mitogenic effect via interaction with a high affinity cell surface receptor. The receptor is a transmembrane glycoprotein of 185 kDa and contains in its cytoplasmic domain a tyrosine kinase which is activated upon ligand binding. The mechanism whereby the mitogenic signal is transmitted within the cell is largely unknown; intracellular signals that might be involved include tyrosine phosphorylation of cytoplasmic proteins, elevation of cytoplasmic Ca^{2+}-concentration, activation of protein kinase C, and induction of specific genes.

Analysis of the amino acid sequence of PDGF revealed that the B chain is almost identical to a portion of $p28^{v\text{-}sis}$, the transforming protein of simian sarcoma virus (SSV) (Waterfield et al. 1983; Doolittle et al. 1983); over a stretch of 109 amino acids, only three substitutions were found, which may re-

Oncogenes and Growth Control
Edited by P. Kahn and T. Graf
© Springer-Verlag Berlin Heidelberg 1986

present the species difference between monkey and man (Johnsson et al. 1984). The A chain of PDGF shows a 60% homology to the B chain.

The finding of a structural homology between the v-*sis* product and PDGF strongly suggested that SSV has acquired the gene for one of the two polypeptide chains of PDGF. Cloning of human c-*sis*, the cellular counterpart of v-*sis*, later verified this assumption (Josephs et al. 1984a; Johnsson et al. 1984; Chiu et al. 1984). The human c-*sis* transcript has a size of 3.5–4.2 kb and contains an open reading frame for a 28 kDa precursor of the B chain of PDGF, but does not code for the A chain (Collins et al. 1985). This structural link between PDGF and a transforming protein implicated a functional similarity, i.e., that the transforming capacity of SSV is due to a PDGF-like growth factor activity. This interpretation is consistent with the fact that SSV induces sarcomas and gliomas in vivo (Deinhardt 1980), i.e., transforms cell types that contain PDGF receptors. Further support came from the subsequent finding by several groups that SSV-transformed cells produce a growth factor which binds to and activates the PDGF receptor (see references in Heldin et al. 1986a).

SSV-Transformation is Mediated by an Externalized PDGF Agonist

The primary translation product of v-*sis*, p28$^{\text{v-}sis}$, is rapidly dimerized after synthesis and then proteolytically cleaved both in the N-terminus and in the C-terminus (Robbins et al. 1983). The apparently stable end product has a molecular weight of 24,000 and is structurally similar to a PDGF B chain homodimer. Evidence that this is the biologically active molecule comes from experiments in which the v-*sis* gene was systematically modified. The N- and C-terminal sequences that are cleaved off during maturation of the product were found to be dispensable for the activity, whereas, as expected, the transforming activity was lost when the middle portion, which codes for the mature protein, was deleted (King et al. 1985). Furthermore, a dimeric form of a 109 amino acid long v-*sis* peptide, corresponding to the mature PDGF B chain, showed PDGF agonist activity when expressed in yeast (Kelly et al. 1985). Interestingly, the transforming activity of v-*sis* was lost when the 5′ flanking helper virus-derived *env* sequence was deleted (Hannink and Donoghue 1984; King et al. 1985). This latter sequence encodes a hydrophobic stretch of amino acids which probably serves as a signal sequence that allows the transport of the gene product into secretory vesicles. It therefore appears that the v-*sis* product has to be compartmentalized with the ligand-binding domain of the PDGF receptor in order to be active, which is compatible with the suggested autocrine mechanism of action of the transforming product. Interestingly, the 5′ region of the normal counterpart, c-*sis*, also encodes a signal sequence (Josephs et al. 1984b) that is required for transforming activity when the gene is transfected into NIH 3T3 cells (Gazit et al. 1984).

The finding that the v-*sis* product has to be externalized in order to be active implies that exogenously added neutralizing antibodies should block its activity. Acutely SSV-transformed human fibroblasts were used for such experiments, since these cells have not undergone such extensive secondary alterations as established cell lines. Addition of anti-PDGF IgG to SSV-transformed fibroblasts led to a reversion of the transformed morphology and to a decrease in the rate of cell proliferation (Johnsson et al. 1985). The functional similarities between the v-*sis* product and PDGF, and the similarities in morphology and growth behavior between SSV-transformed cells and PDGF-stimulated cells, indicate that SSV provides an efficient growth stimulus to cells in vitro, but that the mitogenic signal does not show qualitative differences compared to that obtained via exogenously added PDGF. It is notable, however, that SSV induces malignant tumors, i.e., fibrosarcomas and glioblastomas in newborn marmosets (Deinhardt 1980), a property which cannot be explained by an efficient growth stimulus only. It is possible that SSV can induce a truly malignant phenotype in certain highly susceptible target cells at certain stages of development or that a constitutive growth signal may contribute to malignant transformation by significantly increasing the risk of additional genetic alterations.

PDGF-Like Growth Factors from Cell Lines of Human Tumors

Cell lines derived from tumors of a large variety of histogenetic origins produce growth factors that bind to the PDGF receptor (reviewed in Heldin et al. 1986a). One example is U2 OS, a human osteosarcoma that secretes a growth factor, osteosarcoma-derived growth factor (ODGF), with structural, immunological, and functional properties in common with PDGF (Heldin et al. 1980). A component of 31 kDa which is converted to a doublet of 16.5–17 kDa after reduction, can be immunoprecipitated from the conditioned medium of metabolically labelled U2 OS cells with an antiserum against PDGF (Betsholtz et al. 1983). U2 OS cells have a low number of functional PDGF receptors and there is evidence that the endogenously produced ODGF causes autocrine receptor activation (Betsholtz et al. 1984). However, neutralizing antibodies added to U2 OS cells had no effect on the growth rate of the cells. This indicates either that the production of ODGF is of no importance for stimulation of DNA synthesis in these cells, despite the demonstrated autocrine receptor activation, or alternatively, that ODGF interacts with newly synthesized receptors present in secretory vesicles before their insertion in the cell membrane. It is also possible that the endogenous production of ODGF was important in the early stages of tumorigenesis, but that further growth of the tumor cells in vivo or in cell culture led to additional alterations which rendered them independent of the autocrine growth stimulus.

The osteosarcoma-derived growth factor was recently purified to homogeneity from U2 OS cell-conditioned medium (Heldin et al. 1986b). Analysis of the purified product revealed that it is a homodimer of PDGF A chains. Apparently homodimers of either of the two polypeptide chains of PDGF can bind to and activate the PDGF receptor (A-A in the case of ODGF and B-B in the case of the v-*sis* gene product). Interestingly, pig PDGF has recently been found to contain only B chain sequences and is therefore likely to be a B chain homodimer (Stroobant and Waterfield 1984). However, human PDGF may contain heterodimers, a possibility which is supported by the observations that purified PDGF has stoichiometric amounts of A and B chains, and that it has not been possible to separate homodimers from purified human PDGF (unpublished observations).

The human glioma cell line, U343 MGa Cl 2, also produces a PDGF-like growth factor (Nistér et al. 1984). Furthermore, clones of this cell line show co-variation in the amount of factor they produce and the ability to grow in serum-free medium, providing indirect evidence for an autocrine mechanism (Nistér et al. 1986).

Using mRNA from a clone of the glioma cell line with a high production of PDGF-like growth factor, cDNA for the A chain of PDGF was cloned and sequenced (Betsholtz et al. 1986). The A and B chains of PDGF show extensive homology. The A chain is synthesized as a 211 amino acid precursor with an N-terminal 20 amino acid hydrophobic signal sequence, indicating that, like the B chain, it is secreted from the cell. The precursor is proteolytically cleaved in the N-terminus after a stretch of four basic amino acids, yielding a 125 amino acid long peptide which may undergo additional processing in the C-terminus. The A and B chains show a perfect conservation of all eight cysteine residues, indicating that the chains have similar three-dimensional structures. The amino acid homology is highest (60%) in the middle portions of the two precursors, which corresponds to the mature products (Fig. 1) (Betsholtz et al. 1986).

An intriguing, and yet unresolved, question is why two different PDGF chains have evolved. They are functionally conserved in the sense that homodimers of either one interacts with the PDGF receptor. It remains to be determined whether they possess also functional differences.

The A chain probe hybridizes to three major transcripts (1.9, 2.3 and 2.8 kB) in U343 MGa Cl 2 cells (Betsholtz et al. 1986). Expression of these transcripts is not uncommon among human tumor cells and is independent of B chain mRNA expression. The expression of the A chain transcript correlates perfectly with the secretion of a 31 kDa component which after reduction is converted to 17 kDa (Betsholtz et al. 1986), indicating that all secreted PDGF-like factors in these cell lines are A chain homodimers. This does not rule out the possibility that the B chain is also produced in cells expressing the B chain transcript. Although there is clear evidence that the v-*sis* product is secreted (Johnsson et al. 1985), there are indications that the product remains associated with the cell membrane after externalization and therefore is not seen in the conditioned medium (Robbins et al. 1985)

Production of PDGF-like growth factors by cells carrying PDGF receptors may thus lead to an autocrine loop that causes a constitutive growth signal.

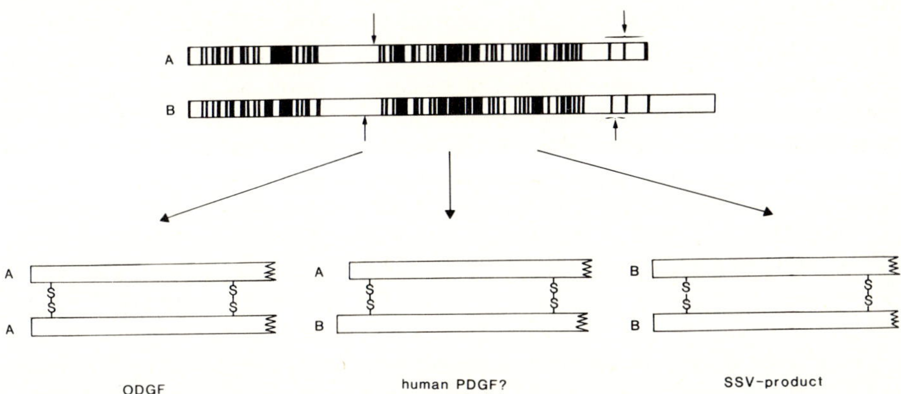

Fig. 1. Schematic illustration of the amino acid homology between the *A* and *B* chain precursors of PDGF, their processing and assembly into heterodimers and homodimers. *Filled areas* represent identical amino acid sequence. Processing sites are indicated by *arrows*; in the N-termini these sites are known, in the C-termini the likely regions where processing occurs are indicated. It should be noted that the exact localization of the disulfide bonds in the dimers are not known. Data are taken from references mentioned in the text

On the other hand, PDGF-like growth factors are also produced by several cell types which do not express detectable levels of PDGF receptors, e.g., a hepatoma cell line (Bowen-Pope et al. 1984), a breast carcinoma cell line (Rozengurt et al. 1985) and a neuroblastoma cell line (van Zoelen et al. 1985), and in these cases probably do not induce autocrine growth. Since many tumors in vivo are surrounded and invaded by proliferating stromal cells, it is an interesting possibility that PDGF-like growth factors produced and released by tumor cells in vivo cause proliferation of connective tissue cells.

PDGF-Like Growth Factors from Normal Cells

Recent studies have indicated that autocrine and paracrine mechanisms may operate not only in cell transformation but also in the stimulation of normal, controlled cell growth under certain conditions.

One example is cytotrophoblasts of the early human placenta, which express a c-*sis* transcript. The c-*sis* expression, as determined by in situ hybridization, correlates with the expression of c-*myc* (Goustin et al. 1985). Furthermore, cultured cytotrophoblasts secrete a PDGF-like growth factor, have PDGF receptors, and respond to PDGF stimulation by showing increased c-*myc* expression. These observations suggest that PDGF may stimulate placental growth during a defined phase of embryogenesis.

PDGF-like growth factors are also produced by smooth muscle cells from newborn rats, but not from adult rats (Seifert et al. 1984), and by cultured rat smooth muscle cells during the transition from a contractile to a synthetic phase (Nilsson et al. 1985). Since smooth muscle cells are responsive to PDGF, the endogenous production of a PDGF-like growth factor may be of significance in autocrine stimulation of smooth muscle cell growth in the vessel wall, e.g., in conjunction with certain stages of the development or in response to certain stimuli.

Human endothelial cells and activated macrophages have been known for some time to produce growth factors. Substantial portions of these growth factor activities seem to be accounted for by PDGF-like growth factors, which is consistent with the fact that both cell types express c-*sis* (Barrett et al. 1984; Shimokado et al. 1985; Martinet et al. 1985). Neither of these cell types respond mitogenically to PDGF, but the growth factor production may be significant in paracrine stimulation of neighboring PDGF responsive cells. Thus, the PDGF-like growth factors released by activated macrophages and endothelial cells may be involved in the connective tissue cell proliferation that accompanies many chronic inflammatory processes, or that occurs in conjunction with injuries to the vessel wall.

As tumor cell-derived homodimers of either of the two chains of PDGF are mitogenically active, and authentic human PDGF might be a heterodimer; it will be of particular interest to determine the subunit structure of the PDGF-like growth factors that are produced by these normal cells. Is there a coordinated expression of the two chains and a formation of heterodimers, or is there a preferential production of either of the homodimers? It is an interesting possibility, which remains to be elucidated, that the various dimers have different functions due to differences in their secretory behavior (see above) and possibly also in their intrinsic activities.

Concluding Remarks

The phenotypic characteristics of SSV-transformed cells agree with a simple model of SSV-transformation which postulates an autocrine stimulation brought about by a growth factor that mimicks PDGF in its functions but has no additional activity. In this context it is pertinent to ask whether there is any functional difference between autocrine growth of transformed cells and normal cells that also secrete PDGF-like growth factors and express functional PDGF receptors. First, it should be noted that, whereas a powerful growth stimulus provided by an endogenous production of growth factors may be important in cell transformation, the acquisition of the fully malignant phenotype involving immortalization, capacity for invasiveness, and formation of metastases clearly require additional genetic alterations. Second, it is likely that autocrine loops in normal cells are subjected to control, e.g., on the level of expression of the genes for the growth factor or its receptor; these genes may be expressed only transiently in normal cells, during defined stages of the development, or as a response to specific stimuli to the cell.

References

Barrett TB, Gajdusek C, Schwartz SM, McDougall JK, Benditt EP (1984) Expression of the *sis* gene by endothelial cells in culture and in vitro. Proc Natl Acad Sci USA 81:6772–6774

Betsholtz C, Heldin C-H, Nistér M, Ek B, Wasteson Å, Westermark B (1983) Synthesis of a PDGF-like growth factor in human glioma and sarcoma cells suggests the expression of the cellular homologue of the transforming protein of simian sarcoma virus. Biochem Biophys Res Commun 117:176–182

Betsholtz C, Westermark B, Ek B, Heldin C-H (1984) Co-expression of a PDGF-like growth factor and PDGF receptors in a human osteosarcoma cell line: implications for autocrine receptor activation. Cell 39:447–457

Betsholtz C, Johnsson A, Heldin C-H, Westermark B, Lind P, Urdea MS, Eddy R, Shows TB, Philpott K, Mellor A, Knott TJ, Scott J (1986) cDNA sequence and chromosomal localization of human platelet-derived growth factor A chain and its expression in tumor cell lines. Nature 320:695–699

Bowen-Pope DF, Vogel A, Ross R (1984) Production of platelet-derived growth factor-like molecules and reduced expression of platelet-derived growth factor receptors accompany transformation by a wide spectrum of agents. Proc Natl Acad Sci USA 81:2396–2400

Chiu I-M, Reddy EP, Givol D, Robbins KC, Tronick SR, Aaronson SA (1984) Nucleotide sequence analysis identifies the human c-*sis* proto-oncogene as a structural gene for platelet-derived growth factor. Cell 37:123–129

Collins T, Ginsburg D, Boss JM, Orkin SH, Pober JS (1985) Cultured human endothelial cells express platelet-derived growth factor B chains: cDNA cloning and structural analysis. Nature 316:748–750

Deinhardt F (1980) Biology of primate retroviruses. In: Klein G (ed) Viral oncology. Raven, New York, pp 357–398

Doolittle RF, Hunkapiller MW, Hood LE, Devare SG, Robbins KC, Aaronson SA, Antoniades HN (1983) Simian sarcoma virus *onc*-gene, v-*sis*, is derived from the gene (or genes) encoding a platelet-derived growth factor. Science 221:275–277

Gazit A, Igarashi H, Chiu I-M, Srinivasan A, Yaniv A, Tronick SR, Robbins KC, Aaronson SA (1984) Expression of the normal human *sis*/PDGF-2 coding sequence induces cellular transformation. Cell 39:80–97

Goustin AS, Betsholtz C, Pfeiffer-Ohlsson S, Persson H, Rydnert J, Bywater M, Holmgren G, Heldin C-H, Westermark B, Ohlsson R (1985) Co-expression of the *sis* and *myc* protooncogenes in human placenta suggests autocrine control of trophoblast growth. Cell 41:301–312

Hannink M, Donoghue DJ (1984) Requirement for a signal sequence in biological expression of the v-*sis* oncogene. Science 226:1197–1199

Heldin C-H, Westermark B, Wasteson Å (1980) Chemical and biological properties of a growth factor from human cultured osteosarcoma cells: resemblance with platelet-derived growth factor. J Cell Physiol 105:235–246

Heldin C-H, Wasteson Å, Westermark B (1985) Platelet-derived growth factor. Mol Cell Endocr 39:169–187

Heldin C-H, Betsholtz C, Johnsson A, Westermark B (1986a) Role of PDGF-like growth factors in malignant transformation. Cancer Rev (in press)

Heldin C-H, Johnsson A, Wennergren S, Wernstedt C, Betsholtz C, Westermark B (1986b) A human osteosarcoma cell line secretes a growth factor structurally related to a homodimer of PDGF A chains. Nature 319:511–514

Johnsson A, Heldin C-H, Wasteson Å, Westermark B, Deuel TF, Huang JS, Seeburg PH, Gray E, Ullrich A, Scarce G, Stroobant P, Waterfield MD (1984) The c-*sis* gene encodes a precursor of the B chain of platelet-derived growth factor. EMBO J 3:921–928

Johnsson A, Betsholtz C, Heldin C-H, Westermark B (1985) Antibodies to platelet-derived growth factor inhibit acute transformation by simian sarcoma virus. Nature 317:438–440

Josephs SF, Guo C, Ratner L, Wong-Staal F (1984a) Human proto-oncogene nucleotide sequences corresponding to the transforming region of simian sarcoma virus. Science 223:487–491

Josephs SF, Ratner L, Clarke MF, Westin EH, Reitz MS, Wong-Staal F (1984b) Transforming potential of human c-*sis* nucleotide sequences encoding platelet-derived growth factor. Science 225:636 – 639

Kelly JD, Raines EW, Ross R, Murray MJ (1985) The B chain of PDGF alone is sufficient for mitogenesis. EMBO J 4:3399 – 3405

King CR, Giese NA, Robbins KC, Aaronson SA (1985) In vitro mutagenesis of the v-*sis* transforming gene defines functional domains of its growth factor-related product. Proc Natl Acad Sci USA 82:5295 – 5299

Martinet Y, Bitterman PB, Mornex J-F, Grotendorst G, Martin GR, Crystal RG (1985) Activated human monocytes express the c-*sis* proto-oncogene and release a mediator showing PDGF-like activity. Nature 319:158 – 160

Nilsson J, Sjölund M, Palmberg L, Thyberg J, Heldin C-H (1985) Arterial smooth muscle cells in primary culture produce a platelet-derived growth factor-like protein. Proc Natl Acad Sci USA 82:4418 – 4422

Nistér M, Heldin C-H, Wasteson Å, Westermark B (1984) A glioma-derived analog to platelet-derived growth factor: demonstration of receptor competing activity and immunological cross-reactivity. Proc Natl Acad Sci USA 81:926 – 930

Nistér M, Heldin C-H, Westermark B (1986) Clonal variation in the production of a PDGF-like growth factor and expression of PDGF receptors in a human malignant glioma. Cancer Res 46:332 – 340

Robbins KC, Antoniades HN, Devare SG, Hunkapiller MW, Aaronson SA (1983) Structural and immunological similarities between simian sarcoma virus gene product(s) and human platelet-derived growth factor. Nature 305:605 – 608

Robbins KC, Leal F, Pierce JH, Aaronson SA (1985) The v-*sis*/PDGF-2 transforming gene product localizes to cell membranes but is not a secretory protein. EMBO J 4:1783 – 1792

Rozengurt E, Sinnett-Smith J, Taylor-Papadimitriou J (1985) Production of PDGF-like growth factor by breast cancer cell lines. Int J Cancer 36:247 – 252

Seifert RA, Schwartz SM, Bowen-Pope DF (1984) Developmental regulation of production of platelet-derived growth factor-like molecules. Nature 311:669 – 671

Shimokado K, Raines ER, Madtes DK, Barrett TB, Benditt E, Ross R (1985) A significant part of macrophage-derived growth factor consists of at least two forms of PDGF. Cell 43:277 – 286

Stroobant P, Waterfield MD (1984) Purification and properties of porcine platelet-derived growth factor. EMBO J 2:2963 – 2967

van Zoelen EJJ, van de Ven WJM, Franssen HJ, van Oostwaard TMJ, van der Saag PT, Heldin C-H, de Laat SW (1985) Neuroblastoma cells express c-*sis* and produce a transforming growth factor antigenitally related to the platelet-derived growth factor. Mol Cell Biol 5:2289 – 2297

Waterfield MD, Scrace G, Whittle N, Stroobant P, Johnsson A, Wasteson Å, Westermark B, Heldin C-H, Huang JS, Deuel TF (1983) Platelet-derived growth factor is structurally related to the putative transforming protein of p28sis of simian sarcoma virus. Nature 304:35 – 39

Transforming Growth Factor-β

HAROLD L. MOSES and EDWARD B. LEOF

Transforming growth factor, type β (TGF-β) was first described by its ability to stimulate mouse embryo-derived AKR-2B cells (Moses et al. 1981) and rat NRK cells to grow in soft agar (Roberts et al. 1981). (The latter cells also required the addition of epidermal growth factor, EGF.) Subsequent studies have shown that TGF-β functions as a growth stimulator only for certain fibroblastic cells, possibly through an indirect mechanism involving the induction of endogenous growth-factor synthesis resulting in autocrine growth (Leof et al. 1986). In fact, TGF-β is a growth inhibitor for most cell types tested (Moses et al. 1985a and unpublished observations). Its growth-inhibitory properties, or those of a closely related molecule, were described by Holley and co-workers several years before its growth-stimulatory effects were discovered (Holley et al. 1978; Tucker et al. 1984a). This review discusses the possible mechanism of growth stimulation and growth inhibition by TGF-β and the possible role of this factor in neoplastic transformation.

Purification and Cloning of TGF-β

TGF-β is a highly ubiquitous molecule (reviewed by Goustin et al. 1986). It has been purified from several normal tissues, the most important of which is human platelets (Childs et al. 1982; Assoian et al. 1983). Regardless of the source, the intact molecule has a molecular weight of 25,000 and is composed of two apparently identical subunits of 12,500; the dissociated subunit is biologically inactive (Assoian et al. 1983).

Derynck et al. (1985) have cloned the gene for TGF-β from a human genomic library and from cDNA libraries derived from human term placenta and the human fibrosarcoma line HT-1080. The amino acid sequence deduced from sequencing of overlapping cDNA fragments suggests a subunit of 112 amino acids, contained in residues 280 – 391 of a hypothetical precursor molecule. Amino acid sequencing of reduced human platelet-derived TGF-β confirmed the notion that the two chains are identical and that the native molecule is a homodimer of disulfide-linked chains (Derynck et al. 1985, this Vol.). The only molecule showing significant sequence and structural homology with TGF-β is inhibin, a potent inhibitor of follicle-stimulating hormone secretion (Mason et al. 1985). The murine TGF-β gene has recently also been cloned and its cDNA sequence determined (R. Derynck, personal communication). The

Oncogenes and Growth Control
Edited by P. Kahn and T. Graf
© Springer-Verlag Berlin Heidelberg 1986

C-terminal precursor cDNA sequence representing the TGF-β coding region is identical in murine and human clones except for one amino acid at position 354 (serine in mouse and alanine in human TGF-β). This high degree of evolutionary conservation suggests that most parts of the TGF-β molecule are necessary for biological activity and that TGF-β probably plays an essential role in normal growth and development.

TGF-β Receptor

TGF-β, unlike TGF-α, has its own specific cell membrane receptors which, like the TGF-β molecule itself, are highly ubiquitous. Specific binding of ^{125}I-TGF-β to various mesenchymal and epithelial cells in primary and secondary cultures as well as to both normal and neoplastic cell lines has been reported (Tucker et al. 1984b). The development of radioreceptor asssays for TGF-β has made it possible to calculate dissociation constants (25 – 140 pM) and receptor number per cell (10,000 – 40,000) (Tucker et al. 1984b; Frolik et al. 1984). The TGF-β receptor is apparently quite different from either the EGF or platelet-derived growth factor (PDGF) receptors. Affinity labelling of the receptor in mouse cells has identified a 565-kDa complex that is apparently a glycoprotein which dissociates in the presence of disulfide reagents into two subunits of 280 – 290 kDa (Massague 1985). Thus far, no kinase activity has been reported for the TGF-β receptor.

Mechanism of Growth Stimulation by TGF-β

TGF-β has been reported to stimulate growth in soft agar of murine AKR-2B and 3T3 cells, rat NRK cells and secondary cultures of human foreskin fibroblasts (Moses et al. 1981, 1985a; Roberts et al. 1981). In other words, these fibroblasts or fibroblast-like cells grow in soft agar only if the serum-containing medium is supplemented with another growth factor, TGF-β. Studies of requirements other than TGF-β have, in general, demonstrated that the same factors required for monolayer growth are also required for proliferation in soft agar (Rizzino 1984; Massague et al. 1985).

TGF-β stimulates DNA synthesis in quiescent monolayer cultures of mouse AKR-2B cells in a completely defined medium without other added growth factors but with delayed kinetics relative to stimulation with other growth factors (Shipley et al. 1985). Stimulation with EGF and insulin, PDGF, fibroblast growth factor, or serum resulted in a 12 – 14-h lag before the onset of DNA synthesis, which then peaked at 20 – 25 h following stimulation. With TGF-β, the lag phase was increased to 24 h with a peak of DNA synthesis between 30 and 35 h. In an effort to determine why TGF-β stimulated DNA synthesis with such delayed kinetics the possibility that it acts as an indirect mitogen through induction of synthesis of endogenous growth factors was examined. Treatment of quiescent AKR-2B cell cultures with TGF-β resulted in an early induction (within 20 min) of c-*sis* mRNA (which encodes one chain of PDGF; Leof et al. 1986). The rise in c-*sis* mRNA was followed by a corresponding increase in PDGF-like protein in the culture medium. In addition, PDGF-regulated genes (c-*fos* and c-*myc*) were stimulated by TGF-β with delayed kinetics relative to that seen with direct PDGF stimulation. The data suggest that the mitogenicity of TGF-β in monolayer cells is mediated by induction of c-*sis* and PDGF which in turn induce the expression of c-*fos*, c-*myc* and other PDGF-inducible genes, culminating in DNA synthesis. Studies with neutralizing PDGF antibodies and anti-sense genes are in progress to more definitely test this model of autocrine growth induced by TGF-β.

Inhibition of Cell Proliferation by TGF-β

Even under circumstances where TGF-β has been demonstrated to be stimulatory for fibroblastic AKR-2B cells, it inhibits the early S-phase induced by EGF and insulin or PDGF (Tucker et al. 1984a; Shipley et al. 1985). In addition, we have recently shown that the growth inhibitor originally described by Holley et al. (1978) from African green monkey (BSC-1) cells is similar, if not identical, to human platelet-derived TGF-β (Tucker et al. 1984a). Growth inhibitor purified from medium conditioned by BSC-1 cells (Holley et al. 1983) and TGF-β purified from human platelets were shown to have almost identical biological activities in stimulating growth of AKR-2B cells in soft agar and inhibiting DNA synthesis in BSC-1 and CCL-64 (mink lung) cells. The growth inhibitor was also shown to compete for binding with ^{125}I-labelled TGF-β to membrane receptors nearly as effectively as the native platelet-derived TGF-β (Tucker et al. 1984a). In addition, both the growth inhibitor and TGF-β have apparent molecular weights of 25,000 and exhibit a single band at half this molecular weight under reducing conditions (Assoian et al. 1983; Holley et al. 1983).

TGF-β inhibits the ability of several human carcinoma cell lines to grow in soft agar (Moses et al. 1985a). No epithelial cell type, either neoplastic or nonneoplastic, has been demonstrated to be stimulated to proliferate by TGF-β; the epithelial cells that have been tested so far were either inhibited or showed no response to TGF-β under standard cell culture conditions. Recently, we demonstrated that TGF-β is a potent growth inhibitor of secondary cultures of human foreskin keratinocytes, with most of the inhibited cells being in the G1 phase of the cell cycle (Moses et al. 1985a; Shipley et al. 1986). The growth inhibition of keratinocytes was reversible. TGF-β also strongly inhibits EGF-induced simulation of DNA synthesis in primary rat hepatocyte cultures (Carr et al. 1986) as well as the growth of primary cultures of human megakaryocytic and erythroid precursors (L. A. Solberg, Jr. and H. L. Moses, unpublished observations).

The mechanisms by which TGF-β inhibits cell proliferation are largely unknown. It is possible that TGF-β is primarily an inhibitor for all cell types and that stimulation of fibroblastic cells occurs through the fortuitous induction of c-*sis*/PDGF, which leads to autocrine growth stimulation.

Potential Role of TGF-β in Neoplasia and Other Disease States

Using mouse embryo-derived cell culture model systems for neoplastic transformation (AKR-2B and C3H/10T1/2 cell lines), we have demonstrated that the chemically transformed derivatives of these cell lines both produce and respond to TGF-β (Moses et al. 1981, 1985a). The nontransformed parent cell lines were found to release as much TGF-β into serum-free conditioned medium as the chemically transformed derivatives. The major change observed in the chemically transformed cells relative to their nontransformed counterparts is the development of a markedly increased sensitivity to growth stimulation in soft agr by TGF-β (Moses et al. 1985a). The TGF-β released by both the nontransformed and transformed lines was in an inactive form that was irreversibly activated by acid treatment. These studies are in agreement with those of Lawrence et al. (1984), who demonstrated that many cell types release TGF-β in an inactive form. The physiological mechanism of activation of TGF-β is not known. The inactive precursor of TGF-β released by cells in culture appears to be in a higher molecular form than the active molecule, perhaps reflecting association with a binding protein (Moses et al. 1985b; Lawrence et al. 1985).

Studies on the TGF-β receptor revealed very slightly reduced numbers of receptors on the chemically transformed cells relative to their nontransformed counterparts, with no detectable change in affinity. This suggested that a post-receptor mechanism was responsible for the increased sensitivity observed in the chemically transformed cells. Using the C3H/10T1/2 cells, which are completely unresponsive to TGF-β with respect to stimulation of growth in soft agar, transfection was carried out with a mouse c-*myc* gene linked to an SV40 promoter and/or with an activated H-*ras* gene, both of which were co-transfected with the dominant neomycin-resistance marker (E. B. Leof and H. L. Moses, unpublished observations). The c-*myc*-transfected cells became highly responsive to stimulation of growth in soft agar by TGF-β, suggesting that c-*myc* expression, at least in part, mediates responsiveness to TGF-β. The *ras*-transfected cells demonstrated marked morphologic transformation in monolayer culture and grew in soft agar in the absence of TGF-β; this growth was only slightly enhanced by the addition of TGF-β. These data demonstrate that transfection with an activated H-*ras* gene induces a phenotype similar to that induced by TGF-β, but without the requirement for added TGF-β, suggesting that *ras* p21 may enhance autocrine stimulation by endogenous TGF-β or may be involved in the transduction of the TGF-β signal.

The potential role of TGF-β in neoplastic transformation of epithelial and other nonfibroblastic cell types may be entirely different from that involved in fibroblastic cells. We have demonstrated that a squamous carcinoma cell line has lost the inhibitory response to TGF-β exhibited by normal keratinocytes (Shipley et al. 1986). The loss of the normal inhibitory response to TGF-β in epithelial cells could result in an enhanced proliferative potential, producing the same effect as the activation of a stimulatory response.

Recent studies by Roberts and Sporn (1985) demonstrated that TGF-β induces a marked desmoplastic reaction when injected subcutaneously into mice. These and other studies indicate that TGF-β could be involved in wound healing and further suggest that TGF-β released by carcinoma cells could contribute to the proliferation of stromal elements, a necessary event in the formation of large tumors. The data further suggest that TGF-β could play a major role in the many diseased states involving fibroblastic proliferation and collagen deposition. A potential mechanism for the increased collagen deposition stimulated by TGF-β has been suggested by studies demonstrating that TGF-β induces an inhibitor of plasminogen activator. This inhibition may result in decreased proteolysis of matrix proteins and thereby lead to a net increase in matrix deposition (Laiho et al. 1986).

Concluding Remarks

TGF-β is a highly ubiquitous molecule produced by a variety of cell types, normal and neoplastic, mesenchymal and epithelial. Since most cells release TGF-β in an inactive form and have receptors for TGF-β, major regulatory steps in TGF-β action are probably at the post-receptor level or at the level of activation of the precursor. TGF-β is primarily a growth inhibitor and not a classical growth factor. Growth stimulatory effects with TGF-β have been observed only in fibroblastic cells. In at least one circumstance, evidence has been presented that the stimulatory effect of TGF-β acts indirectly through induction of c-*sis* and autocrine stimulation by a PDGF-like growth factor. Thus, either the autocrine stimulation by endogenous TGF-β in fibroblastic cells or the loss of the inhibitory effect of TGF-β in epithelial (or other cells normally inhibited by TGF-β) could lead to an increased proliferative potential and thereby contribute to the neoplastic phenotype.

References

Assoian RK, Komoriya A, Meyers CA, Miller DM, Sporn MB (1983) Transforming growth factor-β in human platelets: Identification of a major storage site, purification and characterization. J Biol Chem 258:7155−7160

Carr BI, Hayashi I, Branum EL, Moses HL (1986) Inhibition of DNA synthesis in rat hepatocytes by platelet-derived type-β-transforming growth factor. Cancer Res 46:2330−2334

Childs C, Proper JA, Tucker RF, Moses HL (1982) Serum contains a platelet-derived transforming growth factor. Proc Natl Acad Sci USA 79:5312 – 5316

Derynck R, Jarrett JA, Chen EY, Eaton DH, Bell JR, Assoian RK, Roberts AB, Sporn MB, Goeddel DV (1985) Human transforming growth factor-β cDNA sequence and expression in tumor cell lines. Nature 316:701 – 705

Frolik CA, Wakefield LM, Smith DM, Sporn MB (1984) Characterization of a membrane receptor for transforming growth factor-β in normal rat kidney cells. J Biol Chem 259:10995 – 11000

Goustin AS, Leof EB, Shipley GD, Moses HL (1986) Perspectives in Cancer Research: Growth factors and cancer. Cancer Res 46:1015 – 1029

Holley RW, Armour R, Baldwin JH (1978) Density-dependent regulation of growth of BSC-1 cells in cell culture: Growth inhibitors formed by the cells. Proc Natl Acad Sci USA 75:1864 – 1866

Holley RW, Armour R, Baldwin JH, Greenfield S (1983) Preparation and properties of a growth inhibitor produced by kidney epithelial cells. Cell Biol Int Rep 7:525 – 526

Laiho M, Saksela D, Keski-Oja J (1986) Transforming factor-β alters plasminogen activator activity in human skin fibroblasts. Exp Cell Res 164:399 – 407

Lawrence DA, Pircher R, Kryceve-Martinerie C, Jullien P (1984) Normal embryo fibroblasts release transforming growth factors in a latent form. J Cell Physiol 121:184 – 188

Lawrence DA, Pircher R, Jullien P (1985) Conversion of a high molecular weight latent β-TGF from chicken embryo fibroblasts into a low molecular weight active β-TGF under acidic conditions. Biochem Biophys Res Commun 133:1016 – 1034

Leof EB, Proper JA, Goustin AS, Shipley GD, DiCorleto PE, Moses HL (1986) Induction of c-*sis* mRNA and activity similar to platelet-derived growth factor-like activity by transforming growth factor, type-β: A proposed model for indirect mitogenesis involving autocrine activity. Proc Natl Acad Sci USA 83:2453 – 2457

Mason AJ, Hayflick JS, Ling N, Esch F, Ueno N, Ying S-Y, Guillemin R, Niall H, Seeburg PH (1985) Complementary DNA sequences of ovarian follicular fluid inhibin show precursor structure and homology with transforming growth factor-β. Nature 318:659 – 663

Massague J (1985) Subunit structure of a high-affinity receptor for type β-transforming growth factor: evidence for a disulfide-linked glycosylated receptor complex. J Biol Chem 260:7059 – 7066

Massague J, Kelly B, Mottola C (1985) Stimulation by insulin-like growth factors is required for cellular transformation by type-β-transforming growth factor. J Biol Chem 260:4551 – 4554

Moses HL, Branum EB, Proper JA, Robinson RA (1981) Transforming growth factor production by chemically transformed cells. Cancer Res 41:2842 – 2848

Moses HL, Tucker RF, Leof EB, Coffey RJ Jr, Halper J, Shipley GD (1985a) Type-β transforming growth factor is a growth stimulator and a growth inhibitor. In: Feramisco J, Ozanne B, Stiles C (eds) Cancer cells: Growth factors and transformation. Cold Spring Harbor Press, New York, pp 65 – 71

Moses HL, Shipley GD, Leof EB, Halper J, Coffey RJ Jr, Tucker RF (1985b) Transforming growth factors. In: Boynton AL, Leffert HL (eds) Control of animal cell proliferation. Academic Press, New York, (in press)

Rizzino A (1984) Behavior of transforming growth factors in serum-free media: An improved assay for transforming growth factors. In Vitro 20:815 – 822

Roberts AB, Sporn MB (1985) Transforming growth factors. In: Cancer surveys, vol. 4. Oxford Univ Press, Oxford, England, pp 683 – 705

Roberts AB, Anzano MA, Lamb LC, Smith JM, Sporn MB (1981) New class of transforming growth factors potentiated by epidermal growth factor: Isolation from non-neoplastic tissues. Proc Natl Acad Sci USA 78:5339 – 5343

Shipley GD, Tucker RF, Moses HL (1985) Type-β-transforming growth factor/growth inhibitor stimulates entry of monolayer cultures of AKR-2B cells into S phase after a prolonged prereplicative interval. Proc Natl Acad Sci USA 82:4147 – 4151

Shipley CD, Pittelkow MR, Wille JJ Jr, Scott RE, Moses HL (1986) Reversible inhibition of normal human prokeratinocyte proliferation by type-β-transforming growth factor/growth inhibitor in serum-free medium. Cancer Res 46:2068 – 2071

Tucker RF, Shipley GD, Moses HL, Holley RW (1984a) Growth inhibitor from BSC-1 cells closely related to the platelet type-β-transforming growth factor. Science 226:705 – 707
Tucker RF, Branum EL, Shipley GD, Ryan RJ, Moses HL (1984b) Specific binding to cultured cells of ^{125}I-labelled type-β-transforming growth factor from human platelets. Proc Natl Acad Sci USA 81:6757 – 6761

Transforming Growth Factor-α

RIK DERYNCK

Transforming growth factor-α (TGF-α) was first detected in the culture medium of certain retrovirus-transformed cell lines as an activity which binds to the receptor for epidermal growth factor (EGF) and thereby inhibits the binding of EGF. Subsequent examination showed that this factor is made by many other transformed cell lines but not by adult normal cells in culture (Todaro et al. 1985). Addition of sarcoma growth factor, as it was first called, to rat fibroblasts of the NRK cell line reversibly induced profound morphological changes and colony-forming ability in soft agar (De Larco and Todaro 1978; Todaro et al. 1985). The capacity of this factor to confer a transformed phenotype upon NRK cells, together with the fact that it is synthesized by transformed cells, led to the name transforming growth factor.

Extensive biochemical purification of sarcoma growth factor preparations showed that it contained two TGF peptides, termed TGF-α and -β. TGF-α is related to the epidermal growth factor (EGF) and binds to the EGF receptor. TGF-β is a structurally unrelated protein, which by itself is unable to induce a transformed phenotype and which binds to a distinct receptor (see Moses and Leof, this Vol.). In pure form, TGF-α has only minimal effects on the phenotype of NRK cells. Anchorage-independent growth occurs in this system only when both TGF-α and -β are present (Anzano et al. 1983), although the requirement for both growth factors is not seen in certain other cell systems.

TGF-α and Its Precursor

Since their original discovery, TGF-αs have been detected in culture supernatants and extracts from several transformed rodent and human cells (De Larco and Todaro 1978; Todaro et al. 1980; Marquardt et al. 1983). These TGF-αs all bind to the EGF receptor and display an apparent heterogeneity ranging from a 6-kDa species secreted by several tumor cell lines (Marquardt et al. 1984) to the 34-kDa TGF-α species detected in the urine of cancer patients (Sherwin et al. 1983). The sequences of human and rat TGF-α cDNAs corresponding to the low molecular weight species indicate that the 50 amino acid TGF-α is initially translated as an internal part of a 160 amino acid precursor (see Fig. 1) from which it is then derived by proteolytic cleavage (Derynck et al. 1984; Lee et al. 1985b). The initiator ATG is followed by a short hydrophobic sequence, suggesting the presence of an N-terminal signal se-

Oncogenes and Growth Control
Edited by P. Kahn and T. Graf
© Springer-Verlag Berlin Heidelberg 1986

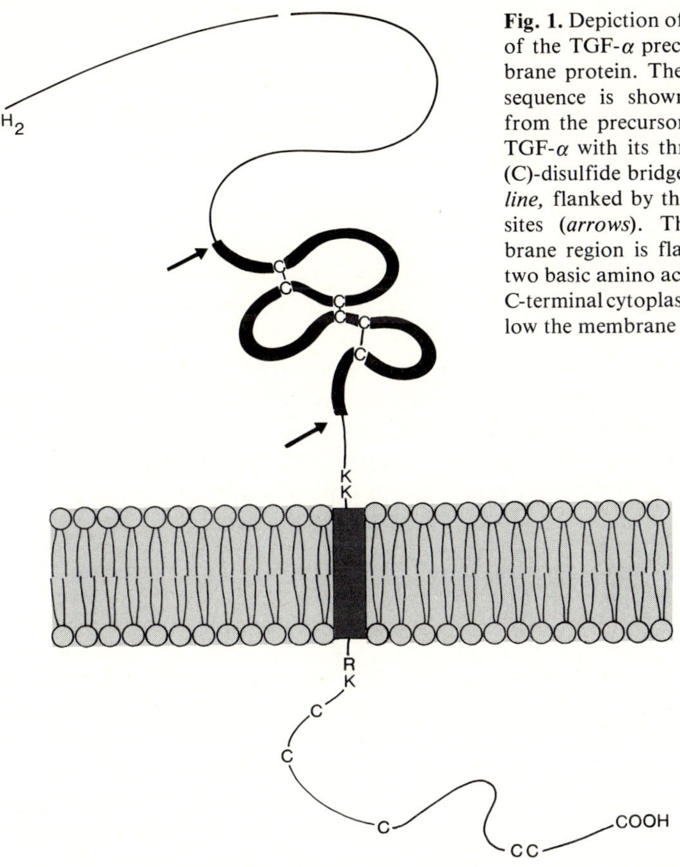

Fig. 1. Depiction of a hypothetical model of the TGF-α precursor as a transmembrane protein. The NH₂-terminal signal sequence is shown as already cleaved from the precursor. The 50 amino acid TGF-α with its three proposed cysteine (C)-disulfide bridges is shown as a *heavy line,* flanked by the proteolytic cleavage sites (*arrows*). The *boxed* transmembrane region is flanked at each side by two basic amino acids (*KK* and *RK*). The C-terminal cytoplasmic domain shown below the membrane is rich in cysteines (*C*)

quence of probably 20 to 22 residues. An Asn-Ser-Thr triplet present at positions 25 – 27 could be a site for N-glycosylation. To generate the 50 amino acid TGF-α, proteolytic cleavage must occur at both the N- and C-terminus between an alanine residue and a valine dipeptide. This Ala-Val-Val trimer is located within the sequence Val-Ala-Ala-*Ala-Val-Val* at the amino terminus of the 50 amino acid TGF-α and within *Ala-Val-Val*-Ala-Ala at its carboxyl end. Processing of a precursor protein by a protease with such specificity has not been described before; cleavage of polypeptide hormone precursors most often takes place at dibasic residues.

An extremely hydrophobic domain begins 9 residues downstream of the carboxy terminus of the 50 amino acid TGF-α. This region is 23 amino acids long and is flanked by paired basic amino acids at positions 96 – 97 and 127 – 128. The length of the hydrophobic domain and the basic character of the flanking amino acids is characteristic of transmembrane domains, suggesting that the TGF-α precursor is inserted into the membrane following the removal of the NH₂-terminal signal peptide. The Ala-Val-Val-specific protease would then cleave the external segment of the precursor and in this way release the 50 amino acid TGF-α. Cells lacking this protease would retain the TGF-α precursor in the membrane and fail to release TGF-α into the medium. Downstream of the 23 amino acid hydrophobic domain is the cysteine-rich region encoding the C-terminus of the precursor. If the TGF-α precursor constitutes a transmembrane protein, then this C-terminal portion would remain at the cytoplasmic side of the membrane and probably not undergo any disulfide bond formation.

Comparison of the deduced rat and human precursor sequences indicates a very strong homology (Derynck et al. 1984; Lee et al. 1985b). Only four differences are observed in the 50 amino acid long TGF-α sequence, but the sequences of the putative transmembrane and cytoplasmic regions are even more conserved, suggesting a potentially important biological function for the C-terminus. Since the cytoplasmic peptide probably remains covalently attached to the transmembrane region (due to the fact that proteolytic processing enzymes are usually found outside but not inside the cell), one hypothesis is that this peptide could play a role in signal transduction, possibly in a way similar to a receptor. This would suggest that the high number of cysteine residues has biological significance. However, it is not known if the TGF-α precursor (perhaps in its unprocessed form) can function as a receptor molecule, as has been proposed for the much larger EGF precursor (see Pfeffer and Ullrich, this Vol.). Whatever the function of the C-terminal peptide may be, it is important to recognize that synthesis and secretion of TGF-α go together with the synthesis of the C-terminal precursor segment, which may have separate activities in the physiology of the cell.

Several lines of evidence indicate that the different-sized species of TGF-α are probably encoded by a single gene (Derynck et al. 1984; Linsley et al. 1985). The larger forms could arise either by proteolytic processing at different sites of the precursor, e.g., at the dibasic residues, by the absence of processing, by the formation of aggregates with other proteins, or by dimerization or oligomerization.

Many of the structural features described here for TGF-α are shared by EGF and by a vaccinia virus-encoded protein, including synthesis as part of a large precursor which contains a potential transmembrane domain. The three peptides also contain extensive amino acid homology (see Pfeffer and Ullrich, this Vol.).

Cellular Sources of TGF-α

Cells transformed with a variety of oncogene-carrying retroviruses, by SV40, or by polyoma secrete TGF-α (Kaplan and Ozanne 1980; Kaplan et al. 1981; Todaro et al. 1985). In a survey of tumor biopsies for the presence of TGF-α mRNA, none of ten hematopoietic tumor cell lines expressed detectable levels. In contrast, TGF-α mRNA was most consistently synthesized in renal and squamous carcinomas, and frequently also in many mammary carcinomas and tumors of neuronal origin. The synthesis of TGF-α by these tumors could explain the presence of a characteristic TGF-α species in the urine of some cancer patients but not in normal controls (Sherwin et al. 1983; Twardzik et al. 1985).

As expected, TGF-α is not detectable in medium from normal cells in vitro or in normal fully developed tissues. However, the possibility that this factor could play a role in the normal physiology of the adult organism cannot yet be excluded. Recent evidence indicates that TGF-α is synthesized during early fetal development (Lee et al. 1985a; Twardzik et al. 1985). TGF-α expression in the murine fetus peaks around day 9 and then drops to undetectable levels by birth (day 21), suggesting that TGF-α may function as a normal embryonic version of a family of EGF-related growth factors. If so, it could be considered as an oncodevelopmental antigen (see also Jakobovits, this Vol.).

Role of TGF-α in Cell Transformation

It has been postulated that TGFs act via an autocrine mechanism during the transformation process, that is, they help sustain the transformed character of the same cells from which they are secreted (Sporn and Todaro 1980; Sporn and Roberts 1985). Such an autocrine mechanism could explain the initial observation that retrovirus-transformed cells have a lower number of EGF binding sites at their surface (Todaro et al. 1976) as a result of endogenous TGF-α secretion and subsequent downregulation of the ligand-receptor complex.

Autocrine growth stimulation also appears to play a role in the induction of malignant transformation in other systems. Examples include the simian sarcoma virus, in which the v-*sis* oncogene encodes a protein related to the β chain of platelet-derived growth factor (PDGF) and transforms cells carrying the PDGF receptor (see Heldin and Westermark, this Vol.), as well as various hematopoietic systems (see Gough and Kahn et al., this Vol.). The fact that TGF-α mRNA is most frequently produced in squamous carcinomas and renal carcinomas (which all contain relatively high levels of EGF receptor mRNA; Derynck et al. 1986), is consistent with the autocrine hypothesis, although there is as yet no direct proof for it.

TGF-α might also contribute to the malignant phenotype by influencing neighboring cells via paracrine mechanisms (see below).

Receptor for TGF-α

It is generally accepted that the biological actions of TGF-α, like the actions of other polypeptide hormones, are mediated through the binding to specific cell surface receptors. Earlier studies indicated that TGF-α interacts not only with the EGF receptor but also with a 60-kDa membrane component which does not bind EGF. It was proposed that this 60-kDa protein was a specific TGF-α receptor species (Massague et al. 1982). However, the induction of anchorage independence can be neutralized by blocking antibodies raised against the EGF receptor (Carpenter et al. 1983), suggesting that this is the functionally important receptor for TGF-α and making it unlikely that the 60-kDa protein is TGF-α-specific.

Both murine EGF and the 50 amino acid TGF-α secreted by transformed rat fibroblasts exhibit a remarkably similar interaction with the EGF receptor in that both peptides compete for receptor binding with the same potency and to the same extent (Massague 1983), although the pH optima for binding are somewhat different. Continued exposure of A431 cells to either TGF-α or EGF induces downregulation of the receptors according to similar kinetics (Massague 1983). Furthermore, TGF-α mimics the action of EGF in activating the EGF receptor-associated tyrosine kinase (Pike et al. 1982). While these results indicate that TGF-α may act through the EGF receptor, it is still unknown how high or low affinity EGF receptors behave in vivo with respect to their TGF-α or EGF binding properties.

Biological Activities of TGF-α: Comparison with EGF

TGF-α and EGF show a number of functional similarities, as well as some differences. These can be summarized as follows. (1) The two peptides are equally potent in cooperating with TGF-β to induce anchorage independence in NRK cells. (2) Injection of TGF-α or EGF into newborn mice accelerates eyelid opening in a very similar concentration-dependent manner (Smith et al. 1985). (3) TGF-α and EGF induce accelerated tooth eruption, retard the growth rate, and inhibit hair growth, although the relative potency of the two factors has not been clearly established. (4) Both TGF-α and EGF induce rapid and transient ruffling responses in cultured cells. At lower doses the response to either factor is similar, but at high doses the maximal responses are higher with TGF-α than with EGF. Pretreatment of the cells with TGF-β greatly enhances the ruffling response to TGF-α but antagonizes EGF-induced ruffling (Myrdal 1985). (5) TGF-α and EGF both induce human epidermal cells to form colonies, with TGF-α being somewhat more potent (Barrandon and Green, personal communication). (6) TGF-α induces Ca^{2+} release in bone organ cultures, while EGF is either less potent or unable to do so, depending on the culture system used. This activity, which is a measure of bone resorption, could be relevant to hypercalcemia in vivo, a condition often associated with advanced stages of malignancies. It is striking that tumor types such as melanomas and squamous, renal, and mammary carcinomas, which are the most frequent expressors of TGF-α mRNA, often exhibit malignancy-associated hypercalcemia (Mundy et al. 1985). However, factors other than TGF-α have been implicated as playing a role in this process. If TGF-α released by the tumor cells indeed participates in the induction of hypercalcemia in vivo, then these activities must be exerted via a paracrine or even an endocrine mechanism. (7) At low concentrations TGF-α, but not EGF, induces neovascularization, although the two factors are equally mitogenic on cultured endothelial cells (Schreiber et al. 1986). The production of TGF-α by tumors could thus stimulate tumor cell proliferation in an autocrine manner and concomitantly contribute to the induction of tumor-derived angiogenesis via a paracrine mechanism. The latter activity of TGF-α could also be physiologically important during early fetal development. (8) TGF-α and EGF induce increased regional arterial blood flow in vascular beds (M. D. Hollenberg, personal communication). The two factors are equally potent in this system, but the maximal response obtained with TGF-α is much higher than with EGF. In addition, prior exposure of the vascular tissue to TGF-α markedly desensitizes the arterial system to EGF but not to TGF-α. The synthesis of TGF-α by tumor cells and the fact that TGF-α does not cause desensitization to its own action may suggest that it plays a role in the local vascular hyperdynamic state associated with malignancy. Taken together, these results indicate that TGF-α and EGF should not be considered as mere analogs, despite their many similarities.

How can two factors which appear to utilize the same receptor exert different biological effects? If both high and low affinity EGF receptors exist (see Schlessinger, this Vol.), then TGF-α and EGF could bind differently to each of these. The behavior of the ligand:receptor complex during internalization could also differ. Detailed studies on receptor:ligand interactions and on the physiological events which they trigger will be needed to resolve this question.

Acknowledgments. I thank all colleagues at Genentech who contributed to the determination of the human TGF-α cDNA sequence and its expression in bacteria: E. Chen, D. Eaton, D. Goeddel and M. Winkler. I thank M. D. Hollenberg (University of Calgary) and H. Green (Harvard University) for allowing the communication of data prior to submission of a manuscript. J. Arch is acknowledged for her skillful assistance in the preparation of this manuscript.

References

Anzano MA, Roberts AB, Smith JM, Sporn MB, De Larco JE (1983) Sarcoma growth factor from conditioned medium of virally transformed cells is composed of both type-α and type-β transforming growth factors. Proc Natl Acad Sci USA 80:6264–6268

Carpenter G, Stoscheck CM, Preston YA, De Larco JE (1983) Antibodies to the epidermal growth factor receptor block the biological activities of sarcoma growth factor. Proc Natl Acad Sci USA 80:5627–5630

De Larco J, Todaro GJ (1978) Growth factors from murine sarcoma virus-transformed cells. Proc Natl Acad Sci USA 75:4001–4005

Derynck R, Roberts AB, Winkler ME, Chen EY, Goeddel DV (1984) Human transforming growth factor-α: Precursor structure and expression in *E. coli.* Cell 38:287–297

Derynck R, Goeddel DV, Ullrich A, Gutterman JU, Williams R, Bringman TS, Berger WH (1986) Synthesis of mRNAs for the transforming growth factors-α and -β and the epidermal growth factor receptor by human tumors. Submitted

Kaplan PL, Ozanne B (1980) Polyoma virus transformed cells produce transforming growth factor(s) and grow in serum free medium. Virology 123:372–380

Kaplan DL, Topp WC, Ozanne B (1981) Simian virus 40 induces the production of a polypeptide transforming growth factor(s). Virology 108:484–490

Lee DC, Rochford RM, Todaro GJ, Villareal LP (1985a) Developmental expression of rat transforming growth factor-α mRNA. Mol Cell Biol 5:3644–3646

Lee DC, Rose TM, Webb NR, Todaro GJ (1985b) Cloning and sequence analysis of a cDNA for rat transforming growth factor-α. Nature 313:489–491

Linsley D, Hargreaves W, Twardzik D, Todaro GJ (1985) Detection of larger polypeptides structurally and functionally related to type I transforming growth factor. Proc Natl Acad Sci USA 82:356–360

Marquardt H, Hunkapiller MW, Hood LE, Twardzik DR, De Larco JE, Stephenson JR, Todaro GJ (1983) Transforming growth factors produced by retrovirus-transformed rodent fibroblasts and human melanoma cells: Amino acid sequence homology with epidermal growth factor. Proc Natl Acad Sci USA 80:4684–4688

Marquardt H, Hunkapiller MW, Hood LE, Todaro GJ (1984) Rat transforming growth factor type 1: Structure and relation to epidermal growth factor. Science 223:1079–1082

Massague J (1983) Epidermal growth factor-like transforming growth factor. II. Interactions with epidermal growth factor receptors in human placenta membranes and A431 cells. J Biol Chem 258:13614–13620

Massague J, Czech MP, Iwata K, De Larco JE, Todaro GJ (1982) Affinity labelling of a transforming growth factor receptor that does not interact with epidermal growth factor. Proc Natl Acad Sci USA 79:6822–6826

Mundy GR, Ibbotson KJ, D'Souza SM (1985) Tumor products and the hypercalcemia of malignancy. J Clin Invest 76:391–394

Myrdal S (1985) Differences in early cellular responses to epidermal growth factor and transforming growth factor, type-alpha. J Cell Biol 101:2440

Pike LJ, Marquardt H, Todaro GJ, Gallis B, Casnellie JE, Bornstein PE, Krebs G (1982) Transforming growth factor and epidermal growth factor stimulate the phosphorylation of a synthetic tyrosine-containing peptide in a similar manner. J Biol Chem 257:14628–14631

Schreiber AB, Winkler ME, Derynck R (1986) Transforming growth factor-α is a more potent angiogenic mediator than epidermal growth factor. Science 232:1250–1253

Sherwin SA, Twardzik DR, Bohn WH, Cockley KD, Todaro GJ (1983) High molecular weight transforming growth factor activity in the urine of patients with disseminated cancer. Cancer Res 43:403–407

Smith JM, Sporn MB, Roberts AB, Derynck R, Winkler ME, Gregory H (1985) Human transforming growth factor-α causes precocious eyelid opening in newborn mice. Nature 315:515–516

Sporn MB, Roberts AB (1985) Autocrine growth factors and cancer. Nature 313:745–747

Sporn MB, Todaro GJ (1980) Autocrine secretion and malignant transformation of cells. N Engl J Med 303:878–880

Todaro GJ, De Larco JE, Cohen S (1976) Transformation by murine and feline sarcoma viruses specifically blocks binding of epidermal growth factor to cells. Nature 264:26–31

Todaro GJ, Fryling C, De Larco JE (1980) Transforming growth factors produced by certain human tumors: Polypeptides that interact with epidermal growth factor receptors. Proc Natl Acad Sci USA 77:5258–5262

Todaro GJ, Lee DC, Webb NR, Rose TM, Brown JP (1985) Rat type-α transforming growth factor: Structure and possible function as a membrane receptor: In: Feramisco J, Ozanne B, Stiles C (eds) Cancer cells 3. Cold Spring Harbor Lab, Cold Spring Harbor, pp 51–58

Twardzik DR, Kimball ES, Sherwin SA, Ranchalis JE, Todaro GJ (1985) Comparison of growth factors functionally related to epidermal growth factor in the urine of normal and human tumor-bearing athymic mice. Cancer Res 45:1934–1939

The Physiology of Epidermal Growth Factor

GRAHAM CARPENTER, LINDA GOODMAN, and LYNN SHAVER

Epidermal growth factor (EGF), a small polypeptide of 53 amino acid residues and a molecular mass of approximately 6000 daltons, was identified and isolated nearly 25 years ago (Cohen 1962) and is presently the best characterized epithelial cell mitogen. Although a great deal is known about the structure of both EGF and its receptor, including the primary sequences, several important questions remain unanswered. First, what is the physiological function of EGF? Does it play a role in epithelial cell renewal, which in the human body requires a constantly high rate of cell proliferation (approximately 4×10^6 cell divisions per second)? Second, by what mechanism does EGF induce cell proliferation? Third, does EGF play a role in the induction and/or maintenance of malignant transformation? The discoveries that many types of malignant cells produce EGF-related proteins (transforming growth factor-α; TGF-α) which have the capacity to induce certain transformed phenotypes (see Derynck, this Vol.) and that molecules related to the EGF receptor can function as oncogenes (v-*erbB, neu*) suggest that parts of the EGF mechanism may have a role in oncogenesis. The importance of understanding this role is underscored by the fact that over 90% of all malignancies arise from epithelial cells (Wright and Alison 1984). This review addresses various aspects of the first two questions. These issues are discussed further in the articles which follow by Derynck, by Schlessinger and by Beug et al.

Distribution of EGF in Human Body Fluids

It is well established that EGF is present in most human body fluids at concentrations ranging from 1 to 800 nanograms (ng) per ml. However, changes in these levels have not, as yet, been correlated with any pathological processes. Interestingly, the concentration of EGF in the urine of newborns is low and increases about fivefold in the first 2 years of life (Mattila et al. 1985). This change represents the largest described alteration of EGF levels in vivo; however, its physiological meaning is not known.

The concentration of EGF in serum or plasma has been difficult to determine accurately. Most authors seem to agree that circulating levels of EGF are quite low − less than 1 ng per ml of plasma. Based on a sensitive bioassay (which also detects TGF-α activity), we estimate that commercial fetal calf serum contains less than 300 picograms (pg) of EGF activity per ml (Carpenter

Oncogenes and Growth Control
Edited by P. Kahn and T. Graf
© Springer-Verlag Berlin Heidelberg 1986

and Zendegui 1986). Cell culture medium with 10% serum therefore contains less than 30 pg per ml, far less than the 1 ng per ml at which most cultured cells respond to EGF, suggesting that EGF is unlikely to contribute significantly to the mitogenic activity of serum in most cell culture media. Furthermore, if EGF serves a mitogenic function in vivo, then the low concentration of this factor in serum suggests that the circulatory system is unlikely to be an important source for target cell populations. The more physiologically relevant source is likely to be localized production, leading to stimulation of neighboring cells in a paracrine fashion. It is also unlikely that filtration and/or concentration of EGF from plasma accounts for the relatively high EGF levels in fluids such as urine, milk, and saliva.

Effects of Exogenous EGF in Animals and Cultured Cells

The administration of EGF to animals produces a number of dramatic effects associated with enhanced proliferation of epithelial tissues. In some cases, such as skin, this stimulation of cell proliferation leads to a more rapid rate of differentiation (Cohen and Elliott 1963), while in other tissues, such as the trachea (Sundell et al. 1980), a metaplastic effect is produced.

Interestingly, two studies report that EGF can have stimulatory and inhibitory effects on different cell populations within the same organ. In the wool follicle, EGF was found to stimulate mitosis of peripheral cells of the acini but to inhibit mitosis in the bulb cells (Moore et al. 1985). Organ culture studies of tooth morphogenesis indicated that EGF stimulates cell proliferation in the dental epithelium, but inhibits proliferation in the dental mesenchyme (Partanen et al. 1985). Surprisingly, dissociated cells of the dental mesenchyme were stimulated by EGF. These findings suggest that the effect of EGF on these cells may be modulated by the interaction of the target cells with nearby cells in the tissue; that is, that the cells may respond positively or negatively to EGF depending on the associations with nearby cells and on the constraints imposed by tissue organization. The capacity of EGF to induce biological effects other than growth stimulation is further demonstrated by the fact that EGF rapidly inhibits gastric acid secretion (Bower et al. 1975), an activity which seems to be dissociated from its mitogenic properties. Most hormones are capable of producing different responses under different physiological circumstances, and if EGF is regarded as a hormone, its potential to induce different responses in different cells or in the same cells under varying conditions is not surprising.

The A431 human squamous cell carcinoma line is a valuable tool in research on the EGF receptor, as it contains EGF receptor levels which are at least 20-fold higher than those found in most other cell lines. EGF inhibits the growth of these cells, an effect which is not understood, but which is generally viewed as an anomaly of little physiological relevance and ascribed to the excessive numbers of receptors present on the cell surface. However, Kawamoto et al. (1983) reported that very low levels of EGF (0.1 ng ml^{-1}) actually stimulate the growth of A431 cells, a result which has been reproduced in several laboratories. In view of the findings (described above) that apparently normal cells appear to show diverse growth responses to EGF, perhaps the response of A431 cells to EGF is not as physiologically irrelevant as is generally assumed.

A number of other nonmitogenic effects of EGF have been reported. These include increased prolactin secretion by the rat pituitary tumor cell line GH$_3$, accompanied by inhibition of cell proliferation (Schonbrunn et al. 1980), and increased chorionic gonadotropin secretion by choriocarcinoma cells (Beneviste et al. 1978). Other growth factors have also been shown to induce nonmitogenic effects. For example, platelet-derived growth factor (PDGF) can exert chemotactic effects (Seppa et al. 1982), while fibroblast growth factor (FGF) has been reported to modulate hormone secretion in pituitary cells (Baird et al. 1985). The capacity to act as both growth-inhibitory and -stimulatory agents has been described for both transforming growth factor-β (Tucker et al. 1984 and Moses and Leof, this Vol.) and for tumor necrosis factor (Sugarman et al. 1985). Taken together, these observations suggest that growth factors may need to be viewed from a somewhat wider perspective as hormones that can elicit different responses from different cells.

Considerably more information is available concerning the biochemical action of EGF on isolated cell populations grown in culture. EGF receptors were found on most nonhematopoietic cell types, and nearly all receptor-positive cells respond to EGF. The most frequently observed biological response of cultured cells to EGF is a mitogenic one, with a few exceptions in which EGF increases the synthesis and secretion of peptide hormones without any measurable mitogenic response.

The EGF Receptor and Its Metabolism

Tyrosine kinase activity associated with the EGF receptor has been detected in all cell types which have been tested; however, due to the technical difficulty in detecting tyrosine phosphorylation in cell populations with a low concentration of receptor, not many cell types have been examined for this enzymatic activity. Unfortunately, the untested cell types includes true epithelial cells which, based on the in vivo activity of EGF, should be the most responsive to EGF as well as the most physiologically significant target cells. Technical difficulties have also made it difficult to detect physiologically relevant in vivo substrates of the receptor kinase.

When EGF binds to its receptor on the cell surface the complex process of hormone internalization and degradation takes place (Carpenter and Cohen 1976) in essentially all EGF-responsive cell types; that is, no cell type has been reported that binds ^{125}I-EGF at the cell surface but does not internalize and degrade the ligand. During this endocytic process, intracellular vesicles containing both EGF and its receptor are formed. The ultimate intracellular destination of these internalized molecules is the lysosome, where both the hormone and its receptor are degraded.

While the fate of the internalized growth factor has been known for some time, the degradation of the internalized receptor has only recently been detected (Stoscheck and Carpenter 1984; Decker 1984; Beguinot et al. 1984; Dunn et al. 1986). The method which has been employed to assess the effect of EGF on the metabolism of its receptor is metabolic labeling of cultured cells followed by a "chase" in the presence or absence of EGF. Subsequently, the radiolabeled receptor is precipitated with specific antibodies and the amount of radioactive material analyzed by SDS gels and fluorography. The results from all laboratories indicate that the presence of EGF in the chase medium leads to as much as a tenfold increase in the degradation rate of the EGF receptor. However, these studies have not been carried out with sufficient precision to support the conclusion that occupied receptors are subject to degradation each time they enter the cell; that is, some recycling may occur.

Despite these uncertainties, it is clear that the fate of the internalized EGF receptor is different from that of many other internalized receptors which are known to be recycled (e.g., LDL, transferrin, asialoglycoprotein). In the case of the insulin receptor, both recycling and degradation routes have been proposed; however, many of the studies suggesting that the internalized insulin receptor is degraded have been performed by chemically crosslinking radiolabeled insulin to its receptor, a technique which may introduce artifacts into the system. It will be of interest to determine whether the internalized receptors for other growth factors such as PDGF, interleukin-2, FGF and insulin-like growth factor-I are subject to rapid degradation.

None of the experiments reported so far has provided evidence for processing of the EGF receptor after internalization and prior to lysosomal degradation. This seems to rule out a mechanism in which the tyrosine kinase domain of the receptor is proteolytically separated from the internalized receptor. However, the metabolic labeling technique is not sufficiently sensitive to detect the processing of a minor fraction of the receptor molecules.

While the biological necessity for internalization of receptors such as LDL and transferrin is clear, the physiological role of the internalization of growth factor receptors is not so obvious. One hypothesis is that internalization represents a mechanism which functions to turn off the proliferative signal at the cell surface. Alternatively, internalization may be coupled to signaling mechanisms for mitogenesis. At present there is no definitive evidence to support either possibility.

It is possible that several intracellular signaling mechanisms may mediate the biological response of cells to EGF (see Hunter, this Vol.). In addition to tyrosine phosphorylation, other possible mechanisms might include phosphoinositide turnover and cyclic nucleotide metabolism. These signaling systems could exist in series or in parallel with tyrosine phosphorylation and, depending on the ultimate biological response, may be present only in specific cell types.

Acknowledgments. The authors received support from the National Cancer Institute (CA24071) and American Cancer Society (BC 294) during the preparation of this manuscript. G. C. is a recipient of an Established Investigator Award from the American Heart Association.

References

Baird A, Mormede P, Ying S-Y, Wehrenberg WB, Ueno N, Ling N, Guillemin R (1985) A nonmitogenic pituitary function of fibroblast growth factor: regulation of thyrotropin and prolactin secretion. Proc Natl Acad Sci USA 82:5545 – 5549

Beguinot L, Lyall RM, Willingham MC, Pastan I (1984) Down-regulation of the epidermal growth factor receptor in KB cells is due to receptor internalization and subsequent degradation in lysosomes. Proc Natl Acad Sci USA 81:2384 – 2388

Beneviste R, Speeg KV, Carpenter G, Cohen S, Lindner J, Rabinowitz D (1978) Epidermal growth factor stimulates secretion of human chorionic gonadotropin by cultured human choriocarcinoma cells. J Clin Endocrinol Metab 46:169 – 172

Bower JM, Camble R, Gregory H, Gerring EL, Willshire IR (1975) The inhibition of gastric acid secretion by epidermal growth factor. Experientia 31:825 – 826

Carpenter G, Cohen S (1976) ^{125}I-labeled human epidermal growth factor (hEGF): binding, internalization, and degradation in human fibroblasts. J Cell Biol 71:159 – 171

Carpenter G, Zendegui J (1986) A biological assay for epidermal growth factor/urogastrone and related polypeptides. Anal Biochem 153: in press

Cohen S (1962) Isolation of a mouse submaxillary gland protein accelerating incisor eruption and eyelid opening in the new-born animal. J Biol Chem 237:1555 – 1562

Cohen S, Elliott GA (1963) The stimulation of epidermal keratinization by a protein isolated from the submaxillary gland of the mouse. J Invest Dermatol 40:1 – 5

Decker SJ (1984) Aspects of the metabolism of the epidermal growth factor receptor in A431 human epidermoid carcinoma cells. Mol Cell Biol 4:571 – 575

Dunn WA, Connolly TP, Hubbard AL (1986) Receptor-mediated endocytosis of epidermal growth factor by rat hepatocytes: receptor pathway. J Cell Biol 102:24 – 36

Kawamoto T, Sato JD, Le A, Polikoff J, Sato GH, Mendelsohn J (1983) Growth stimulation of A431 cells by epidermal growth factor: identification of high-affinity receptors for epidermal growth factor by an anti-receptor monoclonal antibody. Proc Natl Acad Sci USA 80:1337 – 1341

Mattila A-L, Perheentupa J, Pesonen K, Viinikka L (1985) Epidermal growth factor in human urine from birth to puberty. J Clin Endocrinol Metab 61:997 – 1000

Moore GPM, Panaretto BA, Carter NB (1985) Epidermal hyperplasia and wool follicle regression in sheep infused with epidermal growth factor. J Invest Dermatol 84:172 – 175

Partanen A-M, Ekblom P, Thesleff I (1985) Epidermal growth factor inhibits morphogenesis and cell differentiation in cultured mouse embryonic teeth. Dev Biol 111:84 – 94

Schonbrunn A, Krasnoff M, Westendorf JM, Tashjian AH (1980) Epidermal growth factor and thyrotropin-releasing hormone act similarly on a clonal pituitary cell strain. Modulation of hormone production and inhibition of cell proliferation. J Cell Biol 85:786 – 797

Seppa H, Grotendorst G, Seppa S, Schiffman E, Martin GR (1982) Platelet-derived growth factor is chemotactic for fibroblasts. J Cell Biol 92:584 – 588

Stoscheck CM, Carpenter G (1984) Down regulation of epidermal growth factor receptors: direct demonstration of receptor degradation in human fibroblasts. J Cell Biol 98:1048 – 1053

Sugarman BJ, Aggarwal BB, Hass PE, Figari IS, Palladino MA, Shepard HM (1985) Recombinant human tumor necrosis factor-α: effects on proliferation of normal and transformed cells in vitro. Science 230:943 – 945

Sundell HW, Gray ME, Serenius FS, Escobedo MB, Stahlman MT (1980) Effects of epidermal growth factor on lung maturation in fetal lambs. Am J Pathol 100:707 – 726

Tucker RF, Shipley GD, Moses HL, Holley RW (1984) Growth inhibitor from BSC-1 cells closely related to platelet type B transforming growth factor. Science 226:705 – 707

Wright N, Alison M (1984) The biology of epithelial cell populations, vol 1. Oxford Univ Press, Oxford, p 3

Structural Relationships Between Growth Factor Precursors and Cell Surface Receptors

SUZANNE PFEFFER and AXEL ULLRICH

Analysis of the DNA-derived amino acid sequences of several growth factor precursors has revealed the presence of hydrophobic sequences possessing all the characteristics of membrane-spanning domains found in most cell surface receptors. This class of proteins includes the precursors for epidermal growth factor (EGF), transforming growth factor α (TGF-α), and vaccinia virus p19 (p19vacc). The presence of potential transmembrane domains in these growth factor precursors offers new insight into the evolution of growth factors and their cell surface receptors. The subsequent finding that the EGF precursor shares extensive homology with the low density lipoprotein (LDL) receptor and may not be processed in certain tissues has led to the notion that the EGF precursor may play a dual role as a precursor for a secreted growth factor and, in its unprocessed form, as a cell surface receptor of unknown biological function. Additional distinctive structural features shared by growth factor precursors and growth factor receptors point to a common evolutionary origin for these classes of proteins.

The EGF Precursor and Its Relatives

The 53-amino-acid mature EGF molecule is encoded within a 1217-amino-acid EGF precursor (Gray et al. 1983; Scott et al. 1983). It contains an internal stretch of 20 hydrophobic amino acids flanked by polar residues which could anchor the precursor in the membrane and would divide it into a 159-amino-acid cytoplasmic domain and an about 1010-amino-acid extracellular domain (Fig. 1). Why would a secreted growth factor be produced as part of a membrane-bound precursor, and why is the precursor more than 20 times larger than the mature growth factor? In addition to EGF coding sequences, the precursor contains eight EGF-related repeats which are found in two cysteine-rich clusters. Each unit encodes approximately 40 amino acids, including 6 cysteine residues (except repeat 1). These units may represent other biologically active EGF-like polypeptides but, unlike the polypeptide units in precursors such as pro-opiomelanocortin (Nakanishi et al. 1980), they are not flanked by obvious proteolytic cleavage sites in the form of pairs of dibasic amino acids (Arg Arg; Lys Arg). Since the 53 amino acid EGF sequence is not flanked by such endoprotease target sites, a processing enzyme of unknown specificity must cleave the 1217 amino acid prepro-EGF to liberate EGF; other EGF-like

Oncogenes and Growth Control
Edited by P. Kahn and T. Graf
© Springer-Verlag Berlin Heidelberg 1986

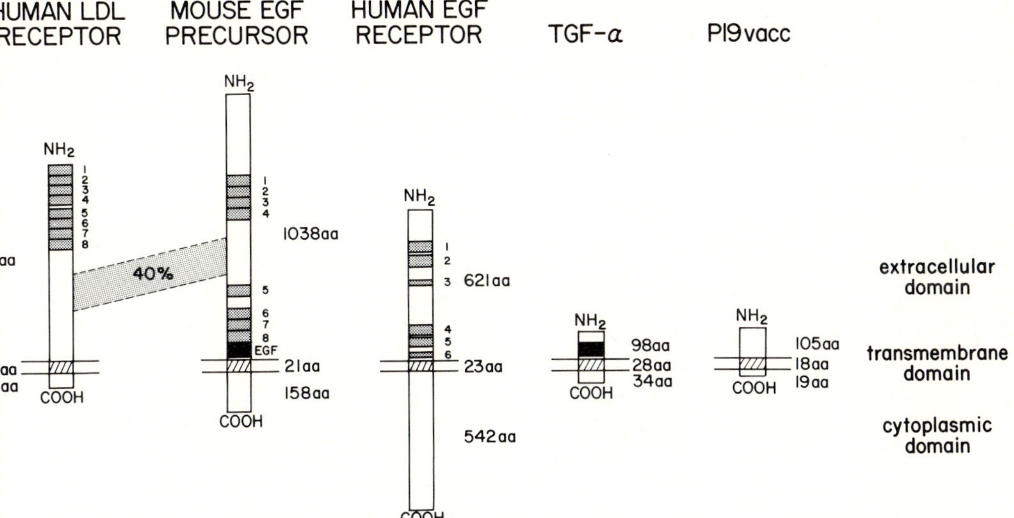

Fig. 1. Comparison of growth factor precursor and cell surface receptor structures. Mouse EGF precursor, TGF-α precursor and p19vacc are shown with putative signal sequences included since the amino-termini of their pro-forms are unknown. Cysteine-rich repeats are *shaded* and molecules are placed in a membrane bilayer with their putative (*hatched*) transmembrane domains. Mature growth factor sequences within their precursors are shown as *black boxes*. No evidence exists that any of the EGF-like repeats (*shaded*) in the EGF precursor actually represent biologically active entities

factors of related or different biological function might also be concomitantly released.

Another well-characterized mitogenic factor, TGF-α (see Derynck, this Vol.) is highly homologous to EGF, especially with respect to its content of cysteines and their specific position within the amino acid sequence. Like EGF, TFG-α is also synthesized as part of a potentially membrane-bound precursor polypeptide, which is much smaller (160 amino acids) than the EGF precursor and does not contain repeated units of TGF-α-related sequences (Fig. 1). The vaccinia virus p19 protein (p19vacc) contains a core region with homology to EGF and TGF-α. Like the TGF-α precursor, it contains only one EGF-related sequence, as well as a potential membrane anchor domain. Thus, EGF, TGF-α and p19vacc share related mature polypeptide sequences as well as potential membrane association. The EGF precursor differs in its repeat structure, most likely a result of sequence duplication.

Interestingly, the recently characterized product of the *Drosophila* notch gene is very similar to the EGF precursor in its overall structure (Fig. 2). The notch locus plays a role in the process by which ventral ectodermal cells differentiate into either epidermal or neural cell precursors. Mutations in this locus lead to differentiation of the ventral ectoderm into neuronal tissue at the expense of epidermal structures. The sequence of the notch gene product (Whar-

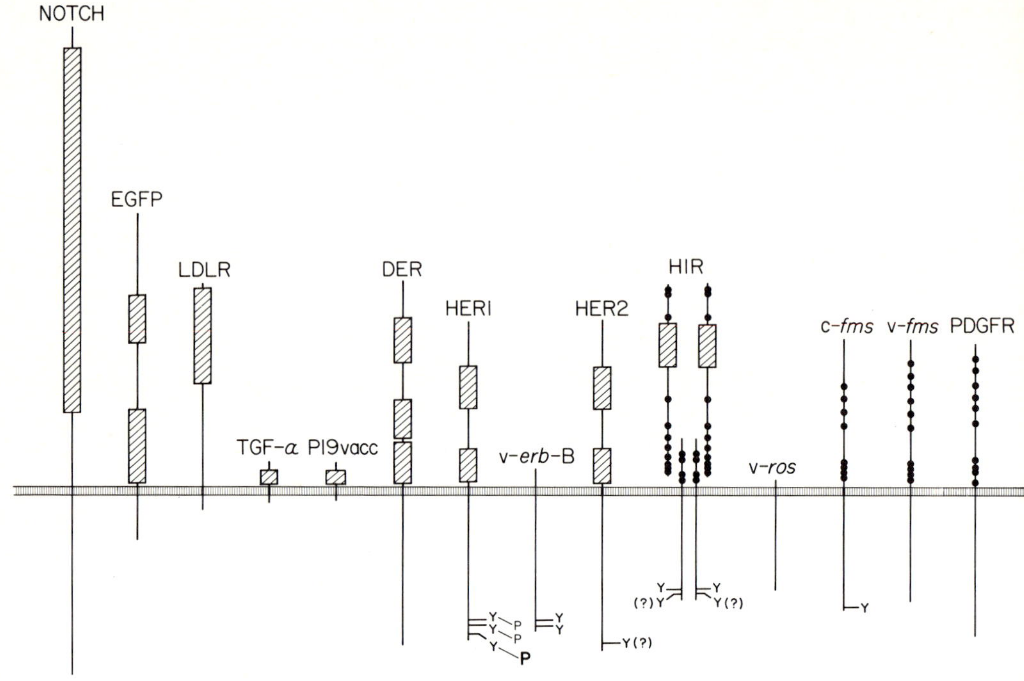

Fig. 2. Overlapping families of cysteine-rich cluster-containing molecules, tyrosine kinase receptors and oncogene products. Presumed structures of actual and putative cell surface polypeptides are shown. Cytoplasmic domains are shown *below* and extracellular regions *above* the plasma membrane bilayer. Cysteine-rich clusters are represented by *hatched boxes* and single cysteine residues by *filled circles*. Confirmed autophosphorylation substrate tyrosine residues of the human EGF receptor are indicated (*Y-P*), in addition to potential substrate sites (*Y*) in homologous locations of other tyrosine kinases. *EGFP* mouse EGF precursor; *LDLR* LDL receptor; *DER Drosophila* EGF receptor related sequence; *HER1* human EGF receptor; *HER2* human EGF receptor and *neu* oncogene related sequence; *HIR* human insulin receptor; *PDGFR* PDGF receptor

ton et al. 1985) predicts a 2703-amino-acid protein that contains a probable signal sequence as well as a potential transmembrane domain. The putative extracellular domain of the protein contains 36 of the prototypic cysteine-rich repeats that were first found in the EGF precursor; they are each about 40 amino acids in length and display the same conserved spacing of cysteine residues as seen in the EGF precursor.

What is the significance of the *notch*/EGF precursor homology? The presence of an amino-terminal signal sequence makes it likely that the *notch* protein is either a membrane or secreted protein. It is very striking that the *notch* protein has such a large putative intracellular domain (almost twice the size of the intracellular domain of the EGF receptor). The sequence is generally hydrophilic and proline-rich, suggestive of an extended structure of functional significance. The presence of the large intracellular domain supports the notion that *notch* is in fact a receptor, rather than a precursor for secreted hormones. However, it is still possible that the extracellular domain is proteolytically processed to yield biologically active peptides derived from the repeat sequences. In the case of the EGF precursor, it appears that the protein may remain intact in some tissues (Rall et al. 1985), yet be processed to generate the mature growth factor in other tissues. Such differential proteolytic processing could also regulate the function of the *notch* gene product in different tissues. The nematode *lin*-12 homeotic gene product also contains at least 11 cysteine-rich, EGF-homologous repeat units, the full significance of which awaits completion of the *lin*-12 sequence (Greenwald 1985).

Cell Surface Receptors Contain Cysteine-Rich Sequence Repeats

The structural motif of cysteine-rich clusters characteristic of the EGF precursor has recently also been found in a number of bona fide cell surface receptors, including those for LDL (Russell et al. 1984), EGF (Ullrich et al. 1984), insulin (Ullrich et al. 1985; Ebina et al. 1985), and a putative receptor with extensive homology to the EGF receptor (HER2; Coussens et al. 1985; Figs. 1, 2). In the LDL receptor, additional sequence homology outside the repeats has also been detected (Yamamoto et al. 1984; Russell et al. 1984). The most significant similarity (about 40%) is between EGF precursor amino acid residues 565–701 and residues 457–595 in the LDL receptor extracellular domain (Fig. 1). Sequence homology decreases in regions surrounding this core, but remains statistically significant for several hundred more amino acids. In addition to direct sequence homology, the extracellular domain of the LDL receptor also contains eight repeated, cysteine-rich units of about 40 amino acids in length. The spatial distribution of the cysteine residues is conserved in each of the repeat units in both molecules, but the relative positions of the cysteines differ somewhat between the LDL receptor and the EGF precursor repeat units. In both cases, other amino acid residues in these cysteine-rich repeats seem also to have been conserved. The repeat units of both proteins are distributed into two larger, distinguishable units of similar size which are arranged in tandem in the LDL receptor but are separated in the EGF precursor (Fig. 1).

The structural theme of repeated cysteine-rich sequences reappears as part of the extracellular ligand-binding region of the EGF receptor, which is the single cell surface receptor known to bind EGF, TGF-α and p19vacc. The EGF receptor extracellular domain contains two cysteine-rich clusters that can be divided into related sequence repeats based upon cysteine residue spacing and sequence homology (Fig. 1). The EGF receptor repeat units are enclosed within a larger duplicated sequence which involves the entire extracellular domain of the receptor. Beyond these repeat units, however, the EGF receptor does not display significant sequence homology with either the EGF precursor or the LDL receptor.

The recently isolated human and rat gene sequences that encode 1255 amino acid polypeptides with extensive regional homology to the human EGF receptor and v-*erbB* have been termed HER2/*neu*, and display all the structural features known to define a cell surface receptor for a polypeptide ligand (Coussens et al. 1985). Examination of the HER2/*neu* primary sequence revealed colinearity with most of the EGF receptor sequence. A hydrophobic N-terminal sequence of HER2/*neu* presumably functions as a signal sequence. As in the EGF receptor, the next 632 amino acids would represent the HER2 receptor extracellular domain that forms the binding pocket for a specific polypeptide ligand. The most conserved structural features in this domain include the repeats of cysteine-rich sequences. Of the 51 cysteine residues contained in the EGF receptor ligand binding domain, 50 are conserved in HER2/*neu*, of which 21 and 26, respectively, are clustered in two subdomains (Fig. 2). Sequences between these conserved, regularly spaced cysteine residues, as well as those flanking the cysteine-rich subdomains, have diverged, resulting in an overall homology of 40%.

Similarly, the structure of the human insulin receptor precursor reveals striking homologies to the human EGF receptor in terms of overall organization and primary sequence (Ullrich et al. 1985; Ebina et al. 1985; Fig. 2). However, only one region of high cysteine content (residues 159–312) is found in the α-subunit coding region, which is thought to be involved in insulin binding. This is in contrast to the EGF receptor, HER2/*neu*, and the LDL receptor, which consist of a single polypeptide chain and contain two clusters of similar cysteine content. It is tempting to speculate that each α-subunit of the heterotetrameric insulin receptor contributes a cysteine-rich region to form a functional domain analogous to the one found in single-chain receptors. Whether the cysteine-rich regions of the two α-subunits constitute the binding pocket for the insulin molecule together with portions of the amino-terminal region of the β-subunits will have to await further experimentation.

While the functional significance of the cysteine-rich repeat units for the EGF precursor and the above-mentioned receptors is presently unknown, they have been postulated to contribute to a multiply-looped, rigid LDL-binding domain in the LDL receptor (Yamamoto et al. 1984). Since cysteine-rich domains are found in a family of true as well as potential cell surface receptors with very diverse functions, they appear to form an essential structural element within the ligand-binding domain but do not define ligand specificity. Ligand specificity may be generated by sequences flanking these cysteine clusters, which may also influence the conformation and orientation of the clusters in the three-dimensional structure of the receptor's ligand-binding domain.

Cysteine-Rich Repeats and the Tyrosine Kinase Family of Receptors

The presence of regularly repeated, extracellular cysteine clusters (Pfeffer and Ullrich 1985) defines a family of bona fide and putative cell surface molecules that overlaps with another gene family: the family of membrane-spanning proteins that use an intrinsic tyrosine-specific kinase activity in the signal transduction process (Hunter and Cooper 1985). As described above, the *notch* gene product, EGF precursor, and LDL receptor may be or are membrane-associated proteins that possess characteristic cysteine-rich repeat units (Fig. 2 left). Human and *Drosophila* EGF receptors, the HER2/*neu* protein, and the insulin receptor each share these features and additionally contain an intracellular tyrosine-kinase domain (Fig. 2, center). A third group of proteins contains a highly homologous intracellular tyrosine kinase domain but lacks the characteristic extracellular cysteine-rich repeat units (Fig. 2, right). Proteins such as v-*fms*, the oncogenic homolog of the c-*fms*/CSF-1 receptor (Sherr et al. 1985; Coussens et al. 1986 and Sherr and Stanley, this Vol.), and the platelet-derived growth factor (PDGF) receptor (Yarden et al. 1986) each possess numerous extracellular cysteine residues, which are conserved in terms of their relative spatial organization (Fig. 2). The first six amino-terminal cysteine residues in both receptors are flanked by related sequences and are most likely the result of a sequence multiplication event. Thus these genes have used a variety of ancestral building blocks to construct a family of diverse cell surface molecules which contribute to the complex communication requirements of cells within highly differentiated multicellular organisms.

The notion that this family of proteins evolved from a few common ancestors is further supported by the following observation (Sudhof et al. 1985). The region of homology between the EGF precursor and LDL receptor is encoded by eight contiguous exons in each respective gene. Of the nine introns that separate these exons, five are located at the same position in the two sequences. This suggests that these portions of the two proteins evolved from a common ancestor by acquisition of these eight exons; the LDL receptor and EGF precursor may then have acquired other introns and the sequences then diverged.

It is apparent that a number of domain-duplication events generated the clusters of cysteine-rich repeat units within this family of proteins. At some point in evolution, a recombination event may have grafted a tyrosine kinase domain onto an ancestor of this family, which then continued to evolve additional cysteine-rich repeat units. One can envision parallel evolution of growth factors, growth-factor-binding proteins and tyrosine kinase domains, portions of which were mixed and matched and then refined according to functional requirements. It is possible, for example, that the *notch* protein, with its large intracellular domain and numerous cysteine-rich repeat units, is an ancestral intermediate between receptor and growth factor in this evolutionary process. Alternatively, the *notch* gene product may be a highly sophisticated signal transducer analogous to the EGF receptor.

References

Coussens L, Yang-Feng TL, Liao Y-C, Chen E, Gray A, McGrath J, Seeburg PH, Libermann TA, Schlessinger J, Francke U, Levinson A, Ullrich A (1985) Tyrosine kinase receptor with extensive homology to EGF receptor shares chromosomal location with *neu* oncogene. Science 230:1132–1139

Coussens L, Van Beveren C, Smith D, Chen E, Mitchell RL, Isacke CM, Verma IM, Ullrich A (1986) Structural alteration of viral homologue of receptor proto-oncogene *fms* at carboxyl terminus. Nature 320:277–280

Ebina Y, Ellis L, Jarnagin K, Edery M, Graf L, Clauser E, Ou J, Masirz F, Kan YW, Goldfine ID, Roth RA, Rutter WJ (1985) The human insulin receptor cDNA: The structural basis for hormone-activated transmembrane signalling. Cell 40:747–758

Gray A, Dull TJ, Ullrich A (1983) Nucleotide sequence of epidermal growth factor cDNA predicts a 128,000-molecular weight protein precursor. Nature 303:722–725

Greenwald I (1985) lin-12, a nematode homeotic gene, is homologous to a set of mammalian proteins that includes epidermal growth factor. Cell 43:583–590

Hunter T, Cooper JA (1985) Protein-tyrosine kinases. Annu Rev Biochem 54:897–930

Nakanishi S, Teranishi Y, Noda M, Notake M, Watanabe Y, Kakidani H, Jingami H, Numa S (1980) The protein-coding sequence of the bovine ACTH-β-LPH precursor gene is split near the signal peptide region. Nature 287:752–755

Pfeffer S, Ullrich A (1985) Epidermal growth factor: Is the precursor a receptor? Nature 313:184

Rall LB, Scott J, Bell GI, Crawford RJ, Penschow JD, Niall HD, Coghlan JP (1985) Mouse pre-pro-epidermal growth factor synthesis by the kidney and other tissues. Nature 313:228–231

Russell DW, Schneider WJ, Yamamoto T, Luskey KL, Brown MS, Goldstein JL (1984) Domain map of the LDL receptor: Sequence homology with the epidermal growth factor precursor. Cell 37:577–585

Scott J, Urdea M, Quiroga M, Sanchez-Pescador R, Fong N, Selby M, Rutter WJ, Bell GI (1983) Structure of a mouse submaxillary messenger RNA encoding epidermal growth factor and seven related proteins. Science 221:236–240

Sherr CJ, Rettenmier CW, Sacca R, Roussel MF, Look AT, Stanley ER (1985) The c-*fms* proto-oncogene product is related to the receptor for the mononuclear phagocyte growth factor, CSF-1. Cell 41:665–676

Sudhof TC, Goldstein JL, Brown MS, Russell DW (1985) The LDL receptor gene: A mosaic of exons shared with different proteins. Science 228:815–822

Ullrich A, Coussens L, Hayflick JS, Dull TJ, Gray A, Tam AW, Lee J, Yarden Y, Libermann TA, Schlessinger J, Downward J, Mayes ELV, Whittle N, Waterfield MD, Seeburg PH (1984) Human epidermal growth factor receptor cDNA sequence and aberrant expression of the amplified gene in A431 epidermoid carcinoma cells. Nature 309:418–425

Ullrich A, Bell JR, Chen EY, Herrera R, Petruzzelli LM, Dull TJ, Gray A, Coussens L, Liao Y-C, Tsubokawa M, Mason A, Seeburg PH, Grunfeld C, Rosen OM, Ramachandran J (1985) Human insulin receptor and its relationship to the tyrosine kinase family of oncogenes. Nature 313:756–761

Wharton KA, Johansen KM, Xu T, Artavanis-Tsakonas S (1985) Nucleotide sequence from the neurogenic locus notch implies a gene product that shares homology with proteins containing EGF-like repeats. Cell 43:567–581

Yamamoto T, Davis CG, Brown MS, Schneider WJ, Casey ML, Goldstein JL, Russell DW (1984) The human LDL receptor: A cysteine-rich protein with multiple Alu sequences in its mRNA. Cell 39:27–38

Yarden Y, Rhee L, Chen E, Ullrich A, Escobedo JA, Williams L (1986) Structure of the mouse PDGF receptor. Nature (in press)

Regulation of Cell Growth by the EGF Receptor

JOSEPH SCHLESSINGER

Epidermal growth factor (EGF) is a small protein of 53 amino acids which acts as a mitogen for various cell types in vitro and in vivo (Carpenter and Cohen 1979 and Carpenter et al., this Vol.). Several lines of evidence suggest that the receptor for this factor can also play a role in the uncontrolled proliferation characteristic of neoplastic cells. First, the v-*erbB* oncogene of avian erythroblastosis virus encodes a truncated EGF receptor (Downward et al. 1984) and we recently proposed that the v-*erbB* protein transforms by functioning as an activated growth factor receptor (see also article by Beug et al.). Second, various animal and human tumor cells produce a growth factor called transforming growth factor-α (TGF-α; Todaro et al. 1980 and article by Derynck). This growth factor is highly related to EGF; both factors bind to the EGF receptor with similar affinities and induce the proliferation of cells bearing the EGF receptor. It has been suggested that TGF-α plays a role in oncogenesis by inducing autocrine growth (Todaro et al. 1980). Finally, the EGF receptor gene is amplified and rearranged in a significant proportion of human brain tumors of glial origin (Libermann et al. 1985). The resultant overexpression of the EGF receptor may play a role in the development or progression of these tumors. Hence, it is clear that the investigation of the structure of the EGF receptor and the mechanism of its activation may provide important clues to fundamental questions underlying the mechanisms of normal growth control and neoplasia.

Structure and Evolution of the EGF Receptor

Following the purification of the human EGF receptor by immunoaffinity chromatography (Yarden et al. 1985) and its partial sequencing (Downward et al. 1984), the complete amino acid sequence of the EGF receptor was deduced from the nucleotide sequence of cDNA clones (Ullrich et al. 1984). The mature receptor is composed of 1186 amino acid residues which are preceded at the N-terminal end by a signal peptide of 24 hydrophobic amino acids (see also Pfeffer and Ullrich, this Vol.). The signal peptide is cleaved following the insertion of the nascent receptor into the membrane of the endoplasmic reticulum. The receptor is cotranslationally glycosylated and transported through the Golgi apparatus to the plasma membrane.

Oncogenes and Growth Control
Edited by P. Kahn and T. Graf
© Springer-Verlag Berlin Heidelberg 1986

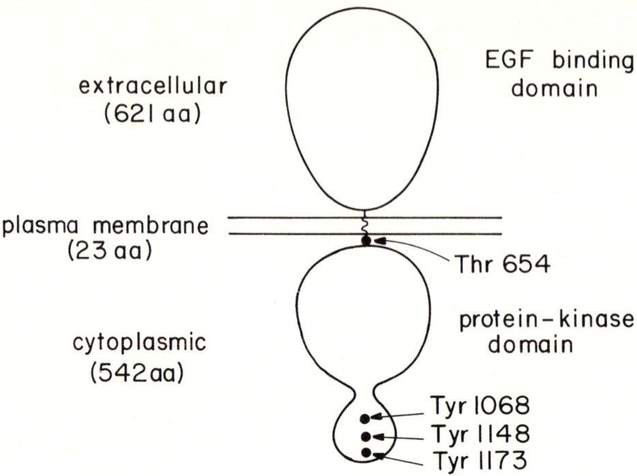

Fig. 1. A model of the EGF receptor. The EGF receptor is composed of three domains. (1) A large extracellular ligand binding domain composed of 621 amino acids (aa). (2) A transmembrane region composed of 23 hydrophobic amino acids. (3) A cytoplasmic region composed of 542 amino acids and containing the protein kinase domain and phosphorylation sites. Threonine (654) is phosphorylated by protein kinase C. The three known autophosphorylation sites are: Tyr(1068), Tyr(1148) and Tyr(1173)

The mature receptor is composed of three major structural elements (Fig. 1). The extracellular EGF-binding domain is composed of 621 amino acid residues and is anchored in the plasma membrane by a single transmembrane region of 23 hydrophobic amino acids. The transmembrane region is followed by a sequence of mostly basic residues, a feature common to many membrane proteins. The cytoplasmic domain of the EGF receptor contains 542 amino acids. We have used immunological probes to demonstrate that the N-terminal end of the receptor is extracellular while its C-terminal end faces the cytoplasm (Kris et al. 1985). The cytoplasmic domain contains a region of approximately 300 amino acid residues which show a high degree of homology to the catalytic domain of the protein tyrosine kinases encoded by the *src*-related oncogenes (Ullrich et al. 1984). Like the other protein-tyrosine kinases, the catalytic domain of the EGF receptor kinase contains a lysine residue which is located 15 residues to the C-terminal side of the consensus sequence Gly – X – Gly – X – phe – Gly – X – Val. The lysine residue, together with the consensus sequence, probably functions as part of the ATP binding site (reviewed by Hunter and Cooper 1985).

The binding of EGF to its receptor leads to the activation of the receptor tyrosine kinase, which phosphorylates various cellular proteins, as well as the EGF receptor itself. In intact cells autophosphorylation occurs mainly on *tyr*(1173), a residue which is deleted in the v-*erbB* protein (Yamamoto et al. 1983). However, at least two additional tyrosine residues are autophosphorylated when EGF is added to solubilized membranes or to the pure receptor (Fig. 1). It is noteworthy that the three known autophosphorylation sites are located at the C-terminal end of EGF receptor (Downward et al. 1985). The mechanism and role of the EGF receptor autophosphorylation are not yet understood.

A remarkable feature of the EGF receptor extracellular domain is the high proportion of cysteine residues. Most cysteines are clustered in two regions each 160 residues long. The two cysteines can be aligned, forming internal repeats with similar spacing between the cysteine residues. Cysteine-rich domains have also been found in the extracellular domains of the LDL receptor, insulin receptor, and a human protein highly homologous to the EGF receptor, termed HER2 (reviewed by Pfeffer and Ullrich 1985 and this Vol., and by Coussens et al. 1985). This latter protein is highly homologous to the EGF receptor, mostly within the tyrosine kinase domain (Coussens et al. 1985). It is possible that HER2 functions as a membrane receptor for an as yet unknown growth factor or hormone. The HER2 protein is probably the human counterpart of the rat *neu* oncogene product, an oncogene associated with neuroblastomas (Schechter et al. 1984). Interestingly the cysteine-rich domains of the EGF receptor and HER2 can be aligned with the cysteine-rich domain of the insulin receptor, suggesting that these domains probably evolved from a common ancestor.

Insight into the evolution of the various EGF receptor domains was obtained from the deduced amino acid sequence of the *Drosophila* homolog of the EGF receptor gene (Livneh et al. 1985; Schejter et al. 1986). The *Drosophila* homolog shows extensive sequence homology to the human EGF receptor and to HER2. Like the human receptor, the *Drosophila* protein shows three distinct domains: an extracellular EGF-binding domain, a hydrophobic transmembrane region and a cytoplasmic kinase domain. The overall amino acid homology is 55% in the kinase domain and 40% in the extracellular domain. Sequence analysis of the *Drosophila* receptor extracellular domain reveals three cysteine-rich clusters (Schejter et al. 1986). The additional cysteine-rich domain in the *Drosophila* protein was probably generated by a duplication of one of the two cysteine-rich domains. The striking conservation of the cysteine-rich domains and the multiple duplication events they have undergone suggests that these regions play an important role in the function of the receptor. However, they probably do not form the ligand-combining site, since receptors with different specificities share the same sequences, and since these clusters are more rigid than would be expected if they are part of a ligand-combining site.

Role of the Functional Domains of the EGF Receptor and v-*erbB*

Binding experiments of ^{125}I-EGF to the EGF receptor on living cells reveal two distinct affinity states (King and Cuatrecasas 1982). "High affinity" sites usually represent 5 – 10% of the total receptor population, depending on cell type. This correlates well with the optimal receptor occupancy required for the initiation of DNA synthesis and it was therefore suggested that the "high affinity" receptors play a role in the mitogenic signaling process. The treatment of fibroblasts and other cells with the tumor promoter phorbol ester (TPA) abolished the "high affinity" state of the EGF receptor (Brown et al. 1979; Shoyab et al. 1979) and reduced its protein-tyrosine kinase activity. TPA activates the Ca^{2+}-sensitive protein kinase C, which in turn phosphorylates the EGF receptor at on several sites. One of these sites is Thr(654), a residue located ten amino within the cytoplasmic domain of EGF-receptor (Hunter et al. 1984; Davis and Czech 1985). Based on these results, it was suggested that the phosphorylation of EGF receptor on Thr(654) regulates the affinity of the extracellular domain towards the ligand and the enzymatic activity of the protein tyrosine kinase domain (see Hunter, this Vol.).

The v-*erbB* oncogene of the avian erythroblastosis virus (AEV) encodes a truncated EGF receptor. Its product, the GP 74^{v-erbB} protein, lacks most of the extracellular EGF-binding domain, as well as the 32 amino acid residues at the C-terminal end of the receptor (containing the major autophosphorylation site of the EGF receptor). It was shown that the v-*erbB* protein possesses intrinsic protein-tyrosine kinase activity toward exogenous substrates and that it also undergoes autophosphorylation (Kris et al. 1985), although the latter sites have not yet been identified.

Erythroblastosis in chickens can also be induced by the chronic avian leukosis virus (ALV), which does not carry an oncogene in its genome. In this case, leukemogenesis appears to result from activation of the c-*erbB*/EGF receptor gene by a promoter insertion mechanism involving integration of the provirus 5' to the region where homology to v-*erbB* starts (Nilsen et al. 1985). Indeed these leukemias express truncated EGF receptor proteins with an intrinsic protein-tyrosine kinase activity and with molecular weights similar to the v-*erbB* protein (Lax et al. 1985). Interestingly, the "v-*erbB*-like" proteins expressed in these leukemias contain the coding information for 34 amino acids homologous to the the human EGF receptor, but missing from v-*erbB* (Nilsen et al. 1985). Hence, it seems that the C-terminal deletion is not required for erythroblastosis. Moreover, recent studies utilizing in vitro site-directed mutagenesis, together with transfection, indicated that a "double truncated" human EGF receptor will transform the rat 1 fibroblast cell line while a receptor mutant lacking the C-terminal deletion fails to do so (Riedel et al. 1986). Although preliminary, these results suggest that the C-terminal end of the EGF receptor may play a role in the tissue specificity of transformation (see Beug et al., this Vol.).

Mechanism of Receptor Activation

An important unanswered question is the mechanism by which binding of the EGF to the extracellular domain of the EGF receptor stimulates the receptor kinase activity in the cytoplasmic portion. Numerous studies have shown that EGF can stimulate the tyrosine kinase activity of its receptor molecule without the involvement of an additional molecule. Hence, two types of model can be proposed for the signal transduction across the plasma membrane: an intramolecular model and an intermolecular model. According to the first view, the binding of EGF induces a conformational change in the extracellular domain and this in turn is transmitted through the membrane-spanning region to the kinase domain which is consequently structurally altered and further activated. Alternatively, the intermolecular model postulates that EGF-induced receptor aggregation activates the kinase activity, and it therefore bypasses the requirement imposed by the first model for a conformational change to be transferred through the transmembrane region. The finding that EGF induces receptor aggregation in intact cells raises the possibility that this plays a role in the cascade of events leading to the mitogenic signal and is consistent with the second model. However, it is not known whether EGF-induced receptor aggregation is an intrinsic property of the receptor molecule or whether additional molecules participate in the in vivo clustering process.

How could receptor aggregation transduce the signal from the extracellular to the intracellular domain? Here I outline the main features of an allosteric aggregation model for EGF receptor. We have proposed that monomeric EGF receptor is in equilibrium with various aggregation states of the receptor (Fig. 2). To simplify the discussion. We describe aggregation as dimer formation (Yarden and Schlessinger 1987a, b). In the illustration below, RC stands for monomeric EGF receptor, R stands for the binding domain of EGF receptor, C stands for the kinase domain and H stands for the ligand.

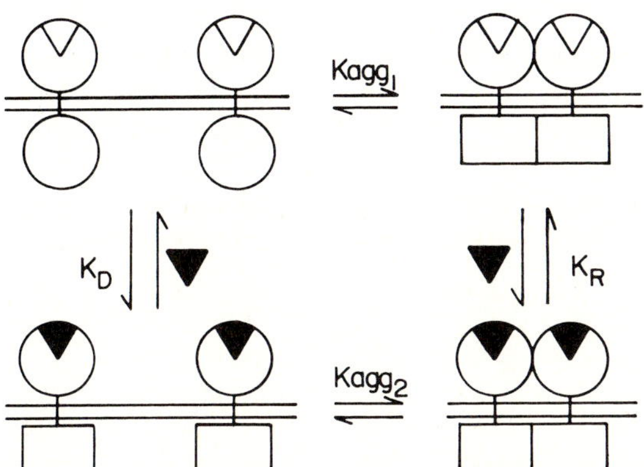

Fig. 2. An allosteric aggregation model for receptor kinase activation by EGF. Monomeric EGF receptor molecules with low kinase activity and low binding affinity for EGF (K_D) exist in equilibrium with aggregated receptor molecules, which have high affinity for EGF (K_R) and with high kinase activity. Kagg$_1$ is the aggregation constant of the free receptor. Kagg$_2$ is the aggregation constant of the occupied receptors. The receptor is schematically illustrated as consisting of three portions which are not drawn to scale

It is assumed that the monomeric receptor (RC) has low kinase activity and low binding affinity toward EGF (K_D) and that the dimeric receptor has stimulated kinase activity ($RC'C'R'$) and high binding affinity toward EGF (K_{R1}, K_{R2}). To further simplify the picture, we assume that the sites on each dimer are identical and that they do not interact, namely: $K_{R1} = K_{R2} = K_R$. Hence

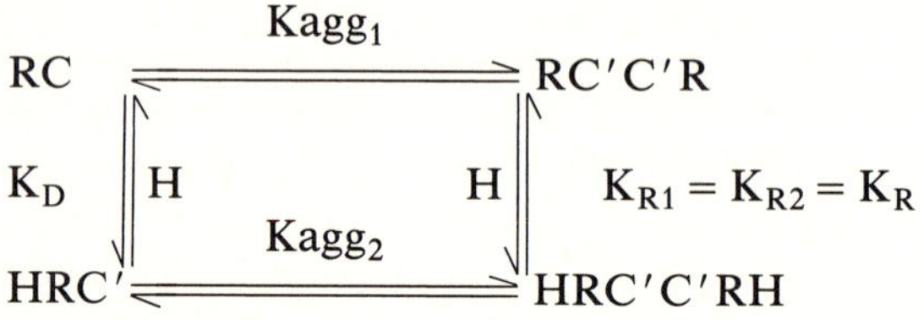

$Kagg_1 = [RC]^2/[RC'C'R]$ is the aggregation constant in the absence of the ligand and $Kagg_2 = [HRC']^2/[HRC'C'RH]$, in its presence $K_D = [RC] \cdot [H]/[RC'H]$; $K_{R1} = [H] \cdot [RC'C'R]/[RC'C'RH]$ and $K_{R2} = ([H] \cdot [RC'C'RH])/[HRC'C'RH]$. Based on microscopic reversibility, $Kagg_2 \cdot K_D^2 = Kagg_1 \cdot K_{R1} \cdot K_{R2}$ or

$$\frac{Kagg_2}{Kagg_1} = \frac{K_R^2}{K_D^2}.$$

If $K_R > K_D$ by a factor of $10-100$, then $Kagg_2 > Kagg_1$ by a factor of $100-10000$. Hence, the binding energy $G = -RT \ln(K_R/K_D)$ confers the aggregated state upon the EGF receptor, which in turn leads to elevated tyrosine kinase activity.

An important feature of the allosteric aggregation model is the existence of at least two aggregation states of EGF receptor with different affinities towards EGF. It was reported that receptor aggregation may be involved in the signal transduction process and that it may be associated with the heterogeneity in the affinity of the EGF binding sites toward the growth factor (reviewed by Schlessinger et al. 1983). It is noteworthy that the pure EGF receptor protein (170 kDa) has a single EGF binding site, which further supports the idea that the heterogeneity in binding sites results from either receptor: receptor or receptor: other molecule interactions. Interconversion of low affinity receptors to high affinity receptors occurs in various cultured cells (King and Cuatrecasas 1982), and it was suggested that the high affinity receptors are those which are required for the mitogenic signal. As already mentioned, TPA inhibits the appearance of the high affinity receptors and also reduces the capacity of the kinase domain of the EGF receptor to induce tyrosine phosphorylation (reviewed by Hunter and Cooper 1985).

The allosteric aggregation model may also provide a simple theoretical framework in which to analyze the effect of the phosphorylation of Thr(654) on the binding and enzymatic activity of the EGF receptor. It postulates that phosphorylation of the EGF receptor by protein kinase C prevents receptor aggregation, that is, the phosphorylated receptor has a reduced aggregation constant ($Kagg_1 \simeq 0$). This in turn leads to the appearance of the low affinity receptors with reduced tyrosine kinase activity.

Obviously, further experiments are required to test the validity of the allosteric aggregation model as a whole. Meanwhile the model provides a useful working hypothesis on which further research can be based.

References

Brown KD, Dicker P, Rozengurt E (1979) Inhibition of epidermal growth factor binding to surface receptors by tumor promoters. Biochem Biophys Res Commun 86:1037–1043

Carpenter G, Cohen S (1979) Epidermal growth factor. Annu Rev Biochem 48:193–216

Coussens L, Yang-Feng TL, Liao Y-C, Chen E, Gray A, McGrath J, Seeburg PH, Libermann TA, Schlessinger J, Francke U, Levinson A, Ullrich A (1985) Tyrosine kinase receptor with extensive homology to EGF receptor shares chromosomal location with *neu* oncogene. Science 230:1132–1139

Davis RJ, Czech MP (1985) Tumor-promoting phorbol diesters cause the phosphorylation of epidermal growth factor receptors in normal human fibroblasts at threonine 654. Proc Natl Acad Sci USA 82:1974–1978

Downward J, Yarden Y, Mayes E, Scrace G, Totty N, Stockwell P, Ullrich A, Schlessinger J, Waterfield MD (1984) Close similarity of epidermal growth factor receptor and v-*erbB* oncogene protein sequences. Nature 307:521–527

Downward J, Parker P, Waterfield MD (1985) Autophosphorylation sites in the epidermal growth factor receptor. Nature 311:483–485

Hunter T, Cooper JA (1985) Protein-tyrosine kinases. Annu Rev Biochem 54:897–930

Hunter T, Ling N, Cooper NA (1984) Protein kinase C phosphorylation of the EGF receptor at a threonine residue close to the cytoplasmic face of the plasma membrane. Nature 314:480–483

King AC, Cuatrecasas P (1982) Resolution of high and low affinity epidermal growth factor receptors: inhibition of high affinity component by low temperature, cycloheximide and phorbol esters. J Biol Chem 257:3053–3060

Kris R, Lax I, Gullick M, Waterfield M, Ullrich A, Fridkin M, Schlessinger J (1985) Antibodies against a synthetic peptide as a probe for the kinase activity of the avian EGF receptor and v-*erbB* proteins. Cell 40:619–625

Lax I, Kris R, Sasson I, Ullrich A, Hayman MJ, Beug H, Schlessinger J (1985) Activation of c-*erbB* in avian leukosis virus-induced erythroblastosis leads to the expression of a truncated EGF receptor kinase. EMBO J 4:3179–3182

Libermann TA, Nussbaum HR, Razon N, Kris R, Lax I, Soreq M, Whittle N, Waterfield MD, Ullrich A, Schlessinger J (1985) Amplification, enhanced expression, and possible rearrangement of the EGF receptor gene in primary human brain tumors of glial origin. Nature 313:144–147

Livneh E, Glaser L, Segal D, Schlessinger J, Shilo BZ (1985) The *Drosophila* EGF receptor gene homolog shows conservation of both hormone binding and kinase domains. Cell 40:599–607

Nilsen TW, Maroney PA, Goodwin RG, Rottman FM, Crittenden LB, Raines MA, Kung H-J (1985) c-*erbB* activation in ALV-induced erythroblastosis: Novel RNA processing and promoter insertion result in expression of an amino-truncated EGF-receptor. Cell 41:719–726

Pfeffer S, Ullrich A (1985) Is the precursor a receptor? Nature 313:184

Riedel H, Lee J, Dull TJ, Kris RM, Waterfield MD, Schlessinger J, Ullrich A (1986) Structural alterations that convert the EGF receptor gene into an oncogene. Submitted

Schechter AL, Stern DF, Vaidyanathan L, Decker SJ, Drebin JA, Greene MI, Weinberg RA (1984) The *neu* oncogene: An *erbB*-related gene encoding a 185,000-M_r tumour antigen. Nature 312:513–516

Schejter ED, Glazer L, Segal D, Ullrich A, Schlessinger J, Shilo BZ (1986) Evolution of the EGF receptor gene family. Submitted

Schlessinger J, Schreiber AB, Levi A, Lax I, Libermann T, Yarden Y (1983) Regulation of cell proliferation by epidermal growth factor. CRC Crit Rev Biochem 14:93−111

Shoyab M, DeLarco JE, Todaro GJ (1979) Biologically active phorbol esters specifically alter affinity of epidermal growth factor receptors. Nature 279:387−391

Todaro GJ, Fryling C, DeLarco JE (1980) Transforming growth factors produced by certain human tumor cells: polypeptides that interact with human EGF receptors. Proc Natl Acad Sci USA 77:5258−5262

Ullrich A, Coussens L, Hayflick JS, Dull TJ, Gray A, Tam AW, Lee J, Yarden Y, Libermann TA, Schlessinger J, Downward J, Mayes ELV, Whittle N, Waterfield MD, Seeburg PH (1984) Human epidermal growth factor receptor cDNA sequence and aberrant expression of the amplified gene in A-431 epidermoid carcinoma cells. Nature 309:418−425

Yamamoto T, Nishida T, Miyajima N, Kawai S, Ooi T, Toyoshima K (1983) The *erbB* gene of avian erythroblastosis virus is a member of the *src* gene family. Cell 35:71−78

Yarden Y, Harari I, Schlessinger J (1985) Purification of an active EGF receptor kinase with monoclonal anti-receptor antibodies. J Biol Chem 260:315−319

Yarden Y, Schlessinger J (1987a) Self-phosphorylation of EGF receptor: Evidence for a model of intermolecular allosteric activation. Submitted

Yarden Y, Schlessinger J (1987b) EGF induces rapid, reversible aggregation of the purified EGF receptor. Submitted

Mutational Analysis of v-*erbB* Oncogene Function

HARTMUT BEUG, MICHAEL J. HAYMAN, and BJÖRN VENNSTRÖM

The v-*erbB* oncogene is contained in two strains of avian erythroblastosis virus (AEV) and encodes their capacity to cause erythroleukemia and sarcomas in chickens. Recent work has revealed that the v-*erbB* gene product represents a truncated and mutated version of the EGF receptor (Downward et al. 1984; Ullrich et al. 1984). This finding was surprising in light of the fact that AEV in hematopoietic cell lineages selectively transforms erythroid cells, which are not known to express the epidermal growth factor (EGF) receptor. Sequence comparisons showed that v-*erbB* is also related to the family of oncogenes encoding tyrosine kinases (Yamamoto et al. 1983).

In this article we discuss the mutational events in the avian EGF receptor gene (c-*erbB*) that lead to its activation as an oncogene. We also focus on the question of how the inappropriate expression of a mutated EGF receptor in erythroid progenitor cells interferes with the control of proliferation and differentiation.

Cell Transformation Capacity of v-*erbB*

V-*erbB*-containing retroviruses replicate in different types of bone marrow cells (Graf et al. 1980) but selectively transform erythroid progenitors which correspond to late BFU-E or early CFU-E stages (Samarut and Gazzolo 1982). In the presence of erythropoietin (EPO), these progenitors undergo four to seven cell divisions while differentiating into erythrocytes, whereas in absence of the hormone they cease dividing and then disintegrate. Transformation of chick erythroid cells by viruses containing the tyrosine kinase encoding oncogenes v-*erbB*, v-*src*, v-*fps*, v-*sea* or other oncogenes such as v-Ha-*ras* and v-*mil* results in the induction of extensive self-renewal capacity, resulting in the outgrowth of essentially immature erythroblasts that can be expanded into mass cultures. However, the transformed cells become committed to erythroid differentiation with a certain frequency and then mature into erythrocytes within 2 – 3 days.

Neither proliferation nor differentiation requires EPO, and no EPO-like activity could be detected in the culture medium from any of the transformed cells (Kahn et al. 1984; Beug et al. 1985, and unpublished observations; see also Kahn et al., this Vol.). This raises the possibility that oncogenes which transform erythroid cells constitutively activate and/or modulate signal trans-

Oncogenes and Growth Control
Edited by P. Kahn and T. Graf
© Springer-Verlag Berlin Heidelberg 1986

ducing pathways normally employed by erythroid-specific receptors such as the EPO receptor (Beug et al. 1985).

In addition to its erythroblast-transforming capacity, v-*erbB* is able to transform several types of mesenchymal cell, again resembling other tyrosine kinase oncogenes such as v-*src*. However, v-*erbB* is more efficient than v-*src* in transforming erythroblasts and induces a less pronounced transformation phenotype in fibroblasts (Kahn et al. 1984; Royer-Pokora et al. 1978). Furthermore, v-*erbB* and v-*src* abolish the strict requirement of chick heart mesenchymal cells for EGF and insulin, although only the latter oncogene is able to induce anchorage-independent growth (Balk et al. 1984). This again suggests that v-*erbB* transforms its target cells by constitutively producing signals which are normally induced only as a consequence of growth-factor : receptor interaction.

Biosynthesis of the v-*erbB* Gene Product and Role of the N-Terminal Truncation

The v-*erbB* gene products of both AEV strains resemble the EGF receptor in that their mature forms represent transmembrane glycoproteins which are synthesized on membrane-bound polyribosomes (Privalsky and Bishop 1984). They are cotranslationally modified by the addition of one or two carbohydrate chains (Schmidt et al. 1985) that are further processed into the complex carbohydrate forms characteristic of many cell surface glycoproteins. During their processing the v-*erbB* proteins are translocated from the rough endoplasmic reticulum (ER) to the plasma membrane, where they become accessible to antibodies that detect extracellular epitopes (Beug and Hayman 1984; Hayman and Beug 1984).

The v-*erbB* proteins of AEV-ES4 and AEV-H exhibit an N-terminal domain of only 61 amino acids as compared to 622 amino acids in the corresponding extracellular domain of the EGF receptor (Fig. 1; see below; also Pfeffer and Ullrich, this Vol., and Schlessinger, this Vol.). The presence of this N-terminal truncation, which includes the signal sequence, might explain why only 10−25% of the v-*erbB* protein synthesized in the rough ER is correctly processed and inserted into the plasma membrane. Consequently, the bulk of the v-*erbB* protein persists in intracellular membranes, where it seems to be degraded (Hayman and Beug 1984; Privalsky and Bishop 1984). The N-terminal deletion presumably also abolishes the ability of the protein to bind EGF.

It is not known whether, as suggested for the EGF receptor (see Schlessinger, this Vol.), clustering and internalization of v-*erbB* proteins is required for mitogenic signaling. A possible function of the extracellular "stump" of the v-*erbB* protein in transformation is suggested by site-directed mutagenesis: of the N-terminus: deletion of the N-terminal half of the extracellular domain abolishes transforming activity, whereas removal of the C-terminal half does

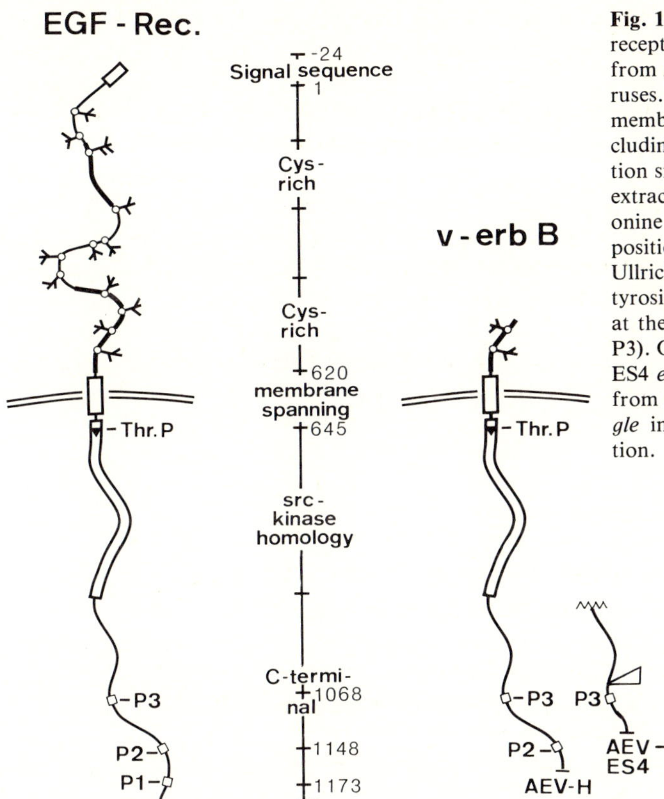

Fig. 1. Comparison of the EGF receptor with the *erbB* proteins from AEV-H and AEV-ES4 viruses. The diagram shows the membrane inserted forms including the potential glycosylation sites (small branches in the extracellular domain), the threonine phosphorylation site in position 654 (see Parker and Ullrich, this Vol.) and the three tyrosine phosphorylation sites at the C-terminal end (P1, P2, P3). Only that portion of AEV-ES4 *erbB* protein which differs from AEV-H is shown. *Triangle* indicates an internal deletion.

not, despite the fact that both mutant genes synthesize similar amounts of v-*erbB* protein (M. Jansson, H. Beug and B. Vennström, unpublished data). It remains to be determined whether or not the former mutation affects the transport of v-*erbB* to the plasma membrane.

Tyrosine Kinase Activity of the v-*erbB* Proteins

The EGF receptor contains three major tyrosine autophosphorylation sites (P1, P2, P3) in its C-terminal domain (see Schlessinger, this Vol.). The v-*erbB* protein of AEV-H lacks 34 C-terminal amino acids, including P1 (Downward et al. 1984), whereas the AEV-ES4 protein exhibits a more extensive truncation encompassing P1 and P2; in addition, it exhibits a deletion of 21 amino acids between the tyrosine kinase domain and P3 as well as several point mutations (Fig. 1; P. Scotting, personal communication).

Both v-*erbB* proteins were found to be autophosphorylated on tyrosine and to phosphorylate on tyrosine in vitro a variety of artificial substrates such as glucose 6-phosphate dehydrogenase and angiotensin (Kris et al. 1985; Hayman et al. 1986; I. Lax and J. Schlessinger, personal communication). In contrast to the chicken EGF receptor, however, neither of the v-*erbB*-kinases could be further stimulated by the addition of EGF (Kris et al. 1985). The AEV-ES4 v-*erbB* protein also differs from chicken EGF receptor in that it exhibits little or no autophosphorylation on tyrosine in virus-transformed fibroblasts metabolically labeled with ^{32}P-orthophosphate (Gilmore et al. 1985; Decker 1985). However, v-*erbB* induces an enhanced tyrosine phosphorylation of a number of cellular proteins such as p36 (which is also phosphorylated in v-*src* transformed cells) and p42 (as in quiescent fibroblasts stimulated with serum, EGF, or phorbol esters; Gilmore et al. 1985). The v-*erbB* protein thus represents a highly mutated and deregulated form of the EGF receptor which is unable to respond to growth factor signals and appears to be constitutively activated.

Role of C-Terminal Truncations

The observation that two AEV strains with similar transforming abilities exhibit different C-terminal truncations suggested that the structure of this domain can vary without grossly altering the transforming capacity of v-*erbB*, provided that the P1 site is deleted. However, analysis of several natural and constructed mutants (*td*130, Yamamoto et al. 1983; *td*359, K. Damm, H. Beug, T. Graf and B. Vennström, unpublished results; dl48, dl83, M. Jansson, unpublished) demonstrated that progressive deletion of C-terminal sequences first weakens and then abolishes their erythroblast-transforming capacity without affecting their fibroblast-transforming capacity. The mutant v-*erbB* proteins retain their plasma membrane localization and, in the case of the *td*359 v-*erbB* protein (which lacks 108 C-terminal amino acids), an apparently unchanged autophosphorylating activity (Hayman et al. 1986; H. Beug and B. Vennström, unpublished observation). These results suggest that a particular configuration of the C-terminus is necessary for the erythroid-transforming capacity of the v-*erbB* oncogene.

Another example in which alterations in the v-*erbB* C-terminal domain uncouple its fibroblast- and erythroblast-transforming capacities comes from studies with newly isolated *erbB*- transducing viruses. Injection of RAV-1, a retrovirus lacking an oncogene, into chickens of the L15 line frequently leads to erythroleukemia as a result of RAV-1 proviral integration into the c-*erbB*/EGF receptor gene. This event leads to a high level of transcription of truncated c-*erbB* mRNAs encoding for proteins which lack the capacity to bind EGF (Nilsen et al. 1985). During this process, c-*erbB* transducing retroviruses are frequently generated. Those viruses which contain a complete c-*erbB* C-terminus induce erythroleukemia only in chickens of the L15 strain and do not transform fibroblasts in culture (H. Robinson, personal communication). Another series of isolates exhibiting C-terminal truncations encompassing P1 and

P2 are leukemogenic in several chicken strains and transform erythroblasts in culture, but still exhibit little or no fibroblast-transforming ability (Tracy et al. 1985; Beug et al. 1986 and B. Vennström and H. Beug, unpublished results). More extensive truncation of c-*erbB* C-terminal sequences and/or other changes generate viruses with full fibroblast and erythroblast transforming ability (B. Vennström and H. Beug, unpublished).

In summary, these results indicate that no simple model accounts for the role of the C-terminus, and suggest that v-*erbB* acts by a pleiotropic mechanism in which different domains and biochemical properties of the protein are responsible for the transformation of different target cells.

Temperature-Sensitive Mutants of *erbB*

Several AEV mutants which are temperature-sensitive (*ts*) for the transformation of erythroblasts have been described (Graf et al. 1978; Palmieri et al. 1982). Two of these mutants (*ts*34 and *ts*167) have been studied in detail and their lesions found to map within v-*erbB*. Mutant-transformed erythroblasts are essentially indistinguishable at the permissive temperature from wild-type transformed cells but can be induced to differentiate synchronously into erythrocyte-like cells if shifted to the nonpermissive temperature (Beug and Hayman 1984). Among the first changes detected after temperature induction is the restoration of the EPO requirement and "normalization" in the pattern of expression of erythroblast- or reticulocyte-specific cell surface glycoproteins, which are aberrantly expressed in the transformed cells (Adkins et al. 1985; Schmidt et al. 1986). Thereafter, temperature-induced differentiation closely resembles terminal differentiation of normal erythroid precursors with respect to many morphological, antigenic, and biochemical markers (Schmidt et al. 1986).

Biochemical characterization of the v-*erbB* proteins synthesized by *ts*34 and *ts*167 revealed that the *ts* lesions in both mutants affect glycoprotein maturation and transport. At the nonpermissive temperature glycosylation of the v-*erbB* protein precursor into the mature plasma membrane form is completely inhibited. Consequently, no newly synthesized v-*erbB* protein is inserted into the plasma membrane, and preexisting cell surface v-*erbB* disappears within 12–24 h (Beug and Hayman 1984). Further studies revealed that plasma membrane localization but not glycosylation is crucial for the transforming function of the v-*erbB* protein (Schmidt et al. 1985). In contrast, the *ts* lesions do not affect the in vitro autophosphorylation of the gp68$^{\text{v-}erbB}$ precursor, suggesting that the kinase of the thermosensitive v-*erbB* protein is still active at the nonpermissive temperature (Hayman et al. 1986). In view of this finding, it was surprising that sequence analysis of the cloned *ts*34 AEV and *ts*167 AEV *erbB* genes revealed only a single amino acid difference within the kinase domain in each of the *ts* mutants relative to the wild-type gene (Choi et al. 1986; P. Scotting, personal communication).

Further work is required to determine whether different assay conditions, for example use of other substrates, will detect thermosensitive kinase activity in the mutant v-*erbB* proteins.

Are *erbB*-Containing Retroviruses Generated by a Common Mechanism?

The v-*erbB* proteins of AEV-H and AEV-ES4 contain an identical stretch of 61 N-terminal amino acids, including 6 amino acids encoded by the retroviral RNA leader sequence spliced onto v-*erbB* (Yamamoto et al. 1983; Debuire et al. 1984; B. Vennström, unpublished data). The following observations suggest that v-*erbB*-containing retroviruses originated by the integration of a provirus next to c-*erbB*, leading to the activation of the gene and its subsequent capture. First, proviral integration in RAV-1-induced erythroleukemias always occurs 5′ to the c-*erbB* exon which encodes the 52 most N-terminal amino acids of the v-*erbB* extracellular domain, suggesting that properties of this exon are important for the transforming function of v-*erbB*. Second, the splice acceptor site for this exon is in the same reading frame as that of the retroviral RNA leader sequence to which the v-*erbB* genes of AEV-H and AEV-ES4 are fused. This leader sequence contains the commonly used AUG, which allows an efficient translation of v-*erbB* (Debuire et al. 1984). The AUG present in the 5′ part of the v-*erbB* gene appears to be nonfunctional, since a frameshift mutation introduced between the RNA leader sequence and v-*erbB* abolishes the production of v-*erbB* protein (M. Jansson and B. Vennström, unpublished). These observations explain the precision of the recombinational event that generated the 5′ junction between the v-*erbB* gene of different AEV isolates and sequences of the ancestral retrovirus vector.

Concluding Remarks

The v-*erbB* protein represents a highly mutated version of the EGF receptor. These mutations abolish or modulate receptor properties as diverse as biosynthesis, plasma membrane localization, and expression and regulation of tyrosine kinase activity. We are still quite far from understanding which of these changes are essential for transformation and how they cooperate in generating a more transforming molecule. In this context, it is interesting that the initial isolate of the AEV-ES4 strain exhibited a low oncogenic potential and long latency, and that the highly oncogenic AEV strains now in use were obtained only after many passages in vivo (Rothe-Meyer and Engelbreth-Holm 1933). This long selection period could account for the various mutations in the ES4 *erbB* gene and might also explain why ES4, but not the less heavily selected AEV-H strain (Hihara et al. 1983), captured v-*erbA*, an oncogene which cooperates with v-*erbB* to induce a more highly transformed phenotype (see Kahn et al., this Vol.).

References

Adkins B, Beug H, Graf T (1985) Protein synthesis in differentiating normal and leukemic erythroid cells. J Cell Physiol 123:269 – 276

Balk SD, Gunther HS, Moisi A (1984) Morphological transformation, autonomous proliferation and colony formation by chicken heart mesenchymal cells infected with avian sarcoma, erythroblastosis and myelocytomatosis virus. Life Sci 35:1157 – 1171

Beug H, Hayman MJ (1984) Temperature-sensitive mutants of avian erythroblastosis virus: surface expression of the v-*erbB* product correlates with transformation. Cell 36:963 – 972

Beug H, Kahn P, Vennström B, Hayman MJ, Graf T (1985) How do retroviral oncogenes induce transformation in avian erythroid cells? Proc R Soc Lond 226:121 – 126

Beug H, Hayman MJ, Raines MB, Kung HJ, Vennström B (1986) RAV-1-induced erythroleukemia cells exhibit a weakly transformed phenotype in vitro and release c-*erbB* containing retroviruses unable to transform fibroblasts. J Virol 57:1127 – 1138

Choi OR, Trainor C, Graf T, Beug H, Engel D (1986) A single amino acid substitution in v-*erbB* confers a thermolabile phenotype to *ts*167 AEV-transformed erythroid cells. Mol Cell Biol 6:1751 – 1759

Debuire B, Henry C, Nenaissa M, Biserte G, Claverie JM, Saule S, Martin P, Stehelin D (1984) Sequencing the *erbA* gene of avian erythroblastosis virus reveals a new type of oncogene. Science 224:1456 – 1459

Decker SJ (1985) Phosphorylation of the *erbB* gene product from an avian erythroblastosis virus-transformed chick fibroblast cell line. J Biol Chem 260:2003 – 2006

Downward J, Parker P, Waterfield MD (1984) Autophosphorylation sites on the epidermal growth factor receptor. Nature 311:483 – 485

Gilmore T, DeClue JE, Martin GS (1985) Protein phosphorylation at tyrosine is induced by the v-*erbB* gene product in vivo and in vitro. Cell 40:609 – 618

Graf T, Ade N, Beug H (1978) Temperature-sensitive mutant of avian erythroblastosis virus suggests a block of differentiation as a mechanism of leukaemogenesis. Nature 275:496 – 501

Graf T, Beug H, Hayman MJ (1980) Target cell specificity of defective avian leukemia viruses: hematopoietic target cells for a given virus type can be infected but not transformed by strains of a different type. Proc Natl Acad Sci USA 77:389 – 393

Hayman MJ, Beug H (1984) Identification of a form of the avian erythroblastosis virus *erbB* gene product at the cell surface. Nature 309:460 – 462

Hayman MJ, Kitchener G, Knight J, McMahon J, Watson R, Beug H (1986) Autophosphorylation of v-*erbB* does not correlate with cell transformation. Virology (in press)

Hihara H, Yamamoto H, Shimohira H, Arai K, Shimizu T (1983) Avian erythroblastosis virus isolated from chick erythroblastosis induced by lymphatic leukemia virus subgroup A. J Nat Cancer Inst 70:891 – 897

Kahn P, Adkins B, Beug H, Graf T (1984) *Src-* and *fps*-containing avian sarcoma viruses transform chicken erythroid cells. Proc Natl Acad Sci USA 81:7122 – 7126

Kahn P, Frykberg L, Brady C, Stanley I, Beug H, Vennström B, Graf T (1986) v-*erbA* cooperates with sarcoma oncogenes in leukaemic cell transformation. Cell 45:349 – 356

Kris RM, Lax I, Gullick W, Waterfield MD, Ullrich A, Fridkin M, Schlessinger J (1985) Antibodies against a synthetic peptide as a probe for the kinase activity of the avian EGF receptor and v-*erbB* protein. Cell 40:619 – 625

Nilsen TW, Maroney PA, Goodwin RG, Rottman FM, Crittenden LB, Paines MA, Kung H (1985) c-*erbB* activation in ALV-induced erythroblastosis: novel RNA processing and promoter insertion result in expression of an amino-truncated EGF receptor. Cell 41:719 – 726

Palmieri S, Beug H, Graf T (1982) Isolation and characterization of four new temperature-sensitive mutants of avian erythroblastosis virus (AEV). Virology 123:293 – 311

Privalsky ML, Bishop JM (1984) Subcellular localization of the v-*erbB* protein, the product of a transforming gene of avian erythroblastosis virus. Virology 135:356 – 368

Rothe-Meyer A, Engelbreth-Holm J (1933) Experimentelle Studien über die Beziehungen zwischen Hühnerleukose und Sarkom an der Hand eines Stammes von übertragbaren Leukose-Sarkom-Kombinationen. Acta Pathol Microbiol Scand 10:380

Royer-Pokora B, Beug H, Claviez M, Winkhardt H-J, Friis R, Graf T (1978) Transformation parameters in chicken fibroblasts transformed by AEV and MC29 avian leukemia viruses. Cell 13:751 – 760

Samarut J, Gazzolo L (1982) Target cells infected by avian erythroblastosis virus differentiate and become transformed. Cell 28:921 – 929

Schmidt JA, Beug H, Hayman MJ (1985) Effects of inhibitors of glycoprotein processing on the synthesis and activity of the *erbB* oncogene. EMBO J 4:105 – 112

Schmidt JA, Marshall J, Hayman MJ, Doederlein G, Beug H (1986) Monoclonal antibodies to novel erythroid differentiation antigens reveal specific effects of oncogenes on the leukaemic cell phenotype. Leukemia Res 10:257 – 272

Tracy SE, Woda BA, Robinson HL (1985) Induction of angiosarcoma by a c-*erbB* transducing virus. J Virol 54:304 – 310

Ullrich A, Coussens L, Hayflick JS, Dull TJ, Gray A, Tam AW, Lee J, Yarden Y, Libermann TA, Schlessinger J, Downward J, Mayes ELV, Whittle N, Waterfield MD, Seeburg PH (1984) Human epidermal growth factor receptor cDNA sequence and aberrant expression of the amplified gene in A431 epidermoid carcinoma cells. Nature 309:418 – 425

Yamamoto T, Hihara H, Nishida T, Kawai S, Toyoshima K (1983) A new avian erythroblastosis virus, AEV-H, carries *erbB* gene responsible for the induction of both erythroblastosis and sarcomas. Cell 34:225 – 232

The c-*fms* Proto-Oncogene and the CSF-1 Receptor

CHARLES J. SHERR and E. RICHARD STANLEY

Cell surface receptors have been traditionally defined through their specific interaction with purified ligands, but the study of oncogene-coded tyrosine kinases and their proto-oncogene homologs has presented the reciprocal problem — namely, can ligands for putative receptors be identified? The demonstration that the v-*erbB* oncogene encodes a truncated form of the epidermal growth factor (EGF) receptor (Downward et al. 1984b) provided the first direct evidence that certain oncogene products could be derived from receptor genes, and underscored the possibility that critical alterations in receptor function might directly contribute to neoplasia. Investigators now suspect that other retroviral oncogenes of the tyrosine kinase gene family (v-*src*, v-*abl*, v-*fes/fps*, v-*yes*, v-*fgr*, and v-*ros*), as well as functionally related oncogenes derived from tumor cells (*neu, trk, met*), could also have arisen from receptor genes. One paradigm involves a member of this gene family, c-*fms*, which encodes a product related, and possibly identical, to the receptor for the macrophage colony-stimulating factor, CSF-1 (Sherr et al. 1985).

Critical Features of the v-*fms* Gene Product

The v-*fms* oncogene of the Susan McDonough strain of feline sarcoma virus (SM-FeSV) was acquired by recombination of feline leukemia virus (FeLV) with c-*fms* proto-oncogene sequences in cat cellular DNA. The v-*fms* gene was inserted into the open reading frame of the viral *gag* gene, so that SM-FeSV encodes a 180 kDa polyprotein specified by 5′ *gag* and 3′ v-*fms* sequences. The polyprotein is translated on membrane-bound polyribosomes and becomes oriented as an integral transmembrane protein with its N-terminal portion in the cisternae of the endoplasmic reticulum (ER) and its C-terminal kinase domain in the cytoplasm. Glycosylation and proteolysis of the *gag*- coded fragment generate a v-*fms*-coded glycoprotein of 120 kDa (gp120$^{\text{v-}fms}$) that remains membrane-associated. The glycoprotein is transported through the ER-Golgi complex to the plasma membrane where it then becomes oriented with its glycosylated N-terminal domain (ca. 450 amino acids) outside the cell and its C-terminal kinase domain (ca. 400 amino acids) in the cytoplasm. Remodeling of N-linked oligosaccharide chains during intracellular transport increases the apparent molecular weight of the glycoprotein to about 140 kDa (gp140$^{\text{v-}fms}$). On the plasma membrane, gp140$^{\text{v-}fms}$ becomes associated with

Oncogenes and Growth Control
Edited by P. Kahn and T. Graf
© Springer-Verlag Berlin Heidelberg 1986

clathrin-coated pits and is recycled into endosomes and subsequently degraded. (For further details, see Rettenmier et al. 1985b, and references therein).

The cytoplasmic domain of the v-*fms*-coded glycoprotein is homologous to prototypic members of the tyrosine kinase gene family (Hampe et al. 1984) and encodes the predicted enzymatic activity (Barbacid and Lauver 1981). When immune complexes containing the v-*fms* gene products are incubated with (γ^{32}P)-ATP and suitable divalent cations, the *fms* proteins act as substrates for an associated kinase activity and are phosphorylated exclusively on tyrosine. Although v-*fms*-coded products in immune complexes will phosphorylate admixed substrates, the "autophosphorylation" reaction predominates. In transformed cells metabolically labeled with [^{32}P]phosphoric acid, only serine phosphorylated forms of the glycoprotein are detected, suggesting that tyrosine phosphorylation is labile and regulated by tyrosine phosphatases. Mutant v-*fms* glycoproteins that retain wild-type kinase activity but are inhibited in their intracellular transport to the cell surface are nontransforming, suggesting that the physiological targets of the enzyme reside at the cell surface (Roussel et al. 1984).

The c-*fms* Proto-Oncogene

The c-*fms* gene is 35 kb in length (as compared to the 3 kb viral oncogene) and maps to the long arm of human chromosome 5 at band q33 – 34 (Roussel et al. 1983; Groffen et al. 1983). Interstitial deletions in this region are associated with a variety of hematopoietic disorders including refractory anemia, myelodysplastic syndromes, and therapy-related acute myelogenous leukemia. In several such cases studied by analysis of the DNA in rodent × human somatic cell hybrids containing the 5q$^-$ chromosome (Nienhuis et al. 1985) or by in situ chromosomal hybridization (Le Beau et al. 1986), the c-*fms* gene was deleted. Most interstitial deletions of chromosome 5 also involved the gene for the granulocyte-macrophage colony-stimulating factor (GM-CSF), suggesting that deletions of one or both of these loci may contribute to disorders of hematopoiesis (Huebner et al. 1985; Le Beau et al. 1986).

Recently, the nucleotide sequence of a genomic 2.5 kb *Eco*RI fragment containing the site of recombination between FeLV and c-*fms* sequences has been determined (Wheeler et al. 1986). This fragment includes a single short open reading frame corresponding to the most 5′ v-*fms* sequences. The first ATG triplet within the c-*fms* coding sequence, located 102 base pairs downstream from the site of FeLV recombination, precedes a sequence encoding 22 hydrophobic amino acids that might serve as the signal peptide of the c-*fms*-coded product. Both the initiator codon and the hydrophobic leader sequence are represented in the v-*fms* gene product, but the hydrophobic leader is interrupted by an arginine residue. Thus, recombination between FeLV and c-*fms* may have occurred within sequences corresponding to the 5′ untranslated region of c-*fms* mRNA, so that the SM-FeSV polyprotein contains "extra" c-*fms*-derived amino acids at the *gag-fms* junction. If the SM-FeSV polyprotein were processed by "signal peptidase", the amino-terminal ends of gp120$^{v\text{-}fms}$ and the c-*fms* product could be similar.

In contrast, analysis of c-*fms* cDNA predicts that the 3′ ends of c-*fms* and v-*fms* differ from each other (Coussens et al. 1986). The C-terminus of the c-*fms* gene product is predicted to be longer than the v-*fms*-coded glycoprotein, and its amino acid sequence distal to the tyrosine kinase domain contains an additional tyrosyl residue not found in the v-*fms* gene product. This residue may be analogous to C-terminal tyrosyl residues in the EGF receptor which are major sites of tyrosine phosphorylation and possibly regulate kinase activity (Downward et al. 1984a).

Expression of c-*fms* mRNA (ca. 4 kb) was first detected in mouse placenta and in human choriocarcinoma cell lines derived from malignant placental trophoblasts (Müller et al. 1983a, b). In a series of different human tumors studied for oncogene expression, c-*fms* transcripts were detected in a variety of different types of cancer; in tumors of any one histologic type, expression of c-*fms* was sporadic (Slamon et al. 1984), possibly reflecting inflammatory infiltration of tumor tissues (see below). In adult cats, the highest levels of 4 kb c-*fms* mRNA were detected in spleen (Rettenmier et al. 1985a), providing the first clue that the c-*fms* gene product might represent a receptor present on hematopoietic cells.

The c-*fms* Gene Product Is Related to the CSF-1 Receptor

Using the immune complex kinase reaction, we screened lysates of cat spleen cells for the presence of a c-*fms* gene product and detected two glycoproteins of 130 and 170 kDa, each of which was phosphorylated on tyrosine. By analogy to the v-*fms*-coded gene products, feline gp130$^{c\text{-}fms}$ and gp170$^{c\text{-}fms}$ represent two differentially glycosylated forms of the same polypeptide, the smaller representing the immature precursor of the latter (Rettenmier et al. 1985a). Expression of these products was limited to a minor population of splenocytes corresponding to inflammatory cells, excluding lymphoid and red cell elements. We therefore inoculated cats with peritoneal irritants and collected inflammatory exudates by lavage 4 days later. When live cells were examined by fluorescence-activated flow cytometry using monoclonal antibodies directed to N-terminal epitopes of gp140$^{v\text{-}fms}$, the fluorescence-positive population corresponded to macrophages (Sherr et al. 1985).

We reasoned that the c-*fms* gene product might correspond to a receptor for a macrophage-specific colony stimulating factor, the most logical candidate being the receptor for CSF-1. CSF-1 acts specifically on cells of the mononuclear phagocyte lineage, whereas other growth factors that regulate cells of this lineage also act on granulocytes and/or erythroid precursors (Stanley et al. 1984). CSF-1 is an acidic glycoprotein composed of two 14 kDa polypeptide chains that are variably glycosylated (Das and Stanley 1982), and binds to a single class of receptors on cells of the mononuclear phagocyte lineage (Byrne et al. 1981; Guilbert and Stanley 1986). The murine receptor has been purified and is a single polypeptide chain of ca. 165 kDa with an associated tyrosine kinase activity that is stimulated by CSF-1 (Y. G. Yeung, P. T. Jubinsky, D. Yeung, A. Sengupta, and E. R. Stanley, unpublished).

Although monoclonal antibodies to the v-*fms* gene product were restricted in their reactivity to the feline c-*fms*-coded glycoprotein, antisera raised against a recombinant v-*fms*-coded polypeptide expressed in bacteria precipitated the murine c-*fms* product. The c-*fms* product was shown to represent the murine CSF-1 receptor by two criteria. First, in assays performed with membrane preparations, the CSF-1 receptor, specifically phosphorylated on tyrosine in the presence of CSF-1, was precipitated with antisera to the recombinant v-*fms*-coded product. Second, complexes formed at the cell surface between ^{125}I-CSF-1 and its receptor were specifically precipitated with these antisera. Taken together, these data indicated that the murine CSF-1 receptor and c-*fms* gene product are closely related, and possibly identical, molecules (Sherr et al. 1985).

The antisera to the recombinant v-*fms*-coded polypeptide also precipitated a c-*fms* gene product from human peripheral blood mononuclear cells. As in animal systems, two forms of the human c-*fms*-coded glycoprotein corresponding to an immature intracellular form (human gp130^{c-fms}) and a mature cell surface form (human gp150^{c-fms}) were detected. The same two glycoproteins, indistinguishable by biochemical and immunological criteria, were identified in human choriocarcinoma cell lines previously shown by Müller et al. (1983b) to express c-*fms* RNA. The latter cells were also found to express high affinity receptors for CSF-1 (Rettenmier et al. 1986). Thus, although CSF-1 has been characterized by its effects on macrophage growth, survival and differentiation, it may play another role in placental development.

The v-*fms*-Coded Glycoprotein Binds CSF-1

Since the v-*fms* gene product did not appear to be truncated in its amino-terminal extracellular domain, we tested whether cells transformed by SM-FeSV could bind CSF-1 (Sacca et al. 1986). Regardless of their species of origin, SM-FeSV transformants expressed specific binding sites for murine CSF-1, whereas parental cell lines or cells transformed by strains of FeSV containing the v-*fes* oncogene did not. The binding of murine ^{125}I-CSF-1 to SM-FeSV-transformed cells was indistinguishable from that observed with feline peritoneal exudate macrophages, but was 2 to 3 orders of magnitude lower in affinity than the binding to murine macrophages. This is consistent with the finding that purified murine CSF-1 is restricted in its biological activity to murine macrophages and their progenitors and does not support the growth and survival of feline macrophages in culture, even at relatively high concentrations. Chemical crosslinking experiments demonstrated that ^{125}I-CSF-1 (ca. 70 kDa) was covalently crosslinked to gp140$^{\text{v-}fms}$ at the cell surface, generating hormone-receptor complexes of ca. 220 kDa that were precipitable with antibodies to the v-*fms* gene product. Thus, the v-*fms*-coded glycoprotein contains a competent CSF-1 binding domain and in this respect, differs from the product of v-*erbB*.

Mechanism of Transformation

The fact that the v-*fms* gene product can bind CSF-1 raises the possibility that it transforms by an autocrine mechanism involving introduction of a competent CSF-1 receptor gene into fibroblasts which produce the growth factor. Indeed, cell lines susceptible to SM-FeSV transformation generally produce CSF-1. However, SM-FeSV-transformed cells do not require an exogenous source of CSF-1 for growth, and antibodies to the v-*fms* product which inhibit CSF-1 binding, or neutralizing antibodies to CSF-1 itself, do not affect the transformed phenotype. These results do not preclude an intracellular interaction between CSF-1 and the v-*fms* product within the secretory compartment of transformed cells.

A major difference between the v-*fms* and c-*fms* products concerns the activity of the receptor kinase. Whereas phosphorylation of the c-*fms* gene product in membranes is greatly enhanced in the presence of CSF-1 (Sherr et al. 1985), phosphorylation of v-*fms*-coded molecules was observed in the absence of the ligand, and the immunoprecipitated polypeptides contained similar amounts of phosphotyrosine, whether or not CSF-1 was added (Sacca et al. 1986). The simplest interpretation is that v-*fms*-coded glycoproteins act constitutively as enzymes because of alterations in their kinase domain. However, these studies were performed with purified murine CSF-1, so that an effect of feline CSF-1 on the v-*fms* gene product cannot be excluded.

As for other members of the tyrosine kinase gene family, no physiologic substrates for the v-*fms* receptor kinase have been identified whose phosphorylation on tyrosine is essential for transformation. However, the turnover of phosphatidylinositides is significantly elevated in SM-FeSV-transformed cells. Using a membrane assay in which exogenous radiolabeled phosphatidylinositol-4,5-diphosphate (PtInsP$_2$) was used as a substrate, the activity of a membrane-bound phospholipase C was found to be significantly elevated in transformed as compared to control cells (Jackowski et al. 1986). This enzyme depended on low concentrations of detergent for optimal activity, was PtInsP$_2$-specific, did not require calcium, and depended on guanine nucleotide triphosphates for activity. The enzyme generates inositol triphosphate and diacylglycerol, two "second messengers" responsible for calcium mobilization and for activation of protein kinase C, respectively (see Berridge, this Vol., Parker and Ullrich, this Vol.). An attractive hypothesis is that the v-*fms*-coded tyrosine kinase is coupled to the activity of this membrane-bound phospho-lipase C, possibly through a G protein intermediate. However, these experiments did not support the concept that the receptor kinase phosphorylates lipid intermediates of the PtdIns cycle, since SM-FeSV transformed cells do not have elevated levels of phosphorylated PtdIns species or increased PtdIns kinase activity.

Some Further Speculations

Since the CSF-1 receptor appears to be restricted in adult animals to circulating monocytes and tissue macrophages, critical mutations or rearrangements in c-*fms* coding sequences might predispose the host animal to certain forms of myelogenous leukemia. Alternatively, inappropriate expression of the c-*fms* gene in fibroblasts or expression of the CSF-1 gene in macrophages could elicit autocrine proliferative responses. These two general mechanisms for oncogenesis are not mutually exclusive and could, in certain circumstances, prove synergistic. The CSF-1 gene has been recently cloned (Kawasaki et al. 1985); hence, surveys for critical alterations at either the c-*fms* or the CSF-1 locus can now be undertaken with presently available molecular probes. Because macrophages function in inflammatory reactions, in antigen processing and presentation, and in the production of other cytokines, alterations in the CSF-1 and c-*fms* genes have implications outside the immediate context of tumor formation. Future studies will most likely yield unanticipated dividends.

Acknowledgments. This work was supported by grants CA 38187 (CJS) and CA 26504 (ERS) from the National Cancer Institute, NIH and by ALSAC of St. Jude Children's Research Hospital.

References

Barbacid M, Lauver AV (1981) Gene products of McDonough feline sarcoma virus have an in vitro-associated protein kinase that phosphorylates tyrosine residues: Lack of detection of this enzymatic activity in vivo. J Virol 40:812−821

Byrne PV, Guilbert LJ, Stanley ER (1981) Distribution of cells bearing receptors for a colony-stimulating factor (CSF-1) in murine tissues. J Cell Biol 91:848−853

Coussens L, Van Beveren C, Smith D, Chen E, Mitchell RL, Isacke CM, Verma IM, Ullrich A (1986) Structural alteration of viral homologue of receptor proto-oncogene *fms* at its carboxyl terminus. Nature 320:277−280

Das SK, Stanley ER (1982) Structure-function studies of a colony-stimulating factor (CSF-1). J Biol Chem 257:13679−13684

Downward J, Parker P, Waterfield MD (1984a) Autophosphorylation sites on the epidermal growth factor receptor. Nature 311:483−485

Downward J, Yarden Y, Mayes E, Scrace G, Totty N, Stockwell P, Ullrich A, Schlessinger J, Waterfield MD (1984b) Close similarity of epidermal growth factor receptor and v-*erbB* oncogene protein sequences. Nature 307:521−527

Groffen J, Heisterkamp N, Spurr N, Dana S, Wasmuth JJ, Stephenson JR (1983) Chromosomal localization of the human c-*fms* oncogene. Nucl Acid Res 11:6331 – 6339

Guilbert LJ, Stanley ER (1986) The interaction of ^{125}I-CSF-1 with bone marrow-derived macrophages. J Biol Chem (in press)

Hampe A, Gobet M, Sherr CJ, Galibert F (1984) The nucleotide sequence of the feline retroviral oncogene v-*fms* shows unexpected homology with oncogenes encoding tyrosine-specific protein kinases. Proc Natl Acad Sci USA 81:85 – 89

Huebner K, Isobe M, Croce CM, Golde DW, Kaufman SE, Gasson JC (1985) The human gene encoding GM-CSF is at 5q21 – q32, the chromosome region deleted in the 5q⁻ anomaly. Science 230:1282 – 1285

Jackowski S, Rettenmier CW, Sherr CJ, Rock CO (1986) A guanine nucleotide-dependent phosphatidylinositol-4,5-diphosphate-specific phospholipase C in cells transformed by the v-*fms* and v-*fes* oncogenes. J Biol Chem 261:4978 – 4985

Kawasaki ES, Ladner MB, Wang AM, Van Arsdell J, Warren MK, Coyne MY, Schweickart VL, Lee MT, Wilson KJ, Boosman A, Stanley ER, Ralph P, Mark DF (1985) Molecular cloning of a complementary DNA encoding human macrophage-specific colony-stimulating factor (CSF-1). Science 230:291 – 296

LeBeau MM, Westbrook CA, Diaz MO, Larson RA, Rowley JD, Gasson JC, Golde DW, Sherr CJ (1986) Evidence for the involvement of GM-CSF and c-*fms* in the deletion (5q) in myeloid disorders. Science 231:984 – 987

Müller R, Slamon DJ, Adamson ED, Tremblay JM, Muller D, Cline MJ, Verma IM (1983a) Transcription of c-*onc* genes c-*ras*^ki and c-*fms* during mouse development. Mol Cell Biol 3:1062 – 1069

Müller R, Tremblay JM, Adamson ED, Verma IM (1983b) Tissue and cell type specific expression of two human c-*onc* genes. Nature 304:454 – 456

Nienhuis AW, Bunn HF, Turner PH, Gopal TV, Nash WG, O'Brien SJ, Sherr CJ (1985) Expression of the human c-*fms* proto-oncogene in hematopoietic cells and its deletion in the 5q⁻ syndrome. Cell 42:421 – 428

Rettenmier CW, Chen JH, Roussel MF, Sherr CJ (1985a) The product of the c-*fms* proto-oncogene: a glycoprotein with associated tyrosine kinase activity. Science 228:320 – 322

Rettenmier CW, Roussel MF, Quinn CO, Kitchingman GR, Look AT, Sherr CJ (1985b) Transmembrane orientation of glycoproteins encoded by the v-*fms* oncogene. Cell 40:971 – 981

Rettenmier CW, Sacca R, Furman WL, Roussel MF, Holt JT, Nienhuis AW, Stanley ER, Sherr CJ (1986) Expression of the human c-*fms* proto-oncogene product (CSF-1 receptor) on peripheral blood mononuclear cells and choriocarcinoma cell lines. J Clin Invest 77:1740 – 1746

Roussel MF, Sherr CJ, Barker PE, Ruddle FH (1983) Molecular cloning of the c-*fms* locus and its assignment to human chromosome 5. J Virol 48:770 – 773

Roussel MF, Rettenmier CW, Look AT, Sherr CJ (1984) Cell surface expression of v-*fms*-coded glycoproteins is required for transformation. Mol Cell Biol 4:1999 – 2009

Sacca R, Stanley ER, Sherr CJ, Rettenmier CW (1986) Specific binding of the mononuclear phagocyte colony stimulating factor, CSF-1, to the product of the v-*fms* oncogene. Proc Natl Acad Sci USA 83:3331 – 3335

Sherr CJ, Rettenmier CW, Sacca R, Roussel MF, Look AT, Stanley ER (1985) The c-*fms* proto-oncogene product is related to the receptor for the mononuclear phagocyte growth factor, CSF-1. Cell 41:665 – 676

Slamon DJ, deKernion JB, Verma IM, Cline MJ (1984) Expression of cellular oncogenes in human malignancies. Science 224:256 – 262

Stanley ER, Guilbert LJ, Tushinski RJ, Bartelmez SH (1984) Growth factors regulating mononuclear phagocyte production. In: Volkman A (ed) Mononuclear phagocyte biology. Dekker, New York, pp 373 – 387

Wheeler EF, Roussel MF, Hampe A, Walker MH, Fried VA, Look AT, Rettenmier CW, Sherr CJ (1986) The amino-terminal domain of the v-*fms* oncogene product includes a functional signal peptide that directs synthesis of a transforming glycoprotein in the absence of feline leukemia virus *gag* sequences. J Virol 59:224 – 233

Activation of the c-*src* Gene

Hidesaburo Hanafusa

Ever since the demonstration that acutely transforming viruses carry onco-genes derived from cellular genes, the basis for the functional differences between the products of these cellular proto-oncogenes (c-*onc* genes) and their viral counterparts (v-*onc* genes) has been an important issue. Comparison of the c-*src* gene product to that of the v-*src* gene encoded by the Rous sarcoma virus (RSV) indicated that the two proteins are similar in size and enzyme activity, but quite different in relative abundance. This finding raised the basic question of whether transformation by RSV is due to the mere overproduction of the c-*src* gene product (as a consequence of placing the gene under the control of viral regulatory elements) or whether qualitative alteration of the coding sequences is involved in "activating" the transformation potential of c-*src*. Early biological studies demonstrated the rescue of transforming sarcoma viruses from chickens infected with RSV mutants which contained deletions in the *src* gene, indicating that c-*src* sequences can replace v-*src* sequences in RSV to restore transforming activity (Hanafusa et al. 1977). However, since this rescue occurred via homologous recombination between the residual v-*src* sequences in the deletion mutants and the c-*src* sequences, these results did not provide a definitive answer to the above question.

Sequence Differences Between c-*src* and v-*src*

Molecular clones containing the chicken genomic c-*src* locus were isolated and analyzed to compare the coding sequences of c-*src* and v-*src* (Takeya and Hanafusa 1983) and to explain the mechanism of transduction of c-*src* by retroviruses (Swanstrom et al. 1983; Takeya and Hanafusa 1983). As expected from the similarity of the protein products, the two genes are almost identical in their coding sequences. A major surprise was the finding that the C-terminal 19 amino acids in $p60^{c\text{-}src}$ are replaced by a new set of 12 amino acids in $p60^{v\text{-}src}$ of the Schmidt-Ruppin strain of RSV (SR-RSV) (Takeya and Hanafusa 1983). Most of the sequence encoding the C-terminus of $p60^{v\text{-}src}$ is present about 1 kb downstream from the termination codon of the c-*src* gene. In this initial comparison with the SR-RSV $p60^{v\text{-}src}$, 8 single amino acid substitutions were found in $p60^{c\text{-}src}$ (in addition to the difference at the C-terminus). More recent analyses (Levy et al. 1986; Mayer, Jove, Hanafusa, unpublished) showed that amino acid 501 in $p60^{c\text{-}src}$ is Lys rather than Arg as originally described (Takeya

Oncogenes and Growth Control
Edited by P. Kahn and T. Graf
© Springer-Verlag Berlin Heidelberg 1986

and Hanafusa 1983), and that amino acid 96 in the SR-RSV p60$^{\text{v-}src}$ is Ile instead of Thr, bringing the total of single amino acid differences to 10. The 12 unique C-terminal amino acids of the SR-RSV p60$^{\text{v-}src}$ are common to the p60$^{\text{v-}src}$ proteins encoded by all other RSV strains examined (Lerner and Hanafusa 1984; Dutta et al. 1985), suggesting that this alteration occurred during the original transduction event. Moreover, the sequence coding for the last two amino acids and the termination codon of p60$^{\text{v-}src}$ is completely homologous to the noncoding sequence upstream of the U_3 region of Rous-associated virus (RAV; Lerner and Hanafusa 1984; Dutta et al. 1985). Therefore, recombination with RAV provided the very 3′ end of the v-*src* coding sequence of RSV.

Biological Acitivity of Overproduced p60$^{\text{c-}src}$

To test whether overproduction of the p60$^{\text{c-}src}$ protein induces cell transformation, the chicken c-*src* gene was ligated to plasmids containing SV40 promoters (Shalloway et al. 1984a; Parker et al. 1984) or avian retrovirus long terminal repeats (LTRs; Iba et al. 1984; Wilhelmsen et al. 1984; Wilkerson et al. 1985) and then transfected into NIH 3T3 cells or chicken embryo fibroblasts (CEF). Successfully transfected cells were found to produce p60$^{\text{c-}src}$ in amounts approximately 50 times that of endogenous p60$^{\text{c-}src}$, that is, at levels equivalent to or greater than those of p60$^{\text{v-}src}$ in RSV-transformed cells. These cells were not morphologically altered and did not grow in soft agarose. Likewise, the overproduction of p60$^{\text{c-}src}$ was unable to stimulate the proliferation of chicken neuroretinal or chondrocyte cells (Iba et al. 1985b). These results demonstrate that overproduction of p60$^{\text{c-}src}$ does not lead to cell transformation. The fact that the half-life of p60$^{\text{c-}src}$ in chicken cells is about three times longer than that of p60$^{\text{v-}src}$ excludes the possibility that the inability of p60$^{\text{c-}src}$ to transform is due to instability of the protein (Iba et al. 1984).

Slightly different results were obtained using a molecular construct in which the c-*src* gene is linked to the Moloney murine leukemia virus LTR. Transfection of NIH 3T3 cells with this construct led to strong overproduction of p60$^{\text{c-}src}$ and the appearance of a few foci containing morphologically altered cells which had a limited growth capacity in soft agarose but were not tumorigenic in vivo (Johnson et al. 1985). These results suggest that p60$^{\text{c-}src}$ can cause some alterations in cells when expressed at extremely high levels.

Transfection of chicken cells with recombinant retrovirus DNA containing c-*src* in place of v-*src* of RSV results in the production of infectious retrovirus (Iba et al. 1984). In this system, about 10^{-3} to 10^{-4} of the total virus population converts to strongly transforming virus by spontaneous mutation.

Alterations Resulting in Activation of the Transforming Potential of c-*src*

To determine the sequence changes that contribute to the activation of p60^{src}, chimeric DNAs were constructed in which various parts of c-*src* were replaced by the corresponding portions of v-*src*. Recombinants containing the N-terminal 431 amino acids from c-*src* and the C-terminal 35 amino acids from v-*src* (the c-v construct) were transforming in CEF and in NIH 3T3 cells (Iba et al. 1984; Shalloway et al. 1984a). The reciprocal construct (the v-c construct) and a recombinant in which the C-terminal 12 amino acids of p60^{v-src} were replaced by c-*src* sequences encoding the 19 C-terminal amino acids were both able to transform CEF (Iba et al. 1984; Wilkerson et al. 1985); however, the former was extremely poor in inducing transformation of NIH 3T3 cells (Shalloway et al. 1984b). The reason for the different results with CEF and NIH 3T3 cells is not known, but in the latter study the v-*src* sequence used for the construction was derived from a different stock of RSV (SR-D as opposed to SR-A) and its sequence has not been determined. In addition, the construct used for transfection of NIH 3T3 cells contains a long noncoding sequence following the termination codon, and it is possible that this sequence influences the mRNA levels. Six and four amino acid changes between SR-RSV v-*src* and c-*src* are located upstream and downstream, respectively, from the Bgl I site used for the construction of the v-c and c-v DNAs. These results indicate that activation can be achieved by mutation in either the N- or C-terminal region. It also appears that the C-terminal 12 amino acids of v-*src* are not essential for transformation as long as other mutations are present.

Construction of additional chimeric DNAs revealed that the substitution of Thr(338) in c-*src* with Ile(338) in v-*src*, or the substitution of an N-terminal fragment of c-*src* encoding Gly(63), Arg(95), and Thr(96) with the equivalent fragment of v-*src* encoding Asp(63), Trp(95) and Ile(96), results in the activation of p60^{c-src} (Takeya, Kato, Grandori, Levy, Iba, Hanafusa, unpublished). In addition, analysis of two transforming mutants derived from the c-*src*-containing virus (Iba et al. 1984) revealed that each *src* gene contains a single mutation, resulting in the substitution of Glu(378) in c-*src* with Gly in one virus, and Ile(441) with Phe in another virus (Levy et al. 1986). Thus, the activation of the c-*src* gene can be achieved by a single mutation at various sites in the coding sequence, a finding which is consistent with the relatively high frequency at which the c-*src* virus mutates to generate transforming virus (Hanafusa et al. 1984; Iba et al. 1984). Since all these transforming viral proteins have elevated protein kinase activity (as described below), these mutations must contribute to conformational alterations that directly or indirectly activate the kinase activity in the C-terminal half of p60^{src}.

All of these studies have been done with constructs that allow elevated expression levels of inserted *src* sequences, and the possibility has not been excluded that weaker expression would still lead to cell transformation. However, it has been demonstrated that a threshold level of p60^{v-src} expression must be attained for cell transformation (Jakobovits et al. 1984). Thus it is probable that elevated levels of expression, together with structural alterations, are required to activate the transformation potential of $p60^{c\text{-}src}$.

Biochemistry of Overproduced $p60^{c\text{-}src}$

What is the basis for the inability of overproduced $p60^{c\text{-}src}$ to induce cell transformation? $p60^{c\text{-}src}$ is not different from $p60^{v\text{-}src}$ in myristylation at the N-terminus (Buss and Sefton 1985; Iba et al. 1985a) or in association with the plasma membranes (Iba et al. 1985a). However, its protein tyrosine kinase activity appears to be different. The conventional in vitro protein kinase assay using immune complexes formed between $p60^{c\text{-}src}$ and polyclonal antibodies gave variable results, probably due to the binding of antibodies to a variety of epitopes (Hanafusa et al. 1984). In contrast, assays performed using a monoclonal antibody that interacts with $p60^{v\text{-}src}$ and $p60^{c\text{-}src}$ in their N-terminal region indicated that the kinase activity of $p60^{c\text{-}src}$ measured with exogenous substrates was consistently much lower (less than 10%) than that of $p60^{v\text{-}src}$ (Iba et al. 1985).

One potentially interesting difference that emerged from these studies concerns the level of autophosphorylation of $p60^{c\text{-}src}$. Iba et al. (1985a) reported that the level of autophosphorylation in $p60^{c\text{-}src}$ is less than 10% that of $p60^{v\text{-}src}$, whereas Coussins et al. (1985) found no difference or only one-half the level of $p60^{v\text{-}src}$ autophosphorylation using the same monoclonal antibody. It is possible that this discrepancy arises from differences in the host cells used; for example, mouse and chicken cells may have different levels of phosphatase activity. The lower kinase activity of $p60^{c\text{-}src}$ correlated with a lower level of phosphorylation on tyrosine residues in total cellular proteins or 34 kDa proteins (Iba et al. 1985a; Coussins et al. 1985).

Another interesting observation is the difference in the sites of tyrosine phosphorylation in $p60^{c\text{-}src}$ and $p60^{v\text{-}src}$ (Iba et al. 1985a). By in vivo labelling, the major phosphorylation in $p60^{v\text{-}src}$ is known to occur at Ser(17) and Tyr(416), whereas $p60^{c\text{-}src}$ is phosphorylated at Ser(17) and tyrosine in the C-terminal half but not at Tyr(416) (Iba et al. 1985a). This difference is correlated with transforming activity because $p60^{src}$ of all transforming viruses, including the spontaneous transformants that have single amino acid substitutions (Levy et al. 1986), contain phosphotyrosine at position 416 (Iba et al. 1985a). Previous studies with in vitro constructed mutants showed that the phosphorylation of Tyr(416) is not essential for $p60^{src}$ to be active in transformation (Snyder et al. 1983; Cross and Hanafusa 1983). Nevertheless, Tyr(416) appears to be available for phosphorylation only in the active form of $p60^{src}$. Therefore, the phosphorylation of Tyr(416) in active $p60^{src}$ may be considered as a marker for the conformational shift from the inactive $p60^{c\text{-}src}$ to the active form. Recently, Cooper et al. (1986) presented evidence that $p60^{c\text{-}src}$ is phosphorylated at Tyr(527) and proposed that this phosphorylation negatively regulates the kinase activity of $p60^{c\text{-}src}$.

The interesting possibility that $p60^{c-src}$ regulates the kinase activity by its own tyrosine phosphorylation deserves further examination. It is also of considerable interest to investigate whether any relationship exists between the mechanisms involved in the activation of $p60^{c-src}$ to a transforming protein and those involved in the activation of $p60^{c-src}$ in certain normal tissues (Brugge et al. 1985; Golden et al. 1986) and in polyoma virus-transformed cells (Bolen et al. 1984 and article by Cheng et al., this Vol.). It remains to be determined whether the pathways involved in transformation by $p60^{v-src}$ are in any way related to the pathways involved in the normal function of $p60^{c-src}$.

References

Bolen JB, Thiele CJ, Israel MA, Yonemoto W, Lipsich LA, Brugge JS (1984) Enhancement of cellular *src* gene product associated tyrosyl kinase activity following polyoma virus infection and transformation. Cell 38:767–777

Brugge JS, Cotton PC, Queral AE, Barrett JN, Nonner D, Keane RW (1985) Neurones express high levels of a structurally modified activated form of $pp60^{c-src}$. Nature 316:554–557

Buss JE, Sefton BM (1985) Myristic acid, a rare fatty acid, is the lipid attached to the transforming protein of Rous sarcoma virus and its cellular homolog. J Virol 53:7–12

Cooper JA, Gould KL, Cartwright CA, Hunter T (1986) Tyr^{527} is phosphorylated in $pp60^{c-src}$: implications for regulation. Science 231:1431–1434

Coussens PM, Cooper JA, Hunter T, Shalloway D (1985) Restriction of the in vitro and in vivo tyrosine protein kinase activities of $pp60^{c-src}$ relative to $pp60^{v-src}$. Mol Cell Biol 5:2753–2763

Cross FR, Hanafusa H (1983) Local mutagenesis of Rous sarcoma virus: the major sites of tyrosine and serine phosphorylation are dispensable for transformation. Cell 34:597–607

Dutta A, Wang LH, Hanafusa T, Hanafusa H (1985) Partial nucleotide sequence of Rous sarcoma virus-29 provides evidence that the original Rous sarcoma virus was replication defective. J Virol 55:728–735

Golden A, Nemeth SP, Brugge JS (1986) Blood platelets express high levels of the $pp60^{c-src}$-specific tyrosine kinase activity. Proc Natl Acad Sci USA 83:852–856

Hanafusa H, Halpern CC, Buchhagen DL, Kawai S (1977) Recovery of avian sarcoma virus from tumors induced by transformation-defective mutants. J Exp Med 146:1735–1747

Hanafusa H, Iba H, Takeya T, Cross FR (1984) Transforming activity of the c-*src* gene. In: Vande Woude GF, Levine AJ, Topp WC, Watson JD (eds) Cancer cells, vol 2. Cold Spring Harbor Lab, Cold Spring Harbor, NY, pp 1–7

Iba H, Takeya T, Cross FR, Hanafusa T, Hanafusa H (1984) Rous sarcoma virus variants that carry the cellular *src* gene instead of the viral *src* gene cannot transform chicken embryo fibroblasts. Proc Natl Acad Sci USA 81:4424–4428

Iba H, Cross FR, Garber EA, Hanafusa H (1985a) Low level of cellular protein phosphorylation by nontransforming overproduced $p60^{c-src}$. Mol Cell Biol 5:1058–1066

Iba H, Jove R, Hanafusa H (1985b) Lack of induction of neuroretinal cell proliferation by Rous sarcoma virus variants that carry the c-*src* gene. Mol Cell Biol 5:2856–2859

Jakobovits EB, Majors JE, Varmus HE (1984) Hormonal regulation of the Rous sarcoma virus *src* gene via a heterologous promoter defines a threshold dose for cellular transformation. Cell 38:757–765

Johnson PJ, Coussens PM, Danko AV, Shalloway D (1985) Overexpressed $pp60^{c-src}$ can induce focus information without complete transformation of NIH 3T3 cells. Mol Cell Biol 5:1073–1083

Lerner TL, Hanafusa H (1984) DNA sequence of the Bryan high-titer strain of Rous sarcoma virus: extent of *env* deletion and possible genealogical relationship with other viral strains. J Virol 49:549–556

Levy JB, Iba H, Hanafusa H (1986) Activation of the transforming potential of p60^{c-src} by a single amino acid change. Proc Natl Acad Sci USA 83:4228 – 4232

Parker RC, Varmus HE, Bishop JM (1984) Expression of v-*src* and chicken c-*src* in rat cells demonstrates qualitative differences between pp60^{v-src} and pp60^{c-src}. Cell 37:131 – 139

Shalloway D, Coussins PM, Yaciuk P (1984a) Overexpression of the c-*src* protein does not induce transformation of NIH 3T3 cells. Proc Natl Acad Sci USA 81:7071 – 7075

Shalloway D, Coussins PM, Yaciuk P (1984b) c-*src* and *src* homolog overexpression in mouse cells. In: Vande Woude GF, Levine AJ, Topp WC, Watson JD (ed) Cancer cells, vol 2. Cold Spring Harbor Lab, Cold Spring Harbor, NY, pp 9 – 17

Snyder MA, Bishop JM, Colby WW, Levinson AD (1983) Phosphorylation of tyrosine-416 is not required for the transforming properties of pp60^{src}. Cell 32:891 – 901

Swanstrom RR, Parker RC, Varmus HE, Bishop JM (1983) Transduction of a cellular and kinase activity oncogene: the genesis of Rous sarcoma virus. Proc Natl Acad Sci USA 80:2519 – 2523

Takeya T (1983) Structure and sequence of the cellular gene homologous to the RSV *src* gene and the mechanism for generating the transforming virus. Cell 32:881 – 890

Wilhelmsen KC, Tarpley WG, Temin HM (1984) Identification of some of the parameters governing transformation by oncogenes in retroviruses. In: Vande Woude GF, Levine AJ, Topp WC, Watson JD (eds) Cancer cells, vol 2. Cold Spring Harbor Lab, Cold Spring Harbor, NY, pp 303 – 308

Wilkerson VW, Bryant DL, Parsons JT (1985) Rous sarcoma virus variants that encode *src* proteins with an altered carboxy terminus are defective for cellular transformation. J Virol 55:314 – 321

Normal and Transforming N-Terminal Variants of c-*abl*

YINON BEN-NERIAH and DAVID BALTIMORE

V-abl was first described as the oncogene contained in Abelson murine leukemia virus (A-MuLV), a virus that transforms both lymphocytes and fibroblast lines. The v-*abl* gene product has a protein tyrosine kinase activity which is essential for the transforming capacity of the virus (Prywes et al. 1983). The normal cellular counterpart, the c-*abl* proto-oncogene, also encodes a protein with tyrosine kinase activity (Konopka and Witte 1985; Ben-Neriah et al. 1986a). In addition to capture by retroviruses, c-*abl* can be activated by chromosomal translocations, such as in human chronic myeloid leukemia. In this review we discuss the structure of the c-*abl* gene, the various transcripts it expresses, and how it is activated to become an oncogene.

Organization of c-*abl*

c-*abl* is a single copy gene located on the long arm of chromosome 9 in humans and on chromosome 2 in mice. It contains at least 15 exons which may be grouped according to their contribution to protein-coding sequences (Fig. 1). A single exon of 3.5 kb encoding the C-terminal 501 amino acids is preceded by 8 exons encoding mainly the kinase domain of the protein (Wang et al. 1984). Further upstream is a group of exons that comprise the N-terminal coding exons, including two exons common to all types of c-*abl* mRNA and at least 4 exons used alternatively in the different forms of c-*abl* mRNA (see below; Ben-Neriah et al. 1986b; Bernards et al., unpublished). The genomic organization of the human c-*abl* locus is similar to that of the mouse: 13 exons have been identified so far (Shtivelman et al. 1985 and personal communication). Recent data indicate that at least one of the human 5′ exons (IV, Fig. 1) is located over 100 kb upstream of the common exons (Bernards et al., unpublished).

Multiple c-*abl* Gene Products

The c-*abl* gene is expressed in every tissue studied so far (Müller et al. 1982; Wang and Baltimore 1983). Its mRNAs separate into two predominant size classes: 5.3 and 6.5 kb in the mouse and 6 and 7 kb in man. The mature mouse testis show an extra band of 4.2 kb (Müller et al. 1982; Ponzetto and Wolge-

Oncogenes and Growth Control
Edited by P. Kahn and T. Graf
© Springer-Verlag Berlin Heidelberg 1986

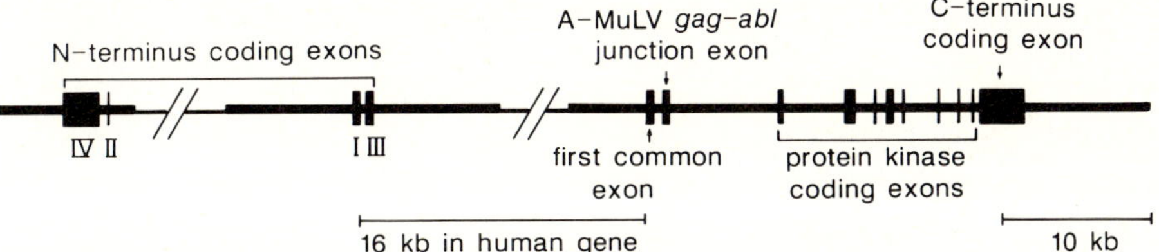

Fig. 1. Organization of the mouse c-*abl* locus. *Vertical bars* individual exons; *horizontal thickened line* introns. The linkage of the N-terminal coding sequences to the rest of the c-*abl* region has not yet been assessed; this is indicated by *breaks* in the horizontal axis. The map was constructed according to Wang et al. (1984) and A. Bernards et al. (unpublished)

muth 1985; Ben-Neriah et al. 1986b). In a mouse lymphoid cell line we identified four types of c-*abl* mRNA, designated I, II, III, and IV. The equivalents of type I and IV mRNAs have also been found in human cells (E. Canaani, personal communication). All these mRNA types have a common core sequence consisting of the eleven 3′ exons but divergent 5′ sequences. S1 nuclease analysis showed that the addition of different 5′ exons by alternative splicing is the origin of the 5′ heterogeneity. The processing of c-*abl* mRNA involves multiple splicing events occurring at a single splice junction flanking the first common c-*abl* axon. The same junction is involved in the 9;22 translocation found in human chronic myelogenous leukemia cells (see below).

The unique splicing pattern of c-*abl* mRNA is not tissue-specific. The two major splice forms, types I and IV, are present in every tissue examined in proportions of about 2:1, respectively. S1 nuclease analysis suggests that they represent about 70% of the total c-*abl* mRNA; the rest consists of types II and III, and perhaps of other types yet to be identified.

The c-*abl* Proteins

The c-*abl* protein has a molecular mass of approximately 150 kDa in the mouse, 140 kDa in man, and 140 kDa in the rat (Witte et al. 1979; Konopka et al. 1984; unpublished results). The different splice products of the c-*abl* locus predict products with 1117 to 1142 amino acids, suggesting that these proteins differ only between 20 – 45 amino acids at the N-terminal end. Nevertheless, we have observed some molecular weight heterogeneity in certain mouse tissues, perhaps due to post-translational modifications.

Comparison of the different N-terminal sequences indicate three unrelated hydrophilic and one hydrophobic amino acid sequence. None of these sequences displays the typical primary structure of a signal sequence or membrane-spanning region characteristic of receptors or transmembrane proteins. The N-terminal sequence predicted from the type IV cDNA starts with met-gly-gln, a sequence that is shared by the *gag* N-terminus of v-*abl*. v-*abl*, v-*src* and c-*src* proteins are all myristylated on a glycine residue found next to the

Receptors and Protein-Tyrosine Kinase Oncogenes

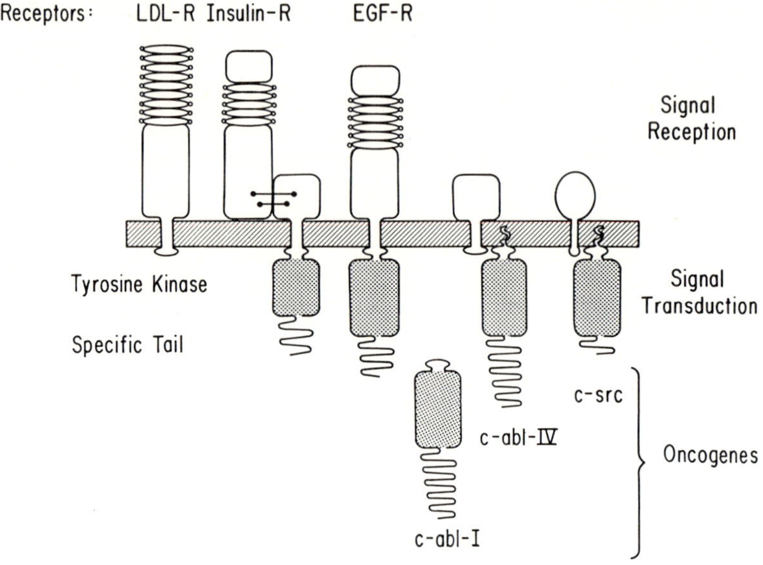

Fig. 2. A schematic illustration of c-*abl* and several other protein tyrosine kinases and receptor proteins. The hatched box represents the plasma membrane separating the extracellular space (*above*) from the cytoplasm (*below*). The protein kinase domain is *stippled*. c-*abl* is represented by two of its N-terminal variants, type *I* and type *IV*. c-*src* and perhaps c-*abl*-IV are attached to the plasma membrane through a myristic acid extension. *LDL-R* low density lipoprotein receptor; *EGF-R* epidermal growth factor receptor; *Insulin R* insulin receptor

N-terminal methionine residue, suggesting that the type IV c-*abl* protein is also myristylated and anchored to the internal side of the plasma membrane. c-*abl* would thereby be expressed as two protein forms, one cytoplasmic and one membrane-bound (see Fig. 2), a prediction supported by cell fractionation experiments (O. N. Witte, personal communication). Antibodies directed to the unique N-termini of the various c-*abl* proteins may aid in the localization of the different forms to particular subcellular compartments.

The N-terminal region of c-*abl* protein might also function to determine the interaction of this protein with other cellular proteins (Fig. 2). Certain receptors, such as the EGF receptor, have protein-tyrosine kinase domains, whose enzyme activity can be activated by the binding of a specific ligand. Other receptors, such as that for insulin, consist of two subunits of which the kinase subunit transduces the signal from the other, ligand-activated subunit. By analogy, the c-*abl* proteins, through an association with another protein determined by the N-terminus, perhaps similarly affect the type of the signal transduced from different receptors.

c-*abl* Is Conserved Among Different Species

c-*abl* homologous sequences have been cloned from mouse, man, and *Drosophila*. The structure of the mouse c-*abl* gene is similar to that of the human, and in their N-terminal third, type I and IV proteins from both species are virtually identical (>99%, Ben-Neriah et al. 1986b; E. Canaani, personal communication). This high degree of conservation suggests functional constraints on the structure of the gene, perhaps due to interaction with other proteins. The *Drosophila* genome has DNA sequences homologous to both c-*abl* and c-*src*. The homology is particularly evident at the amino acid level and covers a large portion (300 amino acids) of the tyrosine kinase domain. The *Drosophila Dash* gene is more homologous to c-*abl* (74%) than to c-*src* (54%) and it is therefore assumed that *Dash* may be functionally related to c-*abl* (Hoffman-Falk et al. 1983). In comparison to c-*abl*, the *Dash* homology extends upstream of the protein kinase domain, as defined by the sequences conserved among all protein-tyrosine kinase genes, and covers the "junction exon" and the first common exon of c-*abl* almost up to the common splice junction (F. M. Hoffman, personal communication). The fact that the homology of *Dash* to the N-terminal region of c-*abl* is far more extensive than to any other member of the protein-tyrosine kinase family further supports the conclusion that *Dash* is functionally related to c-*abl*. In contrast, there is no significant homology between *Dash* and c-*abl* over the C-terminal domain sequences and this region is also relatively poorly conserved between man and mouse (E. Canaani, personal communication). F. M. Hoffman has recently generated a series of deletions and point mutations in the *Dash* locus. Several of these mutants exhibit lethal effects in late embryos, suggesting that they may provide a valuable system in which to investigate the function of c-*abl* in embryonic development.

The Transforming v-*abl* Gene

c-*abl* is present in the genome of two acutely transforming retroviruses, the Abelson murine leukemia virus and Hardy-Zuckerman-2 feline sarcoma virus (HZ-2FeSV). A-MuLV was isolated from a steroid hormone-treated Balb/c mouse infected with a replication competent Moloney murine leukemia virus (Abelson and Rabstein 1970). The infecting virus presumably recombined with the mouse genome at the c-*abl* locus and as a result produced a replication-defective retrovirus containing part of the c-*abl* gene. The structure of the A-MuLV genome is shown schematically in Fig. 3. The virus retains most of the sequences common to all c-*abl* mRNA types and is truncated at the 5'-end. The recombination between the viral *gag* and *abl* genes occurred in the middle of the second common c-*abl* exon (junction exon in Fig. 1). The viral genome shares with c-*abl* a short sequence ("TACA") which is found at the recombination site and which might have facilitated the recombination event

Modes of c-abl Activation

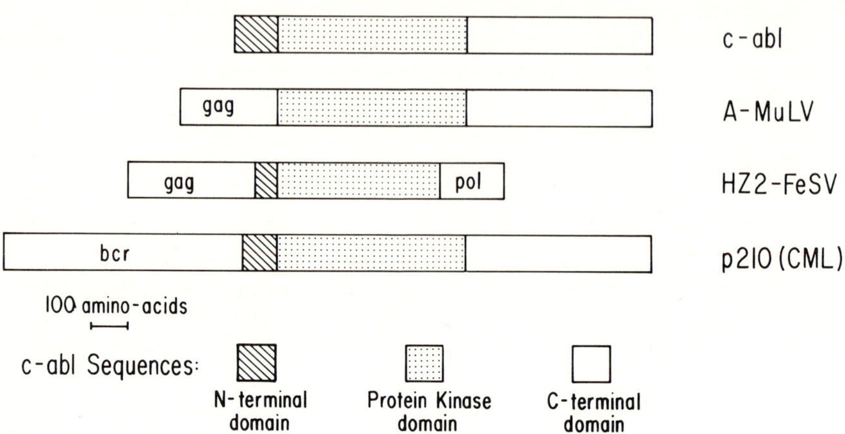

Fig. 3. Schematic illustration of the structure of c-*abl* and its different transforming variants. *gag, pol,* and *bcr* are non-*abl* sequences involved in the transforming activation of c-*abl* by retroviruses or as a consequence of the Philadelphia chromosome translocation in CML = *P210(CML); A-MuLV* Abelson murine leukemia virus; *HZ2-FeSV* Hardy-Zuckerman 2 feline sarcoma virus

(Wang et al. 1984). A-MuLV encodes a 160 kDa *gag-abl* fusion protein in which the N-terminus is derived from Moloney virus *gag* sequence (including p15, p12, and part of P30), replacing about 125 N-terminal amino acids of the predicted normal c-*abl* protein. The viral fusion protein has a strong protein tyrosine kinase activity which can be detected by autophosphorylation (Witte et al. 1980) and by phosphorylation of tyrosine-containing peptides (see Gebhardt and Foulkes, this Vol.).

Mice infected with A-MuLV via an intraperitoneal route develop a typical disease of neurological symptoms, wasting, and lymphoproliferation within a period of 3 to 5 weeks (Rosenberg and Baltimore 1980). The proliferating cells are pre-B lymphocytes that can be propagated in vitro without the addition of lymphocyte-specific growth factors. In contrast, direct intrathymic injection of A-MuLV results in induction of T-cell tumors that can be transplanted and adapted to growth in culture (Cook 1985). A-MuLV is more promiscuous in transformation in vitro; included among its targets are the fibroblast cell lines NIH/3T3 and Rat-1, lymphocytes (Rosenberg and Baltimore 1976), mast cells and myeloid cells (for references, see Gough and Kahn et al., this Vol.). Hematopoietic cells transformed in vitro are released from their growth factor dependence and, upon transfer into syngeneic mice, produce tumors.

Only a small portion of the A-MuLV genome is necessary to transform fibroblasts in vitro. The minimal transforming region determined so far is composed of the 14 first amino acids of *gag* (the p15 region) and the 400 N-terminal amino acids of the transduced *abl*. Somewhat more of the *gag* sequences are needed to transform lymphocytes in vitro. Studies with a lymphoid cell line indicate that the additional p15 sequences stabilize the v-*abl* protein (Prywes et al. 1983).

v-*abl* is among the very few viral oncogenes that transform lymphocytes efficiently both in vitro and in vivo. The specificity for lymphocyte transformation is probably encoded by the kinase domain. Experiments with *src-abl* hybrid retroviruses showed that the *gag* N-terminus of v-*abl* could be replaced by the N-terminus of *src* without abolishing lymphocyte transforming specificity, whereas replacement of the *src* N-terminus with that from v-*abl* resulted in a virus which transformed fibroblasts but not lymphocytes (Mathey-Prevot and Baltimore 1985).

The ability of A-MuLV to transform lymphocytes and fibroblasts is apparently exclusively due to replacement of the normal c-*abl* N-terminus with *gag* sequences. Although sequence comparisons of the region common to v-*abl* and c-*abl* are not complete, so far no differences have been observed between c-*abl* and v-*abl*. This is consistent with the finding that replacement of the c-*abl* N-terminus with A-MuLV *gag* in a position corresponding to the recombination site of A-MuLV generates a highly transforming virus capable of transforming both fibroblasts and lymphocytes (Ben-Neriah and Baltimore, unpublished).

The c-*abl* gene transduced by HZ-2FeSV originated from a cat fibrosarcoma. Its genome (Fig. 3) shares several characteristics with A-MuLV. It is truncated at both the 3′ and 5′ ends relative to c-*abl*. Like the v-*abl* in A-MuLV, HZ-2FeSV codes for a fusion protein in which the N-terminus is coded by *gag* sequences from the helper virus. Recombination in this case occurred in the middle of the first common *abl* exon where a hexanucleotide sequence is shared between c-*abl* and helper virus sequences (Besmer et al. 1986).

Activation of c-*abl* in Chronic Myelogenous Leukemia (CML)

CML is a multistage stem-cell disease which is generally fatal (reviewed by Champlain and Golde 1985). The 9;22 translocation found in the Philadelphia chromosome occurs in over 95% of CML cases. The Philadelphia chromosome is present in all hematopoietic lineages and its origin can be traced by enzymatic marker studies to a single stem cell clone. The chronic stage of the disease is characterized by massive proliferation of granulocytes or sometimes thrombocytes carrying the Philadelphia chromosome. The late stage of the disease, the acute blast crisis, is characterized by a clonal malignant outgrowth of either immature myeloid or B-lymphoid cells. In 75% of the patients, the late stage is associated with the development of additional karyotypic abnormalities superimposed upon the Philadelphia chromosome (Rowley 1975).

The recombination site in the Philadelphia chromosome usually occurs at a variable length 5′ to the first common c-*abl* exon and is clustered around two exons of the *bcr* gene (Heisterkamp et al. 1985). In some cases the *bcr* exons are spliced to c-*abl* from a distance of at least 97 kb (Grosveld et al. 1986). As a consequence of the translocation event, Philadelphia chromosome-positive CML cells express an aberrant c-*abl* transcript of 8.5 kb. This transcript was studied by cDNA cloning and found to be a fusion transcript of *bcr* and c-*abl* (Shtivelman et al. 1985). The fusion most commonly results from an in-frame splicing of two exons of the *bcr* gene to the first common c-*abl* exon, involving the same junction as used for alternative splicing. It is possible that the observed break-point clustering (Groffen et al. 1984) is in part due to selection of productive in frame splice products, resulting in leukemia.

In addition to the 8.5 kb *bcr/abl* transcript, CML cells express a 210-kDa phosphoprotein recognized by c-*abl*-specific antisera (Konopka et al. 1984). With the aid of site-specific antibodies against c-*abl* and *bcr* peptides predicted from the sequence of the 8.5 kb transcript, we have shown that P210 is the product of this transcript (Ben-Neriah et al. 1986a). P210 contains all the predicted C-terminal *abl* sequences up to the common exon junction, after which the normal N-terminus is replaced by the translocated *bcr* sequences (Fig. 3). P210 is overexpressed relative to c-*abl* leukemic cells in acute phase and in variable amounts in cells from the chronic phase of the disease (O. N. Witte, personal communication); the 8.5 kb transcript is clearly overexpressed in all stages of the disease (E. Canaani, personal communication). The expression of P210 is not restricted to myeloid cells, but is also seen in hybrids between CML cells and mouse fibroblasts (Kozbor et al. 1986) and in very low levels in Philadelphia chromosome positive lymphocytes from CML patients (T. Konopka and O. N. Witte, personal communication). This pattern of expression suggests that the presence of the Philadelphia chromosome is not sufficient to induce the malignant phenotype typical for myeloid cells in CML. It will be important to determine whether the expression levels of P210 correlate with the cell type specificity and the aggressiveness of the disease.

It is probable that P210 is expressed from the *bcr* gene promoter; however, the expression of the 8.5 kb *bcr/abl* transcript in the Philadelphia-positive K562 cells greatly exceeds the expression of the normal *bcr* allele (Shtivelman et al. 1985). Furthermore, rearrangement of the *bcr* locus has been seen in CML cells without a 9;22 chromosomal translocation (Bartram 1985), raising the possibility that activation of the *bcr* locus precedes the translocation even in CML. What might be the mechanism by which this translocation contributes to the acquisition of a malignant phenotype? One possibility is that transcriptional activation of the *bcr* locus associated with an open chromatin configuration may promote recombination of *bcr* and *abl* sequences in a manner similar to the mating type switch mechanism in yeast (Klar et al. 1984) and to the immunoglobulin rearrangement process in lymphocytes (Yancopoulos et al. 1986). Under the influence of an activated *bcr* promoter, the translocation event results in overexpression of the *bcr/abl* transcript. Overexpression of p210 may then release the CML granulocytes from their dependence on a certain limiting growth factor, which in turn might lead to their excessive proliferation.

Acknowledgments. We are grateful to Drs. E. Canaani, F. M. Hoffman and O. N. Witte for the communication of valuable unpublished data. We thank A. Bernards, G. Q. Daley, P. Jackson, B. Mathey-Prevot and S. Pillai for critical reading of the manuscript and Ginger Pierce for typing the manuscript. Y. B.-N. was supported by a Fogarty International Fellowship Award. This work was supported by a Program Project Grant (CA 38497) from the National Cancer Institute.

References

Abelson HT, Rabstein LS (1970) Lymphosarcoma: virus-induced thymic-independent disease in mice. Cancer Res 30:2213 – 2222

Bartram CR (1985) *bcr* rearrangement without juxtaposition of c-*abl* in chronic myelocytic leukemia. J Exp Med 162:2175 – 2179

Ben-Neriah Y, Daley GQ, Mes-Masson AM, Witte ON, Baltimore D (1986a) The chronic myelogenous leukemia specific P210 protein is the product of the *bcr/abl* hybrid gene. Science (in press)

Ben-Neriah Y, Bernards A, Paskind M, Daley GQ, Baltimore D (1986b) Alternative 5' exons in c-*abl* mRNA. Cell 44:577 – 586

Besmer P, Murphy JE, George PC, Qiu F, Bergold PJ, Lederman L, Snyder HW Jr, Brodeur D, Zuckerman EE, Hardy WD (1986) A new transforming feline retrovirus and relationship of its oncogene v-*kit* with the protein kinase gene. Nature 320:415 – 421

Champlain RE, Golde DW (1985) Chronic myelogenous leukemia: Recent advances. Blood 65:1039 – 1097

Cook WD (1985) Thymocyte subsets transformed by Abelson murine leukemia virus. Mol Cell Biol 5:390 – 397

Groffen J, Stephenson JR, Heisterkamp N, de Klein A, Bartram CR, Grosveld G (1984) Philadelphia chromosomal breakpoints are clustered within a limited region *bcr* on chromosome 22. Cell 36:93 – 99

Grosveld G, Verwoerd T, Van-Agthoven T, de Klein A, Ramachandran KL, Heisterkamp N, Stam K, Groffen J (1986) The chronic myelocytic cell line K562 contains a breakpoint in *bcr* gene and its role in the Ph' translocation. Nature 315:758 – 761

Heisterkamp N, Stam K, Groffen J, de Klein A, Grosveld G (1985) Structural organization of the *bcr* gene and its role in the Ph' translocation. Nature 315:758 – 761

Hoffman-Falk H, Einat P, Shilo B-Z, Hoffman FM (1983) *Drosophila melanogaster* DNA clones homologous to vertebrate oncogene: evidence for a common ancestor to the *src* and *abl* cellular genes. Cell 32:589 – 598

Klar AJS, Strathern JN, Abraham JA (1984) Involvement of double strand chromosomal breaks for mating type switch in *Saccharomyces cerevisiae*. Cold Spring Harbor Symp Quant Biol 49:77 – 88

Konopka JB, Witte ON (1985) Detection of c-*abl* tyrosine kinase activity in vitro permits direct comparison of normal and altered *abl* gene products. Mol Cell Biol 5:3116 – 3123

Konopka JB, Watanabe SM, Witte ON (1984) An alteration of the human c-*abl* protein in K562 leukemia cells unmasks associated tyrosine kinase activity. Cell 37:1035 – 1042

Kozbor D, Giallongo A, Sierzega ME, Konopka J, Witte O, Showe LC, Croce CM (1986) Expression of a translocated c-*abl* gene in hybrids of mouse fibroblasts and chronic myelogenous leukemia cells. Nature 319:331 – 333

Mathey-Prevot B, Baltimore D (1985) Specific transforming potential of oncogenes encoding protein tyrosine kinases. EMBO J 4:1769 – 1774

Müller R, Slamon DJ, Tremblay JM, Cline MJ, Verma IM (1982) Differential expression of cellular oncogenes during pre- and postnatal development of the mouse. Nature 299:640 – 643

Ponzetto C, Wolgemuth DJ (1985) Haploid expression of a unique c-*abl* transcript in the mouse male germ line. Mol Cell Biol 5:1791 – 1794

Prywes R, Foulkes JG, Rosenberg N, Baltimore D (1983) Sequences of A-MuLV needed for fibroblast and lymphoid cell transformation. Cell 34:569 – 579

Rosenberg N, Baltimore D (1976) A quantitative assay for transformation of bone marrow cells by Abelson murine leukemia virus. J Exp Med 143:1453 – 1463

Rosenberg N, Baltimore D (1980) Abelson virus in viral oncology (Klein G (ed)), Raven Press, New York 1980

Rowley JD (1975) Nonrandom chromosomal abnormalities in hematologic disorders of man. Proc Natl Acad Sci USA 72:152 – 156

Shtivelman E, Lifshitz B, Gale RP, Canaani E (1985) Fused transcript of *abl* and *bcr* genes in chronic myelogenous leukemia.. Nature 315:550 – 554

Wang JYJ, Baltimore D (1983) Cellular RNA homologous to the Abelson murine leukemia virus transforming gene: expression and relationship to the viral sequence. Mol Cell Biol 3:773 – 779

Wang JYJ, Ledley F, Goff S, Lee R, Groner Y, Baltimore D (1984) The mouse c-*abl* locus: molecular cloning and characterization. Cell 36:349 – 356

Witte ON, Rosenberg NE, Baltimore D (1979) A normal cell protein cross reactive to the major Abelson murine leukemia virus gene product. Nature 281:396 – 398

Witte ON, Dasgupta A, Baltimore D (1980) Abelson murine leukemia virus protein is phosphorylated in vitro to form phosphotyrosine. Nature 238:826 – 831

Yancopoulos GD, Blackwell K, Suh H, Hood L, Alt FW (1986) Introduced T cell receptor variable region gene segments recombine in pre-B cell: evidence that B and T cells use a common recombinase. Cell 44:251 – 259

Transformation by the v-*abl* Oncogene

Angelika Gebhardt and J. Gordon Foulkes

The previous chapter by Ben-Neriah and Baltimore described sequences of the v-*abl* gene which are important in cell transformation. In this chapter, we present some thoughts as to the biochemical mechanisms whereby the v-*abl* protein induces the multitude of changes which characterize the transformed phenotype.

The v-*abl* protein is associated with a protein-tyrosine kinase activity (Witte et al. 1980). Cells expressing the v-*abl* kinase contain five to tenfold more protein-bound phosphotyrosine relative to the untransformed parent cell line. The isolation of the v-*abl* gene, its expression in *E. coli* and purification of the corresponding protein to homogeneity has confirmed that the kinase activity is an intrinsic property of the v-*abl* protein (Foulkes et al. 1985). The high specific activity of the purified v-*abl* kinase ($170 \, \mu\mathrm{mol} \, \mathrm{min}^{-1}$) is comparable to the cAMP-dependent protein kinase.

The v-*abl* kinase shares a very significant degree of homology with approximately one-third of all known oncogenes as well as with those growth-factor receptors which encode protein-tyrosine kinase activities (reviewed in Foulkes and Rich-Rosner 1985). These enzymes also demonstrate some homology to serine/threonine specific protein kinases (Hunter and Cooper 1986; and Hunter, this Vol.). Systematic deletions in the cloned v-*abl* gene, combined with a large number of 12 bp linker insertions, has established an excellent correlation between the v-*abl* kinase activity and the ability of the v-*abl* protein to transform cells (see Ben-Neriah and Baltimore, this Vol.).

Together, these studies strongly suggest that the ability of the v-*abl* product to phosphorylate proteins on tyrosine residues is critical for its biological function.

What are the substrates of the v-*abl* kinase? After 6 years of research in numerous laboratories, no physiologically important targets have been firmly established for any of the known protein tyrosine kinases. There are a number of possible reasons for this, which we have discussed in detail previously (Foulkes and Rich-Rosner 1985). They include the inherent complexity of the process under investigation and the low abundance of phosphotyrosine in transformed cells (less than 2% of the total phosphoamino acid pool). In addition, the promiscuous nature of retroviral protein-tyrosine kinases in vivo results in the phosphorylation of many major cytoplasmic proteins which are unrelated to the transformation event. This latter problem is of such magnitude that we estimate that more than 90% of the phosphotyrosine contain-

Oncogenes and Growth Control
Edited by P. Kahn and T. Graf
© Springer-Verlag Berlin Heidelberg 1986

ing proteins in certain transformed cell lines have no role in the transformation process. Hence, it may be these technical problems which have hindered identification of physiological targets.

An alternative possibility is that there are no important phosphotyrosine-containing proteins other than the protein-tyrosine kinases themselves. This idea is supported by the observation that in cells treated with growth hormones whose receptors encode a tyrosine kinase activity, as well as in transformed cells containing the v-*abl* kinase, the major phosphotyrosine-containing proteins are the protein-tyrosine kinases themselves. This is not a result one would expect either by analogy to other signal transduction systems (e.g. the cAMP-dependent protein kinase) or for enzyme-substrate reactions in general. If there are no exogenous substrates, a possible function for phosphotyrosine could be to allow an interaction between the autophosphorylated kinase and the real messenger transduction system.

One could envisage that, following binding of the hormone (e.g. EGF, PDGF), the kinase becomes activated and undergoes rapid autophosphorylation. Once phosphorylated, the receptor might then interact with a signal-generating system, e.g. the catalytic subunit of adenylate cyclase or phosphoinositidase C (see Berridge, this Vol.). In this scheme the autophosphorylated kinase would be analogous to the G proteins which serve to link receptors to signalling mechanisms in a variety of pathways (see Masters and Bourne, this Vol.). Perhaps relevant to this idea is the high energy content of the phosphotyrosine bond in the v-*abl* kinase (Foulkes et al. 1985).

If phosphorylated targets do exist, one approach to search for these proteins is to test candidate targets as substrates in vitro. Such proteins might include G proteins, enzymes of phosphoinositide metabolism, translation/transcriptional factors, and serine/threonine-specific protein kinases and phosphatases (Cohen 1985). If stoichiometric phosphorylation is obtained, one can then look for a corresponding change in the function of the particular substrate. The purified v-*abl* protein, which has a high specific activity, provides an excellent source of tyrosine kinase for such studies. In vitro data alone, however, can be very misleading, as in the report that phosphorylation of topoisomerase I by pp60^{src} resulted in a tenfold decrease in enzyme activity (Tse-Dinh et al. 1984). We have measured topoisomerase I activity in extracts prepared from normal cells and from cells transformed by either the v-*src* or v-*abl* kinase and find no evidence for the inactivation of this enzyme in vivo (Colledge et al. 1986). We have recently detailed the criteria that must be met before establishing a protein as a physiological target (Foulkes and Rich-Rosner 1985).

Although all protein-tyrosine kinases appear to be specific for phosphorylation of tyrosine residues in vitro, the addition of EGF, PDGF, or insulin to responsive cells not only stimulates the tyrosine kinase of the corresponding receptor but also leads to an increases phosphorylation of certain proteins on serine residues. Similarly, cells transformed by either the v-*abl* or v-*src* kinases show an increase in protein-bound phosphoserine. Among these phosphoseryl-proteins, ribosomal protein S6 is of particular interest because its phosphorylation is correlated with growth-promoting stimuli in a variety of systems (see Thomas, this Vol.). The phorbol ester TPA, known to activate protein kinase C (see Parker and Ullrich, this Vol.), also induces S6 phosphorylation. The observation that TPA induces the phosphorylation of the same five S6 phosphopeptides (Maller et al. 1985) suggested that the v-*abl* kinase might regulate S6 phosphorylation via activation of protein kinase C. The most direct mechanism would be for the v-*abl* kinase to phosphorylate protein kinase C on tyrosine. However, protein kinase C isolated from v-*abl*-transformed cells does not contain significant amounts of phosphotyrosine (Fry et al. 1985), suggesting that the v-*abl* kinase activates protein kinase C indirectly.

Several reports claimed that tyrosine kinases phosphorylate phosphatidylinositol (reviewed in Macara 1985). This could lead to an increased production of phosphatidylinositol-4,5 bisphosphate and diacylglycerol, which may in turn activate protein kinase C (see Berridge, this Vol.). If true, this might also explain the absence of important protein targets — perhaps the real substrates are lipids. Recently, however, we have been able to distinguish clearly the major phosphoinositol kinase activity found in Abelson virus-transformed cells from the v-*abl* kinase, suggesting that the direct phosphorylation of phosphatidylinositol by the v-*abl* kinase has no physiological role (Fry et al. 1985). Similar data have now been obtained for several other tyrosine kinases (MacDonald et al. 1985; Sugano and Hanafusa 1985; Thompson et al. 1985).

Another mechanism whereby the v-*abl* kinase might activate protein kinase C would be to stimulate phosphatidylinositol turnover indirectly. This indeed appears to be the case. The addition of serum to serum-deprived normal fibroblasts results in a rapid stimulation of phosphatidylinositide turnover. In contrast, phosphatidylinositol turnover in cells expressing the v-*abl* kinase is unaffected by the addition of serum and the pathway appears to be constitutively active (Fry et al. 1985). These results agree with results obtained from investigations of phosphatidylinositol metabolism in cells transformed by either the *src* or *ros* oncogenes (reviewed in Macara 1985). The step or steps activated by the v-*abl* kinase, which may indicate possible physiological targets for this enzyme, (e.g. phosphoinositidase C or a phosphatidylinositol kinase) await identification.

Besides diacylglycerol and inositol trisphosphate, other phosphoinositide metabolites might act as second messengers in transformed cells. For example, arachidonic acid is a precursor to prostaglandins which might act in an autocrine fashion to alter cAMP levels.

In studying the mechanism of transformation by the v-*abl* kinase it will be necessary to identify physiological substrates of both protein kinase C and the Ca^{2+}-calmodulin-dependent kinases and phosphatases.

A major problem in studying the detailed biochemical mechanisms of cell transformation is to distinguish between primary effects which are a direct consequence of an oncogene product and secondary effects mediated by tumor growth factors (TGFs). Sporn and Todaro (1980) originally proposed that TGFs are produced by transformed cells, are released into the surrounding media and act in an autocrine fashion to stimulate the growth of the cells which produce them. A variety of TGFs have now been described which are related to normal growth hormones (see review by Sporn and Roberts 1985). For example TGF-α has homology to EGF (see Derynck and Pfeffer and Ullrich, both this Vol.), while the product of the *sis* oncogene is homologous to PDGF (see Heldin and Westermark, this Vol.).

Several TGFs, including TGF-α, can be isolated from conditioned medium of Abelson virus-transformed cells (Twardzik et al. 1982; Fry, Gebhardt and Foulkes, unpublished observations). There is no formal proof, however, that TGFs induce autocrine growth in cells transformed by the v-*abl* oncogene. On the contrary, two recent reports provide evidence against an autocrine mechanism in v-*abl* transformed cells. In both cases, cell lines depending on the growth factor IL-3 for growth in tissue culture (normal mast cells and a myeloid line) lose their IL-3 requirement after transformation by the v-*abl* kinase. None of these lines, however, shows any significant change in the number of IL-3 receptors, nor do they it secrete IL-3 like growth factors into the media (Cook et al. 1985; Pierce et al. 1985 and Gough, this Vol.). These results do not exclude an involvement of TGFs. To examine the possible role of TGF-α in Abelson virus-transformed cells we first analyzed the phosphorylation state of the EGF receptor. On the hypothesis that TGF-α is released into the medium, binds to the EGF receptor and induces autophosphorylation of the receptor, one might predict an increased phosphorylation of the EGF receptor in cells expressing the v-*abl* kinase. However, immunoprecipitation of ^{32}P-labelled EGF receptor revealed it to be significantly underphosphorylated in Abelson virus-transformed cells (less than 5% of the level in untransformed cells) (Gebhardt and Foulkes, unpublished observations). The probable explanation for this is an extensive downregulation of the EGF receptor by TGF-α released into the medium. The normal function of downregulation in biological systems is to act as a feedback mechanism to switch off the response to the agonist. If this is also true for the EGF receptor in v-*abl*-transformed cells, one could argue that there was no functional receptor in these cells. In this case it would be difficult to explain how TGF-α could function in an autocrine manner.

To further examine the role of TGF-α in transformation by v-*abl*, we decided to use the Swiss 3T3 variant cell line NR6. These cells were selected from their parental line for a lack of mitotic response to EGF (Pruss and Herschman 1977). In these cells the EGF receptor is absent. Infection of NR6 cells with Abelson virus resulted in the appearance of morphologically transformed cells (Gebhardt et al. 1986). These NR6 transformants express the same level of v-*abl* kinase activity as found in transformed Swiss 3T3 cells. Both transformed NR6 cells, as well as transformed wild-type cells, are capable of anchorage-independent growth in semisolid medium, normally an excellent correlate of tumourigenicity (Shin et al. 1975). However, these cells exhibited striking differences in their ability to form tumours in nude mice. Whereas

10^5 cells of the wild-type line formed tumours in nude mice after only $2-3$ weeks, no tumours were detected even after 3 months following injection of up to 10^7 Abelson virus-transformed NR6 cells. This suggests that the failure of the latter cells to form tumours is due to their capacity to respond to TGF-α. There are a number of possible explanations for this result. First, there might be differences in the availability of exogenous growth factors when cells are injected into inguinal subcutaneous sites in vivo compared to tissue culture cells. In the effective absence of TGF-α, the in vivo milieu could inhibit the growth of v-*abl*-transformed NR6 cells. For example, TGF-β can either stimulate or inhibit cell growth, depending on the concentration of other growth factors (Tucker et al. 1984; Roberts et al. 1985 and Moses, this Vol.). A second, and perhaps more exciting possibility, is that stimulation of the EGF receptor and the subsequent response could be a precondition for tumour formation. In this regard, it is interesting that the ability of certain transformed epidermal cell lines to form tumours in nude mice can be blocked by anti-EGF receptor antibodies (Masui et al. 1984).

Based on these observation one might propose the following scheme. The presence of the v-*abl* kinase induces the production of TGF-α, which acts in an autocrine fashion to stimulate the EGF receptor. The activated receptor may in turn stimulate the production of other factors which could play a role in neovascularization, growth of surrounding normal tissue or changes in extracellular matrix proteins.

Acknowledgments. This work was supported by the Medical Research Council U.K. A. G. is recipient of a postdoctoral grant from the Deutsche Forschungsgemeinschaft.

References

Cohen P (1985) The role of protein phosphorylation in the hormonal control of enzyme activity. Eur J Biochem 151:439−448

Colledge WH, Edge M, Foulkes JG (1986) A comparison of topoisomerase I activity in normal and transformed cells. Biosci Rep 6:301−307

Cook W, Metcalf D, Nicola NA, Burgess AW, Walker F (1985) Malignant transformation of a growth factor-dependent myeloid cell line by Abelson virus without evidence of an autocrine mechanism. Cell 41:677−683

Foulkes JG, Rich-Rosner M (1985) Tyrosine-specific protein kinases as mediators of growth control. In: Cohen P, Houslay MD (eds) Molecular aspects of cellular regulation, vol 4. Elsevier, Amsterdam, pp 217−252

Foulkes JG, Chow M, Gorka C, Frackelton AR, Baltimore D (1985) Purification and characterization of a protein-tyrosine kinase encoded by the Abelson murine leukaemia virus. J Biol Chem 260:8070−8077

Fry MJ, Gebhardt A, Parker PJ, Foulkes JG (1985) Phosphatidylinositol turnover and transformation of cells by Abelson murine leukaemia virus. EMBO J 4:3173−3178

Gebhardt A, Bell JC, Foulkes JG (1986) Abelson transformed fibroblasts lacking the EGF receptor are not tumourigenic in nude mice. EMBO J (in press)

Hunter T, Cooper JA (1986) Viral oncogenes and tyrosine phosphorylation. In: Boyer PD, Krebs EG (eds) Enzyme control by phosphorylation. The Enzymes (in press)

Macara IG (1985) Oncogenes, ions and phospholipids. Am J Physiol 48:C3−C11

MacDonald ML, Kuenzel EA, Glomset JA, Krebs EG (1985) Evidence from two transformed cell lines that the phosphorylations of peptide tyrosine and phosphatidylinositol are catalyzed by different proteins. Proc Natl Acad Sci USA 82:3993–3997

Maller JL, Foulkes JG, Erikson E, Baltimore D (1985) Phosphorylation of ribosomal protein S6 on serine after microinjection of the Abelson murine leukaemia virus tyrosine-specific protein kinase into *Xenopus* oocytes. Proc Natl Acad Sci USA 82:272–276

Masui H, Kawamoto I, Sato JD, Wolf B, Sato G, Mendelsohn J (1984) Growth inhibition of human tumor cells in athymic mice by anti-epidermal growth factor receptor monoclonal antibodies. Cancer Res 44:1002–1007

Pierce JH, Di Fiore PP, Aaronson SA, Potter M, Pumphrey J, Scott A, Ihle JN (1985) Neoplastic transformation of mast cells by Abelson-MuLV: Abrogation of IL-3 dependence by a non-autocrine mechanism. Cell 41:685–693

Pruss RM, Herschman HR (1977) Variants of 3T3 cells lacking mitotic response to epidermal growth factor. Proc Natl Acad Sci USA 74:3917–3921

Roberts AB, Anzano MA, Wakefield LM, Roche NS, Stern DF, Sporn MB (1985) Type β transforming growth factor: A bifunctional regulator of cellular growth. Proc Natl Acad Sci USA 82:119–123

Shin S, Freedman VH, Risser R, Pollack R (1975) Tumorigenicity of virus-transformed cells in *nude* mice is correlated specifically with anchorage independent growth in vitro. Proc Natl Acad Sci USA 72:4435–4439

Sporn MB, Roberts AB (1985) Autocrine growth factors and cancer. Nature 313:745–747

Sporn MB, Todaro GJ (1980) Autocrine secretion and malignant transformation of cells. New Engl J Med 303:878–880

Sugano S, Hanafusa H (1985) Phosphatidylinositol kinase activity in virus-transformed and non-transformed cells. Mol Cell Biol 5:2399–2404

Thompson DM, Cochet C, Chambaz EM, Gill GN (1985) Separation and characterization of a phosphatidylinositol kinase activity that co-purifies with the epidermal growth factor receptor. J Biol Chem 260:8824–8830

Tse-Dinh YC, Wong TW, Goldberg AR (1984) Virus and cell-encoded tyrosine protein kinases inactivate DNA topoisomerases in vitro. Nature 312:785–786

Tucker RF, Shipley GD, Moses HL (1984) Growth inhibitor from BSC-1 cells closely related to platelet type β transforming growth factor. Science 226:705–707

Twardzik DR, Todaro GJ, Marquardt H, Reynolds FH, Stephenson JR (1982) Transformation induced by Abelson murine leukaemia virus involves production of a polypeptide growth factor. Science 216:894–897

Witte ON, Dasgupta A, Baltimore D (1980) Abelson murine leukaemia virus is phosphorylated in vitro to form phosphotyrosine. Nature 283:826–831

mos

Donald G. Blair

The Moloney murine sarcoma virus (Mo-MSV; Moloney 1966), isolated over 20 years ago from a Balb/c rhabdomyosarcoma induced by high multiplicity passage of the Moloney strain of murine leukemia virus (Mo-MuLV), has been one of the most extensively studied retroviral systems. This review focuses on v-*mos*, the sequence responsible for its acute transforming potential, and on its cellular proto-oncogene homolog, c-*mos*. Although the catalog of oncogene-containing retroviruses is large, *mos* and Mo-MSV represent unique members of this community and have unusual and interesting properties. As will become clear, although *mos* is grouped with the tyrosine kinase oncogenes, its structural and functional peculiarities have set it apart from other members of this group and have hampered attempts to understand its normal role and mechanism of action.

The Structure of v-*mos* and c-*mos*

Variations in passage history, coupled with biological clonings of these viruses, led to the isolation of Mo-MSV variants (i.e., ml, HT-1, 124, etc.) with distinct biological properties. Some of these properties, such as the high ratio of transforming to helper virus in Mo-MSV 124 stocks (Ball et al. 1973) facilitate molecular studies of the virus. Still other variants, such as the myeloproliferative strain of MSV (Stocking et al. 1985) have demonstrated the broad oncogenic capacity of v-*mos*. Direct genetic analysis of Mo-MSV has proven difficult (as for all defective mammalian viruses), but one temperature-sensitive mutant has been described (Blair et al. 1979). In this mutant, v-*mos* is expressed as a *gag-mos* fusion protein (Murphy and Arlinghaus 1982) and temperature-sensitive RNA processing is apparently responsible for the *ts* phenotype (Horn et al. 1981).

Advances in our understanding of *mos* at a molecular level, however, began with the application of recombinant DNA techniques to this system. Cloning of the integrated proviruses of the ml and HT-1 strains (Vande Woude et al. 1979), as well as the circular unintegrated provirus of the 124 strain (Verma et al. 1980), allowed a detailed analysis of their molecular structure. The DNA sequence of the Mo-MSV 124 genome (Van Beveren et al. 1981) and the v-*mos* region of the ml (Brow et al. 1984) and HT-1 (Seth and Vande Woude 1985) strains of MSV have been determined. In each case the differences

Oncogenes and Growth Control
Edited by P. Kahn and T. Graf
© Springer-Verlag Berlin Heidelberg 1986

between the viral *mos* sequences are minor, while major differences are found in the MuLV-related sequences. Cloning and sequence analysis of the normal c-*mos* homolog of mouse (Oskarsson et al. 1980; Jones et al. 1980), man (Watson et al. 1982), rat (van der Hoorn et al. 1982), and more recently, chicken (Schmidt et al., in preparation) has allowed one of the most extensive molecular and biological characterizations of an oncogene/proto-oncogene system accomplished to date.

Viral *mos* is, in reality, a gene fusion. The v-*mos* gene of MSV-124 encodes a 374 amino acid protein whose initiating methionine and first five amino acids are provided by the MuLV *env* gene. It differs at 11 positions from that predicted by the mouse c-*mos* sequence, while the ml v-*mos* sequence differs at 4 positions and the HT-1 sequence shows no differences. This is consistent with the fact that the HT-1 strain was derived from a single hamster tumor after inoculation of passage 6 of the original MSV stock (Huebner et al. 1966), while both ml and 124 were isolated after extensive animal passage of the un-cloned MSV stock. The amino acid differences do not appear to affect the oncogenic potential of the viruses, although a precise detailed comparison has not been made.

The mouse c-*mos* sequence is unique in that it is completely colinear with its viral homolog and lacks detectable introns. The major *mos* open reading frame is 75% conserved between mouse and man, both starting with a conserved 5 amino acid sequence and terminating at a conserved opal codon. All other c-*mos* genes also appear to lack introns, but human c-*mos* contains a second reading frame which overlaps the first 90 codons of the major conserved region and extends 20 codons upstream. Sequence analysis also reveal that human and mouse c-*mos* share a second region of conserved homology (75% at the DNA level), a 217 bp region located 353 bp upstream from the conserved ATG of human c-*mos*. The degree of conservation is suggestive of some biological function, although none has yet been identified.

Human c-*mos* has been mapped to chromosome 8, but the exact localization (8q22, Neel et al. 1982; or 8q11, Caubet et al. 1985) is in dispute. No human malignancies have been shown to contain an activated c-*mos* oncogene, although this gene is activated in mouse plasmacytomas as a result of the integration of an intracisternal-A particle genome into the 5′ coding region (Canaani et al. 1983).

Biological Analysis

Both the viral and cellular homologs of *mos* have been subject to extensive biological analysis using DNA transfection techniques (Table 1). Efficient transformation by cloned v-*mos* requires the presence of either upstream or downstream enhancer sequences, which can be provided by the proviral long terminal repeat (LTR) (Blair et al. 1980). Subgenomic fragments containing

Table 1. Transforming activity of v-*mos* and c-*mos*

Transfecting DNA [a]	LTR position	Transformation efficiency (ffu/pmol) [b]
v-*mos*	–	10–40
v-*mos*+LTR	Upstream/downstream	6000–20000
Mouse c-*mos*+LTR	Upstream	3000–6000
	Downstream	3–10
Human c-*mos*+LTR	Upstream	40–300
Chicken c-*mos*+LTR	Upstream	600–1500

[a] All constructs were transfected onto NIH 3T3 cells in the presence of carrier DNA (Blair et al. 1980). The LTR position relative to *mos* is indicated.

[b] ffu, focus-forming units per pmol of DNA.

v-*mos* but lacking an enhancer are able to transform NIH 3T3 mouse cells, but are about 1000-fold less efficient.

The amino acid identity between mouse c-*mos* and HT-1 v-*mos* demonstrates that the cellular sequence has oncogenic potential. The mouse c-*mos* locus lacks both promoters and enhancers which will function in NIH 3T3 mouse fibroblasts. However, the cloned normal mouse c-*mos* locus transforms effectively when an LTR is linked upstream, but is approximately 1000-fold less efficient if the LTR is linked downstream, sharply contrasting to what was observed with v-*mos* (Oskarsson et al. 1980). A sequence termed UMS, which is located approximately 1500 bp upstream to c-*mos*, was shown to be responsible for inhibiting transforming activity. UMS apparently acts as a transcription terminator (McGeady et al. submitted) and thus blocks readthrough from upstream promoters.

Although the human c-*mos* locus contains no upstream sequences homologous to UMS, attempts to activate the transforming potential with LTR sequences were unsuccessful. Recombinant DNA constructs containing human/mouse hybrid *mos* genes indicated that removal of sequences immediately upstream to the human ATG could enhance the transforming activity (Blair et al. submitted). Placement of an upstream LTR within 25 bp of the conserved ATG, which eliminated the human overlapping reading frame, activated the transforming potential of human c-*mos* (Table 1) but the efficiency was 20–50-fold lower than with mouse c-*mos*, and the transformed foci were small and contained fewer refractile cells. It is possible that the human overlapping reading frame acts, by promoter exclusion, as the functional equivalent of the mouse UMS; i.e., it provides a barrier to the inopportune expression of *mos*. However, even when activated, human c-*mos* is an inefficient transforming gene in mouse cells. Cells transformed with human c-*mos* contain high levels of the 33 kDa human c-*mos*-encoded product, suggesting that high levels of an inefficient transforming protein are required to induce the transformed phenotype. Analysis of human/mouse hybrid *mos* constructs suggests that several regions (domains) of *mos* are responsible for these differences (Blair et al. submitted), with the C-terminal domain (amino acid positions 223 – 346) having the most significant effect. This is the region of *mos* which possesses homology to the *src* kinase domain, and, coupled with the recent evidence that the *mos* product is a serine kinase (Maxwell and Arlinghaus 1985), indicates a functional role for this domain in *mos*-induced oncogenesis.

The ability of c-*mos* to function as an oncogenic sequence has been also observed with the rat (van der Hoorn et al. 1982) and chicken (Schmidt et al. in preparation) homologs. In the case of the chicken c-*mos*, efficient activation by an LTR is observed (Table 1) despite the fact that its homology with v-*mos* is markedly less than that of the poorly transforming human homolog (62% vs. 75%). However, it remains to be seen whether the human and chicken c-*mos* genes transform cells of the corresponding species more efficiently than they do mouse cells. Nevertheless, it is clear that both at the sequence level and the functional level, *mos* exhibits a more rapid and more variable divergence than, for example, the highly conserved *ras* family of oncogenes. The significance of this will undoubtedly become clearer as we learn more about the target of *mos* and its mechanism of oncogenesis.

RNA and Protein Expression

At both the protein and RNA levels, the detection of *mos* expression has proven to be a frustrating enigma. Even in virally transformed cells, the structure of the message responsible for *mos* expression remains unclear. C-*mos* RNA expression appears to be both developmentally regulated and tissue-specific, and has only recently been detected for the first time, (Propst and Vande Woude 1985), primarily in mouse ovaries and testes. The level of expression in these tissues is low (10 copies or less per cell) and the size of the transcript varies in a tissue specific fashion (from 1.3 to 6 kb), with the variation apparently due to differences at the 5′ end of the RNA (Propst and Vande Woude 1985). To date, all other attempts to detect c-*mos* expression in tumors or cells in culture have been unsuccessful, although it is possible that newer, more sensitive techniques and renewed interest may uncover either organ-specific or tumor-specific expression in certain cases.

The detection and analysis of either the viral or cellular *mos* gene product has also presented particular difficulties. Attempts to derive *mos*-specific antisera from tumor-bearing animals of various species have been unsuccessful, but antiserum raised against a synthetic 12 amino acid peptide representing the C-terminus of *mos* was able to detect a 37 kDa protein in MSV 124-transformed cells (Papkoff et al. 1981). The v-*mos* product undergoes phosphorylation at serine residues, and is present at extremely low levels in stably transformed cells although cells acutely infected with MSV apparently contain 30 – 100 times more *mos* protein (Papkoff et al. 1982). Since the majority of cells acutely infected with MSV at high multiplicities die, this suggested that high levels of *mos* expression are lethal. The *mos* protein has been localized primarily in the cytoplasm of infected cells.

Biochemical Functions

The mechanism by which *mos* circumvents normal cellular growth control and induces cell transformation remains unknown. The identification of a biochemical activity associated with the *mos* protein has been hampered by the low level of expression, even in transformed cells, and by the lack of effective antisera. Nevertheless, it has been possible to demonstrate that the Mo-MSV 124 *mos* product possesses a serine/threonine kinase activity (Maxwell and Arlinghaus 1985). This activity has been shown to be temperature-sensitive in a *ts* transformation mutant of MSV (Kloetzer et al. 1983). Recently, *mos* derived from the HT-1 strain of MSV has been expressed at high levels in bacteria, and this bacterially expressed protein possessed both ATP binding and ATPase activities (Seth and Vande Woude 1985). These activities are consistent with sequence analysis, since *mos* contains a region homologous to the consensus ATP-binding site of the catalytic subunit of the cyclic AMP-dependent bovine protein kinase (Barker and Dayhoff 1982) as well as to the ATP-binding and kinase domains of the *src* family of tyrosine phosphokinases. Although the *mos* protein is now known to possess several distinct biochemical activities, it should be emphasized that none of these has been shown to be directly involved in the oncogenic function of the protein. It will be of particular interest to determine if mouse c-*mos* (i.e., HT-1 *mos*) exhibits a kinase activity equivalent to that detected in the more heavily mutated Mo-MSV 124 v-*mos* product.

Another open question is whether the proteins expressed from the complex family of messenger RNAs detected in testes and ovaries are functionally equivalent to the viral and activated cellular gene products. If they are, then *mos* may represent a clear case in which the inopportune expression (i.e., in a tissue where it is normally silent) of a normal cell sequence is sufficient to induce the transformed state. This intrinsic oncogenic potential, capable of transforming fibroblastic and hematopoietic cells, may suggest why the expression of *mos* seems to be so tightly controlled and why cis-acting inhibitory elements, such as UMS in mouse, and perhaps the overlapping reading frame in man, have been seen in these two cellular loci. Since these elements must of necessity be specifically overcome in tissues (i.e., testes or ovaries) whose *mos* is expressed, the positive regulation of *mos* expression presents several interesting problems. Clearly, most of the questions pertaining to both regulation and gene function remain to be answered. However, if our present knowledge of *mos* is any indication, the answers are likely to reveal novel facts about how mammalian genes function both normally and in oncogenesis.

References

Ball JK, McCorter JA, Sunderland SM (1973) Evidence for helper independent murine sarcoma virus. I. Segregation of replication-defective and transformation-defective viruses. Virology 56:268 – 284

Barker WC, Dayhoff MO (1982) Viral *src* gene products are related to the catalytic chain of mammalian cAMP-dependent protein kinase. Proc Natl Acad Sci USA 79:2836 – 2839

Blair DG, Hull MA, Finch EA (1979) The isolation and preliminary characterization of temperature-sensitive transformation mutants of Moloney sarcoma virus. Virology 95:303 – 316

Blair DG, McClements WL, Oskarsson MK, Fischinger PJ, Vande Woude GF (1980) Biological activity of cloned Moloney sarcoma virus DNA: Terminally redundant sequences may enhance transformation efficiency. Proc Natl Acad Sci USA 77:3504 – 3508

Brow MAD, Sen A, Sutcliffe JG (1984) Nucleotide sequence of the transforming gene of ml murine sarcome virus. J Virol 49:579 – 582

Canaani E, Dreazen O, Klar Z, Rechavi G, Ram D, Cohen JB, Givol D (1983) Activation of the c-*mos* oncogene in a mouse plasmacytoma by insertion of an endogenous intracisternal A-particle genome. Proc Natl Acad Sci USA 80:7118 – 7122

Caubet JF, Mathieu-Mahul D, Bernheim A, Larsen CJ, Berger R (1985) Human proto-oncogene c-*mos* maps to 8q11. EMBO J 4:2245 – 2248

Huebner RJ, Hartley JW, Rowe WP, Lane WT, Capps WI (1966) Rescue of the defective genome of Moloney sarcoma virus from a noninfectious hamster tumor and the production of pseudotype sarcoma viruses with various murine leukemia viruses. Proc Natl Acad Sci USA 81:5305 – 5309

Horn JP, Wood TG, Murphy EC, Blair DG, Arlinghaus RA (1981) A selective temperature-sensitive defect in viral RNA production in cells infected with a *ts* mutant of murine sarcoma virus. Cell 25:37 – 46

Jones M, Bosselman RA, Houtin VJF, Berub A, Fan H, Verma IM (1980) Identification and molecular cloning of Moloney mouse sarcoma virus-specific sequences from uninfected mouse cells. Proc Natl Acad Sci USA 77:2651 – 2655

Kloetzer WS, Maxwell SA, Arlinghaus RB (1983) p85$^{gag\text{-}mos}$ encoded by *ts*110 Moloney murine sarcoma virus has an associated protein kinase activity. Proc Natl Acad Sci USA 80:412 – 416

Maxwell SA, Arlinghaus RB (1985) Serine kinase activity associated with Moloney murine sarcoma virus-124-encoded p37^{mos}. Virology 143:321 – 333

Moloney JB (1966) A virus induced rhabdomyosarcoma of mice: an Em study of virus induced murine sarcoma. Natl Cancer Inst Monogr 22:139 – 142

Murphy E, Arlinghaus R (1982) Comparative tryptic peptide analysis of candidate p85$^{gag\text{-}mos}$ of *ts*110 Moloney murine sarcoma virus and p38-p23 *mos* gene-related proteins of wild-type virus. Virology 121:372 – 383

Neel BG, Jhanwhar SC, Chaganti RSK, Hayward WS (1982) Two human c-onc genes are located on the long arm of chromosome 8. Proc Natl Acad Sci USA 79:7842 – 7846

Oskarsson M, McClement WL, Blair DG, Maizel JV, Vande Woude G (1980) Properties of a normal mouse cell DNA sequence (*sarc*) homologous to the *src* sequence of Moloney sarcoma virus. Science 207:1222 – 1224

Papkoff J, Lai MHT, Hunter T, Verma IM (1981) Analysis of transforming gene products from Moloney murine sarcoma virus. Cell 27:109 – 119

Papkoff J, Verma IM, Hunter T (1982) Detection of a transforming gene product in cells transformed by Moloney murine sarcoma virus. Cell 29:417 – 426

Propst F, Vande Woude G (1985) c-*mos* proto-oncogene transcripts are expressed in mouse tissues. Nature 315:516 – 518

Seth A, Vande Woude GF (1985) Nucleotide sequence and biochemical activities of the Moloney murine sarcoma virus strain HT-1 *mos* gene. J Virol 56:144 – 152

Stocking C, Kollek R, Ostertaz W (1985) Long terminal repeat sequences impart hematopoietic transformation properties to the myeloproliferative sarcoma virus. Proc Natl Acad Sci USA 82:5746 – 5750

Van Beveren C, van Straaten F, Galleshaw JA, Verma IM (1981) Nucleotide sequence of the genome of a murine sarcoma virus. Cell 27:97–108

Vande Woude GF, Oskarsson M, Enquist LW, Nomura S, Sullivan M, Fischinger PJ (1979) Cloning of integrated Moloney sarcoma proviral DNA sequences in bacteriophage λ. PNAS 76:4464–4468

Van der Hoorn FA, Hulsebos E, Berns AJM, Bloemers HPJ (1982) Molecularly cloned c-*mos* (rat) is biologically active. EMBO J 1:1313–1317

Verma IM, Lai MHT, Bosselman A, McKenneth MA, Fan H, Berns A (1980) Molecular cloning of unintegrated Moloney mouse sarcoma virus DNA in bacteriophage. Proc Natl Acad Sci USA 77:1773–1777

Watson R, Oskarsson M, Vande Woude GF (1982) Human DNA sequence homologous to the transforming gene (*mos*) of Moloney murine sarcoma virus. Proc Natl Acad Sci USA 79:4078–4082

Structure and Function of the Human Interleukin-2 Receptor

MASANORI HATAKEYAMA, SEIJIRO MINAMOTO, HISASHI MORI, and
TADATSUGU TANIGUCHI

Clonal expansion of resting T cells requires the sequential stimulation of two sets of glycoproteins: the T cell receptor (T3-Ti) complex and the interleukin-2 receptor (IL-2R) complex. The human T3-Ti complex is composed of two components: a polymorphic heterodimer of α and β chains, which determine both antigen specificity and major histocompatibility complex (MHC) restriction, and a complex of nonpolymorphic molecules called T3-γ, -δ and -ε, which may play a role in transducing the T-cell activation signal (Meuer et al. 1984). Triggering of the T3-Ti complex induces expression of the genes for interleukin-2 (IL-2), a T cell growth factor, and its receptor (IL-2R). The net effect is therefore that clonal expansion of T cells occurs by converting the signal generated upon specific antigen: MHC/T3-Ti interaction into a mitogenic signal mediated by IL-2 and IL-2R (see Fig. 1). The molecular nature of the IL-2 system has been studied in detail (reviewed by Taniguchi et al. 1986). In this review, we describe the main features of the human IL-2R or IL-2R complex.

IL-2: Receptor Binding

The existence of the IL-2R on activated T cells was first demonstrated by Robb and co-workers using affinity-purified, radiolabeled IL-2 (Robb et al. 1981). Like certain other growth factor receptors, the IL-2R exists in two forms which differ only on the basis of their affinity for the ligand (Robb et al. 1984): the high affinity IL-2R has a dissociation constant (K_d) of $10-100$ pM while the low affinity receptor has a K_d of $1-10$ nM. IL-2-specific signals seem to be delivered solely by the high affinity form; the biological significance of the low affinity IL-2R remains unclear. As described above, resting T cells do not express either high or low affinity IL-2R. Antigen stimulation induces expression of both types of IL-2Rs and of IL-2 itself, and T cell proliferation occurs following interaction of the high affinity IL-2R with IL-2 (Smith 1980; Taniguchi et al. 1983). The IL-2:IL-2R interaction results in down-regulation of the high affinity receptor (i.e. decrease in receptor number) by rapid internalization of the receptor-ligand complexes and/or by a change in the receptor from high to low affinity (Smith and Cantrell 1985). Therefore, the magnitude of the IL-2-dependent T cell response is determined both by the IL-2 concentration and by the number of functional, high-affinity surface IL-2Rs on the responder cells.

Oncogenes and Growth Control
Edited by P. Kahn and T. Graf
© Springer-Verlag Berlin Heidelberg 1986

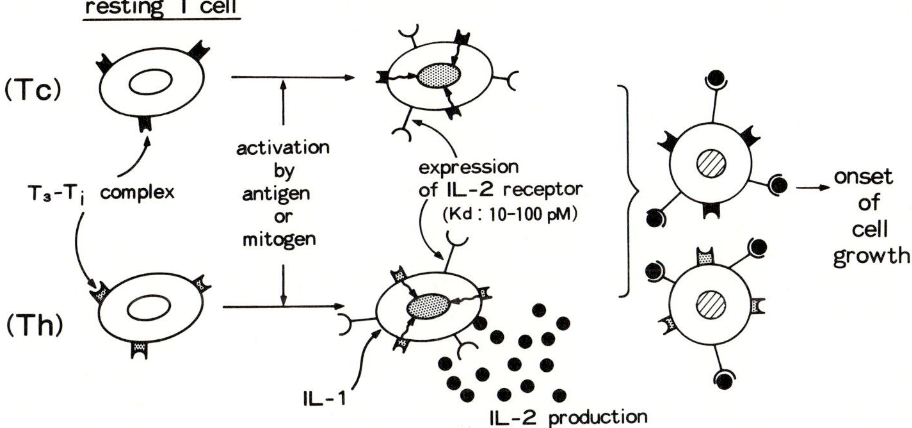

Fig. 1. Model for the induction of T cell growth. Tc, cytotoxic T cell; Th, helper T cell; $T_3 - T_i$, T cell receptor

The demonstration that the IL-2R is present on the surface not only of T cells but also of immature thymocytes (Raulet 1985), B cells (Zubler et al. 1984), and macrophages (Herrmann et al. 1985) suggests that IL-2 might have multiple functions in the immune system.

Analysis of the IL-2 Receptor/Tac Antigen

One fruitful approach to analyzing the IL-2 receptor has involved use of a monoclonal antibody prepared by Uchiyama et al. (1981) which specifically reacts with activated, but not with resting, human T cells and which blocks IL-2-dependent T cell growth. It also blocks the binding of IL-2 to IL-2R-bearing T cells. These results strongly suggest that the antigen recognized by this antibody, which has been designated the Tac antigen, is the human IL-2R or molecules closely associated with it (Uchiyama et al. 1981; Leonard et al. 1982, 1983). The cDNAs encoding the Tac antigen were molecularly cloned from libraries, prepared first from mRNA of T cell lines infected with the human T-cell leukemia virus (HTLV-I) (Leonard et al. 1984; Nikaido et al. 1984; Cosman et al. 1984) and subsequently from normal peripheral blood lymphocytes (Hatakeyama et al. 1985). No differences were detected between the cDNAs derived from these two sources. The amino acid sequence of the Tac antigen, as deduced from the cDNA sequence, is composed of 272 residues, including a 21 amino acid signal peptide. The molecular weight of the Tac protein without the signal sequence is 28,428. This protein is composed of a 219 amino acid extracellular domain, a 19 amino acid transmembrane domain and a 13 amino acid cytoplasmic domain. There are two potential N-linked glycosylation sites and multiple potential O-linked glycosylation sites. The Tac antigen appears to undergo post-translational modifications which include glycosylation, phosphorylation and sulfation, generating a 55 kDa cell surface Tac glycoprotein (Leonard et al. 1985a).

The Tac antigen is encoded by a single gene on human chromosome 10. The gene is divided into eight exons and seven introns and spans more than 25 kb. Exons 2 and 4 have a striking homology and are assumed to have been generated by a gene duplication event (Leonard et al. 1985b). The exon structures show some degree of homology to the Ba fragment of human complement factor B. Northern blot analysis of Tac-positive cells with a Tac antigen cDNA probe revealed the presence of two mRNAs (1.5 kb and 3.5 kb), the size difference being due primarily to the utilization of different polyadenylation signals. Both mRNAs encode molecules that can react with anti-Tac serum. In addition, the existence of an mRNA which deletes exon 4 sequence by alternative splicing was also observed from cDNA analysis; however, this mRNA does not direct expression of cell surface molecules which can react with anti-Tac serum or bind IL-2 (Leonard et al. 1984).

One of the structural features of the Tac antigen is that it contains only a small (13 amino acid) cytoplasmic domain. In contrast, growth factor receptors such as the EGF and insulin receptors, HER-2/c-*erbB*-2 and CSF-1/c-*fms* have large cytoplasmic domains with tyrosine kinase activity which appear to be important in transmembrane signaling (Coussens et al. 1985). This difference raises the interesting and as yet unanswered question of how the unusually small cytoplasmic domain of the IL-2R transmits a mitogenic signal.

Transfection of Tac antigen cDNA into fibroblast cell lines such as mouse L cells and monkey COS cells resulted in the expression of molecule(s) capable of binding with anti-Tac serum and with IL-2. However, the expressed receptor molecule(s) showed low affinity for the ligand and could not transduce the IL-2-specific growth signal (Greene et al. 1985). These observations raised the question as to whether the cloned Tac antigen cDNA directly encodes the functional (high affinity) IL-2R.

To clarify this issue, attempts were made to reconstitute the functional, high-affinity IL-2R by introducing the Tac-encoding cDNA into lymphoid cells (Hatakeyama et al. 1985). An expression vector was constructed in which the Rous sarcoma virus LTR sequence was linked upstream of the Tac antigen cDNA. Transfection of this construct into mouse L929 fibroblasts (used as a control) resulted in the expression of only a nonfunctional, low affinity IL-2R on the cell surface, in agreement with previous observations. However, transfection into the mouse thymoma (T lymphoid) cell line EL-4 (which has a Thy1[+], Ly1[+], Lyt2[−], and L3T4[+] phenotype but does not express IL-2R) resulted in the expression of high as well as low affinity IL-2Rs. These results clearly demonstrated for the first time that the two classes of human IL-2R can be generated from a single gene and that the Tac antigen itself is a constituent of both high and low affinity IL-2Rs in lymphocytes. The fact that expression of the high affinity IL-2R requires a lymphocyte environment strongly suggests that this form arises via association of the Tac antigen with another lymphocyte-specific membrane determinant(s), which is yet to be identified. In other words, the Tac antigen may be only a part of the "IL-2R complex" in which the Tac antigen acts as a ligand-binding component and another molecule(s) transduces the IL-2 specific signal. The hypothesized IL-2R complex would be reminiscent of the T cell receptor complex in which both α and β chains, each containing a very short cytoplasmic region, form a heterodimer that is responsible for antigen/MHC recognition (Meuer et al. 1984). In this case, the signal generated by antigenic stimulation seems to be transduced via T3 molecules that are closely associated with the αβ heterodimer. The cytoplasmic domain of the Tac antigen contains a region with 6 positively charged amino acid residues (Arg and Lys) which may allow interaction with other molecules. The 113KDa and 180KDa peptides which occasionally co-precipitate with Tac antigen in the anti-Tac immunoprecipitation (Leonard et al. 1983) could represent other functional units of the IL-2R complex.

However, it is not ruled out that high affinity IL-2 receptors do not arise via association with another molecule(s) but via a lymphocyte-specific post-translational modification(s). Regardless of how they are generated, the expression of high affinity receptors appears to be suppressed in fibroblasts, since cell hybrids obtained by fusing mouse T lymphoid cells that express both high and low affinity human IL-2Rs with mouse L929 fibroblasts express only low-affinity IL-2R (S. Minamoto and M. Hatakeyama, unpublished observation).

Phosphorylation of IL-2 Receptor

Of the 13 amino acid residues in the cytoplasmic domain of Tac antigen, there are two possible phosphorylation sites, Ser(247) and Thr(250). It has been suggested that treatment of T cells with the phorbolester TPA leads to phosphorylation of these two residues (Schackelford and Trowbridge 1984). In addition, IL-2 has been reported to activate protein kinase C, indicating that one or both of the above-mentioned residues is physiologically phosphorylated following IL-2 stimulation. In the EGF receptor, phosphorylation of Thr (654), which is located in a position (just under the inner membrane surface) corresponding to that of Ser(247) or Thr(250) in the Tac antigen, leads to a decrease in the receptor affinity for EGF (Davis and Czech 1985), raising the possibility that the IL-2R may also be regulated by phosphorylation. Site-directed mutagenesis of the phosphorylation sites in the IL-2R followed by assays of biological function in T lymphoid cells should resolve this issue.

IL-2 As a Growth Inhibitor

Mouse T lymphoid (EL-4) cells expressing the human high-affinity IL-2Rs respond to human, but not mouse, recombinant IL-2. However, IL-2 does not stimulate, but strongly inhibits the proliferation of these cells (Hatakeyama et al. 1985). Such bifunctional properties of growth factors have also been described for EGF, TGF-β and tumor necrosis factor (Sporn and Roberts 1985 and Carpenter et al. and Moses and Leof, this Vol.). IL-2 also induces some HTLV-I-infected human T cell lines, which respond to IL-2 through high affinity IL-2R, to cease proliferation (Sugamura et al. 1985). Such bifunctional effect can perhaps be viewed as one of the general features of growth factors. Whether the cellular response is mitogenic or inhibitory may be determined not only by the ligand-receptor interaction but by the combined activation and/or inactivation of other biochemical reactions and/or gene expressed in the responder cells. Although there is no evidence at present that such a negative growth effect of IL-2 really operates in vivo, one can also speculate that IL-2 functions in a negative feedback regulation of the T cell response under certain physiological conditions.

IL-2 and Oncogene Expression

Recent studies show that treatment of T cells expressing the IL-2 receptor with IL-2 stimulates expression of the c-*myc* and c-*myb* oncogenes (Reed et al. 1985; K. A. Smith, personal communication). The activation of these nuclear oncogenes seems to be important for the $G_0{\to}G_1$ transition in the cell cycle (see Bravo and Müller, this Vol.). Preliminary analysis of the levels of several c-*onc* mRNAs in EL-4 transformants before and after treatment with IL2 suggests that the level of c-*myc* mRNA decreases following IL-2 treatment, whereas that of other oncogenes, including the *ras* gene family, are unaffected (H. Mori et al., unpublished observation). Further work should reveal the molecular nature of the signal pathway between extracellular stimulation by IL-2 and the nuclear gene activation and/or inactivation which ultimately directs the cellular responses.

IL-2 Receptor and Adult T Cell Leukemia

Adult T cell leukemia (ATL) is a unique hematological disease caused by HTLV-I infection. ATL cells constitutively express both high and low affinity IL-2Rs on their cell surface (Uchiyama et al. 1985), which originally led to the hypothesis that the IL-2/IL-2R-mediated autocrine mechanism operates in ATL leukemogenesis. Indeed, autocrine stimulation by IL-2 has been observed in non-ATL-related leukemic T cells (Duprez et al. 1985). However, the finding that ATL cells generally do not produce IL-2 or respond to exogeneous IL-2 (Uchiyama et al. 1985) suggests that they do not require IL-2 for their autonomous proliferation and therefore argues against the autocrine hypothesis.

What mechanisms might induce the aberrant expression of IL-2R in ATL cells? It has been reported that a protein derived from the pX-region of HTLV-I genome (p40^x or tat protein) functions as a strong trans-activator of transcription from the LTR sequence of HTLV-I itself (Sodroski et al. 1984; Fujisawa et al. 1985 and article by Haseltine et al.). The p40^x protein might also have an effect on a cellular gene promoter region that includes IL-2 and IL-2R (W. C. Greene, personal communication). Activation of both ligand and receptor genes might play a role in the initiation and/or promotion of ATL leukemogenesis but then become inoperative and/or unnecessary in the acute phase of ATL. Another possibility is that humoral factors are involved in the induction of IL-2R overexpression. In fact, an ATL-derived factor designated ADF and a factor tentatively designated as IL-4a, which is a natural ligand to the T11 molecule, have been described (Teshigawara et al. 1985; Milanese et al. 1986). The p40^x protein may also activate genes for such humoral factors.

Concluding Remarks

The availability of molecularly cloned IL-2 and IL-2R (or a part of the receptor) and of a transfection system in which functional IL-2R molecules can be assayed have made it possible to begin unraveling the complex mechanism of T cell growth. Further study of the IL-2/IL-2R system promises to provide important information regarding more general molecular mechanisms of cell growth and regulation, particularly the mechanism of signal transduction by receptors with very small cytoplasmic domains.

Acknowledgments. We thank Drs. T. Uchiyama, R. R. Hardy and E. Barsoumian for valuable suggestions. We also thank Ms. M. Nagatsuka for excellent assistance.

References

Davis RJ, Czech MP (1985) Tumor-promoting phorbol diesters cause the phosphorylation of epidermal growth factor receptors in normal human fibroblasts at threonine-654. Proc Natl Acad Sci USA 82:1974 – 1978

Duprez V, Lenoir G, Dautry-Varsat A (1985) Autocrine growth stimulation of a human T-cell lymphoma line by interleukin 2. Proc Natl Acad Sci USA 82:6932 – 6936

Fujisawa J, Seiki M, Kiyokawa T, Yoshida M (1985) Functional activation of the long terminal repeat of human T-cell leukemia virus type I by a transacting factor. Proc Natl Acad Sci USA 82:2277 – 2281

Greene WC, Roob RJ, Svetlik PB, Rusk CM, Depper JM, Leonard WJ (1985) Stable expression of cDNA encoding the human interleukin-2 receptor in eukaryotic cells. J Exp Med 162:363 – 368

Hatakeyama M, Minamoto S, Uchiyama T, Hardy RR, Yamada G, Taniguchi T (1985) Reconstitution of functional receptor for human interleukin-2 in mouse cells. Nature 318:467 – 470

Herrmann F, Cannistra SA, Levine H, Griffin JD (1985) Expression of interleukin-2 receptors and binding of interleukin-2 by gamma interferon-induced human leukemic and normal monocytic cells. J Exp Med 162:1111 – 1116

Leonard WJ, Depper JM, Uchiyama T, Smith KA, Waldmann TA, Greene WC (1982) A monoclonal antibody that appears to recognize the receptor for human T cell growth factor; partial characterization of the receptor. Nature 300:267 – 269

Leonard WJ, Depper JM, Robb RJ, Waldmann TA, Greene WC (1983) Characterization of the human receptor for T cell growth factor. Proc Natl Acad Sci USA 80:6957 – 6961

Leonard WJ, Depper JM, Crabtree GR, Rudikoff S, Pumphrey J, Robb RJ, Svetlik PB, Peffer N, Waldmann TA, Greene WC (1984) Molecular cloning and expression of cDNAs for the human interleukin-2 receptor. Nature 311:626 – 631

Leonard WJ, Depper JM, Kronke M, Robb RJ, Waldmann TA, Greene WC (1985a) The human receptor for T cell growth factor. J Biol Chem 260:1872 – 1880

Leonard WJ, Depper JM, Kanehisa M, Kronke M, Peffer N, Svetlik PB, Sullivan M, Greene WC (1985b) Structure of the human interleukin-2 receptor gene. Science 230:633 – 639

Meuer SC, Acuto O, Hercend T, Sclossman SF, Reinherz EL (1984) The human T cell receptor. Annu Rev Immunol 2:23 – 50

Nikaido T, Shimizu A, Ishida N, Sabe H, Teshigawara K, Maeda M, Uchiyama T, Yodoi J, Honjo T (1984) Molecular cloning of cDNA encoding human interleukin-2 receptor. Nature 311:631 – 635

Raulet DH (1985) Expression and function of interleukin-2 receptors on immature thymocytes. Nature 314:101 – 103

Reed JC, Sabath DE, Hoover RG, Prystowsky MB (1985) Recombinant interleukin-2 regulates levels of c-myc mRNA in a cloned murine T lymphocytes.

Robb RJ, Munck A, Smith KA (1981) T cell growth factor receptors. Quantitation, specificity, and biological relevance. J Exp Med 154:1455 – 1474

Robb RJ, Greene WC, Rusk CM (1984) Low and high affinity cellular receptors for interleukin-2. Implications for the level of Tac antigen. J Exp Med 160:1126 – 1146

Schackelford DA, Trowbridge IS (1984) Induction of expression and phosphorylation of the human interleukin-2 receptor by a phorbol diester. J Biol Chem 259:11706 – 11712

Smith KA (1980) T-cell growth factor. Immunol Rev 51:337 – 357

Smith KA, Cantrell DA (1985) Interleukin-2 regulates its own receptors. Proc Natl Acad Sci USA 82:864 – 868

Sodroski JG, Rosen CA, Haseltine WA (1984) Trans-acting transcriptional activation of the long terminal repeat of human T lymphotropic viruses in infected cells. Science 225:381 – 385

Sporn MB, Roberts AB (1985) Autocrine growth factors and cancer. Nature 313:701 – 705

Sugamura K, Nakai S, Fujii M, Hinuma Y (1985) Interleukin-2 inhibits in vitro growth of human T cell lines carrying retrovirus. J Exp Med 161:1243 – 1248

Taniguchi T, Matsui H, Fujita T, Takaoka C, Kashima N, Yoshimoto R, Hamuro J (1983) Structure and expression of a cloned cDNA for human interleukin-2. Nature 302:305 – 310

Taniguchi T, Matsui H, Fujita T, Hatakeyama M, Kashima N, Fuse A, Hamuro J, Nishi-Takaoka C, Yamada G (1986) Molecular analysis of the interleukin-2 system. Immunol Rev 92 (in press)

Teshigawara K, Maeda M, Nishino K, Nikaido T, Uchiyama T, Tsudo M, Wano M, Yodoi J (1985) Adult T cell leukemic cells produce a lymphokine that arguments interleukin-2 receptor expression. J Mol Cell Immunol 2:17 – 26

Uchiyama T, Broder S, Waldmann TA (1981) A monoclonal antibody (anti-Tac) reactive with activated and functionally matured human T cells. I. Production of anti-Tac monoclonal antibody and distribution of Tac^+ cells. J Immunol 126:1393 – 1397

Uchiyama T, Hori T, Tsudo M, Wano Y, Umadome H, Tamori S, Yodoi J, Maeda M, Sawami H, Uchino H (1985) Interleukin-2 receptor (Tac antigen) expressed on adult T cell leukemia cells. J Clin Invest 76:446 – 453

Zubler RH, Lowenthal JW, Evard F, Hashimoto N, Deuos R, MacDonald HR (1984) Activated B cells express receptors for, and proliferation in response to pure interleukin-2. J Exp Med 160:1170 – 1183

III Signal Transduction and *ras* Oncogenes

Once a growth factor receptor has been activated by the binding of ligand, the mitogenic signal must travel through the cytoplasm to the nucleus, ultimately leading to a round of DNA synthesis and to cell division. The articles in this section describe some of the mechanisms by which this intracellular signalling might occur in the cytoplasm. One principle which is emerging is that interconnected pathways exist, some of which may eventually converge, while others remain separate. These pathways are likely to involve the activation of signal-transducing proteins and "second messengers", the nature of which is another major theme of this section. In addition, the articles illustrate the point that some of the changes which occur in response to growth factor stimulation are also seen in cells transformed with retroviral oncogenes.

The section begins with a discussion of the role of phosphorylation in signal transduction. The fact that many growth factor receptors and oncogene products possess protein-tyrosine kinase activity strongly implicates tyrosine phosphorylation as being important in the mitogenic response. However, these kinases lead to the phosphorylation of a large number of cellular proteins, and therefore, despite intensive efforts in the last several years, it is still unclear which substrates are critical for the transformation process.

Several subsequent articles focus on the recent finding that growth factors and oncogenes stimulate the metabolism of inositol lipids in the membrane, yielding two products which can each act as a second messenger: inositol triphosphate and diacylglycerol. Inositol triphosphate leads to the mobilization from intracellular stores of Ca^{2+}, an ion which is also believed to act as a second messenger. Diacylglycerol stimulates a serine/threonine kinase known as protein kinase C, an enzyme which is increasingly recognized as playing a critical role in the mitogenic response of fibroblasts and is also implicated in the response of hematopoietic progenitor cells to specific growth factors. Among the known effects of protein kinase C is the activation of a sodium/proton exchange system and a consequent increase in intracellular pH, which appears to be necessary but not sufficient for the induction of DNA synthesis. One important substrate for an EGF-regulated serine/threonine kinase which has not yet been identified is the 40S ribosomal protein S6. Multiple phosphorylations of this protein are thought to be associated with the increased rate of protein synthesis observed during the mitogenic response.

Another class of molecules which appears to have a signal transducing function and which is discussed in this section is the so-called G protein family. These membrane-associated proteins transduce signals from hormone or sensory receptors to various effector systems, including adenylate cyclase. The interaction of G proteins with effector enzymes is regulated not only by the receptor molecules but also by guanine nucleotides. Binding of GTP activates G proteins to interact with the appropriate effectors, and the interaction is terminated by a G protein-mediated hydrolysis of GTP. Interestingly, DNA sequence analysis has revealed significant homology between one of the G protein subunits and the *ras* proto-oncogene family, the subject of the last several articles in the section. The *ras* family in mammalian cells consists of three closely related genes which have been detected in an oncogenically activated form in different murine sarcoma virus strains and in a wide variety of human tumors. Like the G proteins, the normal c-*ras* proteins are associated with the plasma membrane, bind GTP and have GTPase activity. These findings have raised the intriguing possibility that the *ras* proteins participate in the transduction of mitogenic signals between unidentified growth factors and effector systems. In contrast, the transforming *ras* proteins bind GTP but hydrolyze it poorly, suggesting that they may be constitutively active.

The section concludes with a description of the yeast homologues of the mammalian *ras* genes. Like their mammalian counterparts, the yeast *RAS* genes encode proteins with GTPase activity. Mutations which abolish *RAS* gene function lead to defects in sporulation and utilization of nutrients. The observation that these mutants can be rescued by mammalian *ras* genes suggests that the homology between yeast and mammalian *ras* genes is functional as well as structural. These findings provided the stimulus to exploit the relative simplicity of the yeast system for the analysis of *RAS* function. Such studies have revealed that *RAS* proteins in baker's yeast stimulate adenylate cyclase and therefore support the notion that they participate in signal transduction. However, mammalian *ras* proteins do not seem to interact with adenylate cyclase, and therefore, their biochemical function and their role in signal transduction are still open.

Phosphorylation in Signal Transmission and Transformation

Tony Hunter

The response of cells to external stimuli requires the transduction of signals across the plasma membrane. Surface receptors with specific ligand binding domains accomplish the task of recognizing individual stimuli. Although one can conceive of many types of signal which could be generated upon ligand binding to a receptor, the obvious need to integrate the wide variety of external inputs encountered by the average cell suggests that there will be a limited number of effector mechanisms, each linked to multiple receptors. Signal systems can be expected to have a number of properties in common. There is a requirement for a high degree of sensitivity to the stimulus, but only a small binding energy is available from receptor: ligand interaction. Thus the generation of a significant internal response demands that signal mechanisms have a considerable degree of gain. Another prerequisite is that the signaling system can be switched on and off effectively and quickly. The necessity for signal integration means that the signal generators must be subject to regulation to allow potentiation and both autologous and heterologous inhibition.

Protein kinases and their complementary phosphoprotein phosphatases make ideal systems for the generation and transmission of signals in the cell. Protein phosphorylation is a well-recognized means for modulating the activity of proteins in a rapid and reversible fashion. In addition, protein kinases and phosphatases act catalytically, thus providing direct signal amplification. They also act pleiotropically, each type of protein kinase or phosphatase being able to use several different protein substrates. In principle, the signal can be transduced either by phosphorylation or by dephosphorylation of a target protein, the off-switch involving the converse reaction. Clearly the generation of such a signal requires that either the protein kinase, or the protein phosphatase, or both, has an activity that is regulated, directly or indirectly, by the stimulus in question. Several different types of protein kinase have activities which can be modulated and thus are potentially suitable for signal transmission systems. For example, there are protein kinases activated by diffusible metabolites such as cyclic nucleotides or diacylglycerol, while others are stimulated by peptide growth factors. Less is known about the regulation of protein phosphatases, although several specific inhibitors have been isolated.

In principle a "phosphorylation" signal generated at the plasma membrane could be propagated in a variety of ways. For instance, the target protein could be an enzyme whose activity is increased by phosphorylation. The catalytic nature of this second step in the pathway would provide further am-

Oncogenes and Growth Control
Edited by P. Kahn and T. Graf
© Springer-Verlag Berlin Heidelberg 1986

plification of the signal. Alternatively, a phosphorylated target protein could regulate the activity or function of another protein in a stoichiometric fashion. In either case, these events could form part of a cascade of steps leading to phenotypic changes. Whatever proteins are phosphorylated by the signal generator, one needs to explain how their phosphorylation causes the induction of secondary signal-dependent events. Events occurring at the plasma membrane can be conceived in terms of either a phosphorylated target protein (or a secondarily activated protein) moving in the plane of the membrane to its site of action. Alternatively, a diffusible compound which interacts with a membrane component could be produced by one of the steps in a cascade. Nuclear events, such as the induction of gene expression, might be triggered in a number of ways. Translocation into the nucleus of a phosphorylated target protein (or a secondarily activated protein) is one possibility. Alternatively, a low molecular weight compound or ion generated at the membrane might diffuse into the nucleus. If the target protein were a cytoskeletal element, one could imagine mechanical signal propagation via the cytoskeleton to the nucleus.

Two types of protein kinase are most germane to a discussion of signal transmission and transformation: the growth factor receptor protein-tyrosine kinases and the Ca^{2+}/phospholipid-dependent diacylglycerol-regulated protein-serine kinase, protein kinase C. Both types of protein kinase are implicated in signals generated at the plasma membrane in response to external mitogenic stimuli. The cAMP-dependent protein kinase system, which clearly acts as a signal pathway, will not be considered here because there is no strong evidence that it is directly involved in cell growth control or transformation.

Growth Factor Receptors that Are Ligand-Stimulated Protein-Tyrosine Kinases

There are at least five growth factor receptors which have protein-tyrosine kinase activities that are stimulated severalfold upon binding their cognate ligand, namely those for epidermal growth factor (EGF), platelet-derived growth factor (PDGF), colony-stimulating factor-1 (CSF-1), insulin, and insulin-like growth factor-1 (IGF-1) (reviewed in Heldin and Westermark 1984; Hunter and Cooper 1985; Sherr et al. 1985 and Sherr and Stanley, this Vol.). The EGF receptor is the best studied of these receptors. The amino acid sequence predicted from the nucleotide sequence of cDNA clones shows that the EGF receptor has a large glycosylated N-terminal extracellular EGF-binding domain, a single membrane-spanning segment, and a cytoplasmic region containing a sequence that is clearly identifiable as the catalytic domain of a protein kinase (Ullrich et al. 1984 and Pfeffer and Ullrich, this Vol.). The C-terminal region beyond the protein kinase domain probably has a negative regulatory function. Binding of EGF to the external domain must induce a change in the intracellular domain that increases protein kinase activity. The exact mechanism by which the effect of EGF binding is transmitted across the membrane is still unclear. Bearing in mind that the only physical connection between the external and internal domains is a single transmembrane segment of 23 amino acids, signal transduction may well require an EGF-dependent change in oligomeric state of the receptor in the membrane. In principle, the inactive form could be monomeric and the active one dimeric, or vice versa. The properties of site-directed mutations in the transmembrane domain and the regions abutting the membrane may answer this question (see also Schlessinger, this Vol.).

The structure of the CSF-1 receptor is similar in many respects to that of the EGF receptor (Sherr et al. 1985), as is that of the PDGF receptor (Ullrich, A. and Williams, L. T. personal communication). In apparent contrast, the insulin and IGF-1 receptors have an $(\alpha\beta)_2$ structure. The two chains in each $\alpha\beta$ subunit of the insulin receptor, however, are clearly derived by proteolysis from a precursor with a structure akin to the EGF receptor (Ullrich et al. 1985; Ebina et al. 1985). The mechanism of transmembrane signaling is not understood for these receptors either, although the dimeric nature of the insulin and IGF-1 receptors may be an important clue.

Are there other growth factor receptors which are protein-tyrosine kinases? The minimal size for such receptors would appear to be about 800 amino acids, and, based on this criterion, there are certainly other uncharacterized growth factor receptors which could be protein-tyrosine kinases (e.g., the G-CSF receptor). In addition, the 185-kDa *neu* gene product, which is a surface protein with protein-tyrosine kinase activity and a structure similar to the EGF receptor, seems likely to be a growth factor receptor, although no ligand has yet been identified. There are also three oncogenes derived from genes that encode growth factor receptor-like protein-tyrosine kinases which are not identical with any known receptor (see below). In toto there could be more than 20 protein-tyrosine kinase growth factor receptor genes.

The common function of these receptors strongly implies that tyrosine phosphorylation is an integral part of a mitogenic signal system. In an attempt to confirm this idea, considerable effort has been put into identifying substrates for the growth factor receptor protein-tyrosine kinases. Several such proteins have been reported, but to date there is no proof that any of them is critical for mitogenesis. The most promising substrate is a 42-kDa protein (p42), which is rapidly phosphorylated in quiescent cells treated with a variety of mitogens such as PDGF and EGF. Little is known about p42 except that it is a scarce cytosolic protein present in a variety of cell types and is rather highly conserved among vertebrates (Cooper and Hunter 1985). In principle one might pinpoint critical phosphorylation targets by selecting cell mutants unable to respond to individual growth factors. The few such nonresponsive cell lines isolated to date, however, have all been receptor mutants rather than substrate mutants (Schneider et al. 1986). In the absence of demonstrable essential substrates, direct proof that activation of tyrosine phosphorylation is necessary for mitogenic signaling by these receptors may ultimately be obtained by comparing the ability of cells lacking a particular receptor to respond following introduction of genes encoding either the wild-type receptor or mutant forms with enfeebled protein kinase activities. Such studies are in progress, but without direct evidence, it is important to bear in mind the possibility that these receptors might have additional functions which are involved in signaling.

A number of general principles apply to the growth factor receptor protein-tyrosine kinases. Enhancement of their phosphotransferase activities is very rapid following ligand binding, and in every case an early event is auto-phosphorylation. The fact that autophosphorylated insulin and EGF receptor molecules have increased activity toward exogenous substrates (Rosen et al. 1983; Bertics and Gill 1985) suggests that this may relieve the effect of negative regulatory domains on protein kinase activity. All receptors of this type cluster rapidly after ligand binding and are then internalized via coated pits into endosomes. Their fate at this stage is not certain. In the long term, internalized receptors are degraded by the lysosomal pathway, but there may be limited recycling to the surface. There are two schools of thought concerning the significance of internalization of liganded receptors. One is that this makes the cell refractory to further stimulation by the same ligand, although the fact that only 10% of surface receptors need to be occupied to provide a mitogenic signal would tend to argue against this idea. The other is that the internalized receptor delivers a signal at a site distant from the cell surface during its transit through the cell. The latter notion is attractive in terms of inducing events other than at the plasma membrane.

Activation of Protein Kinase C and Phosphatidylinositol Turnover by Growth Factors and Other External Stimuli

Many mitogens, including some of the growth factors whose receptors are protein-tyrosine kinases, induce a rapid stimulation in the turnover of phosphatidylinositol (PI) (Berridge and Irvine 1984 and Berridge, this Vol.). This pathway involves the phosphorylation of PI, in two steps, to yield PIP_2, which is then hydrolyzed by phospholipase C to form diacylglycerol (DG) and inositol trisphosphate (IP_3). DG is an activator of the Ca^{2+}/phospholipid-dependent serine/threonine-specific protein kinase, protein kinase C (Nishizuka 1984 and Parker and Ullrich, this Vol.). Thus the binding of most mitogens leads to the activation of protein kinase C. Tumor promoters such as TPA can substitute for DG in the activation of protein kinase C, which is, in fact, the major cellular receptor for tumor promoters (Nishizuka 1984). The ability of TPA to act as a mitogen for some cells suggests that activation of protein kinase C could be a primary event in the initiation of a mitogenic response. This raises the possibility that there might be just a single mitogenic pathway which works through protein kinase C and which can be activated indirectly by the receptor protein-tyrosine kinases. There are reasons for thinking, however, that this is not the case. Cells in which the level of protein kinase C has been reduced by continuous exposure to TPA can still respond mitogenically to appropriate growth factors (Coughlin et al. 1985). In addition, cells which have been selected to be nonresponsive to TPA can still respond to growth factors (Bishop et al. 1985), implying that the two mitogenic pathways are separable. Interestingly, these mutant cells possess normal levels of functional protein kinase C, suggesting that they have a postreceptor defect. It seems likely, then, that there are distinct tyrosine phosphorylation and protein kinase C mitogenic pathways, which are used in parallel. However, there are a number of interconnections between them which offer opportunities for control and integration of signals activating the two pathways.

As mentioned above, one such connection is the activation of PI turnover by the growth factors whose receptors are protein-tyrosine kinases. How this occurs is unknown. The idea that these enzymes might phosphorylate PI directly now seems unlikely (Thompson et al. 1985). Instead, one of the enzymes of the PI cycle or its regulatory proteins could be phosphorylated on tyrosine by the receptor protein-tyrosine kinases. Since phospholipase C seems to be the rate-limiting step in the formation of DG and IP_3, this enzyme is a likely target for phosphorylation. Alternatively, given the recent indications that phospholipase C may be regulated by GTP-binding proteins akin to N_s and N_i, one of these regulatory proteins might be phosphorylated by receptor protein-tyrosine kinases.

The likely involvement of protein kinase C in a mitogenic signal pathway makes a knowledge of its substrates important. The activated form of protein kinase C is membrane-associated, and one expects many of its substrates to be in this cellular environment. A number of proteins which are phosphorylated by protein kinase C have been identified. Several are surface receptors, including the EGF (Cochet et al. 1984), interleukin-2 (Shackelford and Trowbridge 1984), α_1-adrenergic (Leeb-Lundberg et al. 1985), and transferrin receptors (May et al. 1985). In principle, phosphorylation of the EGF and interleukin-2 receptors by protein kinase C could provide ligand-independent activation of their mitogenic pathways. However, phosphorylation of all these receptors by protein kinase C seems to inhibit their functioning. For instance, protein kinase C phosphorylation of the EGF receptor at a single threonine residue reduces its ability to bind EGF and decreases its protein kinase activity (Cochet et al. 1984). Rather than providing a positive mitogenic signal, this seems likely to be part of a feedback system which allows transmodulation between other receptors and the EGF receptor, such as the observed downregulation of EGF binding by PDGF. This negative regulation of the EGF receptor exemplifies another of the interconnections between the protein kinase C and protein-tyrosine kinase pathways.

Among the other known substrates for protein kinase C are pp60$^{c\text{-}src}$ (Gould et al. 1985) and the glucose transporter (Witters et al. 1985). The functional consequences of these phosphorylations are unknown, but the modification of pp60$^{c\text{-}src}$ could explain the observed TPA-induced increase in tyrosine phosphorylation of p42, the growth factor receptor protein-tyrosine substrate, while that of the glucose transporter could be involved in the early increase in glucose uptake which occurs in response to mitogens. Protein kinase C also phosphorylates the GTP-binding subunit of N_i, the inhibitory adenyl cyclase regulatory protein (Katada et al. 1985) and suppresses its activity, which could affect intracellular cAMP levels. In addition, there is indirect evidence that phospholipase C and the Na^+/H^+ antiporter are substrates for protein kinase C.

IP$_3$ is also a second messenger and acts to release Ca^{2+} from intracellular reserves (Berridge and Irvine 1984). Clearly this will stimulate a number of Ca^{2+}-dependent processes, but it is unlikely that this Ca^{2+} flux is required for the activation of protein kinase C, since the resting concentration of Ca^{2+} is probably sufficient for this purpose. However, calmodulin-dependent protein kinases will be activated and could propagate a Ca^{2+}-dependent signal.

Perturbation of Phosphorylation-Dependent Signaling Systems Implicated in Transformation and Unrestricted Growth

There are several oncogenes which are either known or thought to be derived from growth factor receptor protein-tyrosine kinase genes. The v-*erbB* gene of avian erythroblastosis virus (AEV) clearly arose from the EGF receptor gene (Ullrich et al. 1984 and articles by Schlessinger and by Beug et al., this Vol.), while the v-*fms* gene was probably derived from the CSF-1 receptor gene (Sherr et al. 1985 and Sherr and Stanley, this Vol.). As mentioned above, the *neu* gene (also known as c-*erbB2* and HER2) product has a structure very closely related to the EGF receptor (Bargmann et al. 1986). The v-*ros* protein has a structure reminiscent of a growth factor receptor (Neckameyer and Wang 1985), and the same is true for the *trk* gene product (Martin-Zanca et al. 1986) and the *met* gene product (Dean et al. 1985). All of these proteins may transform cells by delivering a continuous and unregulated mitogenic signal which overwhelms the tyrosine phosphorylation pathway.

There seem to be a number of principles at work in the oncogenic activation of these growth factor receptor genes. The loss of the putative N-terminal extracellular binding domain is common to the v-*erbB,* v-*ros* and *trk* proteins. This necessarily makes the receptors unable to respond to ligand, although it is unclear whether the absence of a binding domain will put the receptor protein-tyrosine kinase into an activated or simply an unregulatable state. The loss of C-terminal sequences is observed for the v-*erbB* and v-*fms* proteins. In the case of the v-*erbB* protein, this removes the major autophosphorylation site, and possibly thereby relieves the inhibitory effect of this C-terminal domain. In contrast to the other receptor-related oncogene products, the *neu* oncogene product is very similar in structure to its normal cellular counterpart. In this case a simple point mutation which introduces a charged residue into the transmembrane domain appears to activate the protein-tyrosine kinase, possibly by mimicking ligand occupancy (R. A. Weinberg, personal communication).

Despite the strong indications that these proteins transform through unregulated tyrosine phosphorylation, proof is not yet at hand. Without knowing the critical substrates for the growth-factor receptor protein-tyrosine kinases, we cannot determine whether these proteins are constitutively phosphorylated in the relevant transformed cells as would be expected. The same caveat applies to another series of oncogenic proteins which have protein-tyrosine kinase activity, namely the v-*src*, v-*yes*, v-*fgr*, v-*fps/fes*, and v-*abl* proteins. These all have unregulated activities, which might be able to phosphorylate proteins normally phosphorylated by growth factor receptor protein-tyrosine kinases, but again the paucity of knowledge concerning substrates leaves us unable to substantiate this hypothesis.

Molecular clones of protein kinase C have only recently been isolated (see Parker and Ullrich, this Vol.) and have not yet been thoroughly analyzed, but none of the known oncogenes appears to correspond to the protein kinase C gene. The v-*mos* and v-*mil/raf* genes, however, do encode protein-serine kinases (Maxwell and Arlinghaus 1985; Moelling et al. 1984). In principle, these enzymes might mimic protein kinase C, but in contrast to protein kinase C these oncogenic protein kinases are soluble. Thus if they intervene in the protein kinase C pathway, they probably do not do so at a primary step.

Conclusions

Both of the mitogenic signal pathways discussed have many of the desirable properties listed at the start, and it is evident that phosphorylation must play an important role in mitogenesis. Nevertheless, the whole process of mitogenesis involves cascades of events of which only the first steps will be regulated by the signal. The protein kinase C and tyrosine phosphorylation pathways are initiated via phosphorylation events at the plasma membrane. The identities of the critical substrates are still elusive, but there are several obvious plasma membrane-associated proteins to examine (e.g., phospholipase C).

While it is clear that we are only at a start in understanding the many roles of phosphorylation in signal transduction, the increasing number of oncogenes whose products perturb or mimic signaling systems is compelling evidence that phosphorylation is indeed critical. A study of these oncogenes is not only enhancing our understanding of the molecular basis of transformation, but is also allowing us to unravel the function of phosphorylation in signaling through the unmasking of cognate normal genes involved in this process.

References

Bargmann CI, Hung MC, Weinberg RA (1986) The *neu* oncogene encodes an epidermal growth factor receptor-related protein. Nature 319:226–230

Berridge MJ, Irvine RF (1984) Inositol trisphosphate, a novel second messenger in cellular signal transduction. Nature 312:315–320

Bertics PJ, Gill GN (1985) Self-phosphorylation enhances the protein-tyrosine kinase activity of the epidermal growth factor receptor. J Biol Chem 260:14642–14647

Bishop R, Martinez R, Weber MJ, Blackshear PJ, Beatty S, Lim R, Herschman HR (1985) Protein phosphorylation in a TPA non-proliferative variant of 3T3 cells. Mol Cell Biol 5:2231–2237

Cochet G, Gill GN, Meisenhelder J, Cooper JA, Hunter T (1984) C-kinase phosphorylates the epidermal growth factor (EGF) receptor and reduces its EGF-stimulated tyrosine protein kinase activity. J Biol Chem 259:2553–2558

Coughlin SR, Lee WMF, Williams PW, Giels GM, Williams LT (1985) c-*myc* gene expression is stimulated by agents that activate protein kinase C and does not account for the mitogenic effect of PDGF. Cell 43:243–251

Cooper JA, Hunter T (1985) A major substrate for growth factor activated protein-tyrosine kinases is a low abundance 42,000 dalton protein. Mol Cell Biol 5:3304–3309

Dean M, Park M, Le Beau MM, Robins TS, Diaz MO, Rowley JD, Blair DG, Vande Woude G (1985) The human *met* oncogene is related to the tyrosine kinase oncogenes. Nature 318:385–388

Ebina Y, Ellis L, Jarnagin K, Edery M, Graf L, Clauser E, Ou JH, Masiarz F, Kan YW, Goldfine, Roth RA, Rutter WJ (1985) The human insulin receptor cDNA: the structural basis of hormone-activated transmembrane signalling. Cell 40:747–758

Gould KL, Woodgett JR, Cooper JA, Buss JE, Shalloway D, Hunter T (1985) Protein kinase C phosphorylates pp60^{src} at a novel site. Cell 42:849–857

Heldin CH, Westermark B (1984) Growth factors: mechanism of action and relation to oncogenes. Cell 37:9–20

Hunter T, Cooper JA (1985) Protein-tyrosine kinases. Annu Rev Biochem 54:897–930

Katada T, Gilman AG, Watanabe Y, Bauer S, Jakobs KH (1985) Protein kinase C phosphorylates the inhibitory guanine-nucleotide-binding regulatory component and apparently suppresses its function in hormonal inhibition of adenylate cyclase. Eur J Biochem 151:431–437

Leeb-Lundberg LMF, Cotecchia S, Lomasney JW, DeBernardis JE, Lefkowitz RJ, Caron MG (1985) Phorbol esters promote α_1-adrenergic receptor phosphorylation and receptor uncoupling from inositol phospholipid metabolism. Proc Natl Acad Sci USA 82:5651–5655

Martin-Zanca D, Hughes SH, Barbacid M (1986) A human transforming gene present in a colon carcinoma is a hybrid between a tropomyosin gene and a novel tyrosine protein kinase locus. Nature 319:743–748

May WS, Sahyoun N, Jacobs S, Wolfe M, Cuatrecasas P (1985) Mechanism of phorbol diester-induced regulation of transferrin receptor involves the action of activated protein kinase C and an intact cytoskeleton. J Biol Chem 260:9419–9426

Maxwell SA, Arlinghaus RB (1985) Serine kinase activity associated with Moloney murine sarcoma virus 124-encoded p37^{mos}. Virology 143:321–333

Moelling K, Heimann B, Beimling P, Rapp UR, Sander T (1984) Serine- and threonine-specific protein kinase activities of purified *gag-mil* and *gag-raf* proteins. Nature 312:558–561

Neckameyer WS, Wang LH (1985) Nucleotide sequence of avian sarcoma virus UR2 and comparison of its transforming gene with the other members of the tyrosine protein kinase oncogene family. J Virol 53:879–884

Nishizuka Y (1984) The role of protein kinase C in cell surface signal transduction and tumor promotion. Nature 308:693–698

Rosen OM, Herrera R, Olowe Y, Petruzelli LM, Cobb MH (1983) Phosphorylation activates the insulin receptor tyrosine protein kinase. Proc Natl Acad Sci USA 80:3237–3240

Schneider CA, Lim RW, Terwilliger E, Herschman HR (1986) Epidermal growth factor-non-responsive 3T3 variants do not contain epidermal growth factor receptor-related antigens of mRNA. Proc Natl Acad Sci USA 83:333–336

Shackelford DA, Trowbridge IS (1984) Induction of expression and phosphorylation of the human interleukin-2 receptor by a phorbol diester. J Biol Chem 259:11706–11712

Sherr CJ, Rettenmeier CW, Sacca R, Roussel MF, Look AT, Stanley ER (1985) The c-*fms* proto-oncogene product is related to the receptor for the mononuclear phagocyte growth factor, CSF-1. Cell 41:665–676

Thompson DM, Cochet C, Chambaz EM, Gill GN (1985) Separation and characterization of a phosphatidylinositol kinase activity that copurifies with the epidermal growth factor receptor. J Biol Chem 260:8824–8830

Ullrich A, Coussens L, Hayflick JS, Dull TJ, Gray A, Tam AW, Lee J, Yarden Y, Liberman TA, Schlessinger J, Downward J, Mayes ELV, Waterfield MD, Whittle N, Seeburg PH (1984) Human epidermal growth factor receptor cDNA sequence and aberrant expression of the amplified gene in A431 epidermoid carcinoma cells. Nature 309:418–425

Ullrich A, Bell JR, Chen EY, Herrera R, Petruzelli LM, Dull TJ, Gray A, Coussens L, Liao YC, Tsubokawa M, Mason A, Seeburg PH, Grunfeld C, Rosen O, Ramachandran J (1985) Human insulin receptor and its relationship to the tyrosine kinase family of oncogenes. Nature 313:756–761

Witters LA, Vater CA, Lienhard GE (1985) Phosphorylation of the glucose transporter in vitro and in vivo by protein kinase C. Nature 315:777–778

Inositol Lipids and Cell Proliferation

Michael J. Berridge

One problem which is central to an understanding of both cellular growth control and the mechanism of hormone action is the nature of the signal pathway and, in particular, the identity of the second messenger molecules which trigger DNA synthesis. The main components of the signal pathways by which growth factors stimulate cell growth are shown in Fig. 1. The fact that different growth factors do not all act via the same mechanism suggests that there may be qualitatively different signal pathways. At some stage these pathways merge with each other, but the points of convergence are still obscure. The importance of understanding these pathways and their transduction mechanisms is underscored by recent findings which suggest that certain oncogenes encode proteins which participate in signal pathways.

This chapter describes the sequential flow of information along one pathway in which inositol lipids play an important role, as well as the possible involvement of specific oncogene products in this pathway. As will be discussed further, inositol lipids appear to function as part of a transduction mechanism for generating second messengers such as inositol 1,4,5-trisphosphate (Ins1,4,5P$_3$) and diacylglycerol (DG), which in turn transmit information into the cell (Michell 1982; Berridge and Irvine 1984; Nishizuka 1984). These two second messengers then regulate two separate ionic events, both of which are crucial for the onset of DNA synthesis. Ins1,4,5P$_3$ acts to mobilize calcium from intracellular stores (Berridge and Irvine 1984) whereas DG stimulates protein kinase C (Nishizuka 1984 and Parker and Ullrich, this Vol.), which then activates a Na$^+$/H$^+$ exchanger to raise intracellular pH (see Moolenaar,

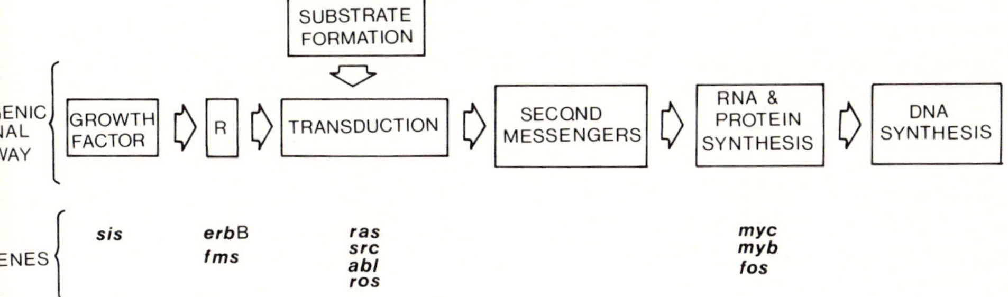

Fig. 1. Major components of mitogenic signal pathways. The position where different oncogenes might participate in signaling is included below

Oncogenes and Growth Control
Edited by P. Kahn and T. Graf
© Springer-Verlag Berlin Heidelberg 1986

this Vol.). The increases in intracellular calcium and pH are responsible for activating the transcription of certain genes (including the *myc* and *fos* proto-oncogenes) and may thus participate in initiating the sequence of events which culminates in DNA synthesis and cell division. While many of the details are still somewhat speculative, it is clear that the hydrolysis of a minor membrane lipid generates two very different intracellular second messengers which are key elements of the mitogenic pathway.

Growth Factors and Receptors

One of the growth factors which appears to act through the inositol lipid pathway is platelet-derived growth factor (PDGF). Recently the v-*sis* oncogene was shown to encode a protein which is almost identical to PDGF, supporting the contention that this oncogene may transform via an autocrine mechanism (see Heldin and Westermark, this Vol.) presumably also utilizing this pathway. Another instance in which the inositol lipid pathway may play an important role in neoplastic growth is small cell lung carcinoma cells, which seem to depend for growth upon an autocrine secretion of bombesin (Cuttitta et al. 1985), a potent stimulator of inositol lipid hydrolysis.

The inositol lipid signal pathway is initiated when an appropriate growth factor binds to its receptor at the cell surface. As yet the nature of most such receptors, which are potential oncogene products, remains to be character-ized. For other signal pathways a link between oncogenes and receptors has been established: *erbB* is a truncated and mutated EGF receptor (Downward et al. 1984 and articles by Beug et al. and by Schlessinger, both this Vol.), while c-*fms* is related to the CSF-1 receptor (Sherr et al. 1985 and Sherr and Stanley, this Vol.).

Synthesis of PtdIns4,5P$_2$, the Substrate for the Inositol Lipid Signal Pathway

Transduction mechanisms act via specific precursors which are converted into second messengers following stimulation of the cell by a growth factor. The growth factor receptors which use tyrosine kinase as a transduction mechanism presumably phosphorylate a target protein which then acts as a messenger molecule, although these remain to be identified. In the case of the inositol lipid signal pathway, the substrate is phosphatidylinositol 4,5-bis-phosphate (PtdIns4,5P$_2$), which generates the second messengers Ins1,4,5P$_3$ and DG (Fig. 2). PtdIns4,5P$_2$ is formed by the stepwise phosphorylation of phosphatidylinositol (PtdIns) via two different kinases. PtdIns kinase con-verts PtdIns into phosphatidylinositol 4-phosphate (PtdIns4P), which is then phosphorylated to PtdIns4,5P$_2$ by a second kinase. In addition to these two kinases, there are corresponding phosphomonoesterases which remove the

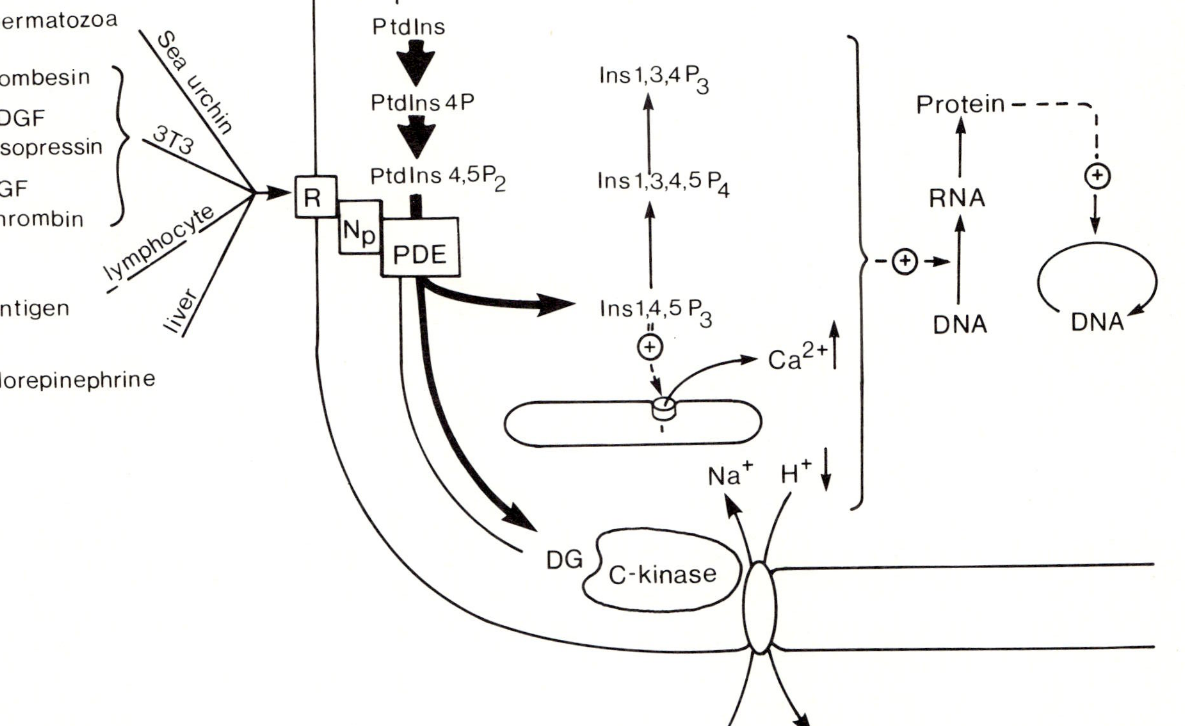

Fig. 2. Summary of the main components of the inositol lipid pathway. The pathway is activated by growth stimuli acting on their specific receptors (*R*) in many different cell types. Substrate for the pathway is formed by the conversion of PtdIns into PtdIns4,5P$_2$. The transducing mechanism acts via a G-protein (N$_p$) which couples the receptor to the phosphodiesterase (PDE) that cleaves PtdIns4,5P$_2$ to the two second messengers DG and Ins1,4,5P$_3$. The latter is converted into additional inositol phosphates (Ins1,3,4,5P$_4$ and Ins1,3,4P$_3$) which may also have messenger functions. DG acts through protein kinase C to activate the Na$^+$/H$^+$ exchanger, which reduces the intracellular hydrogen ion concentration, while Ins1,4,5P$_3$ mobilizes intracellular calcium. The increase in pH and calcium contribute to the onset of gene transcription and protein synthesis which culminate in DNA synthesis

4- and 5-phosphates and thereby convert PtdIns4,5P$_2$ back to PtdIns. The three inositol lipids are thus in a dynamic equilibrium with each other. The kinases ensure that the receptor mechanism is constantly supplied with the PtdIns4,5P$_2$ required to generate second messengers.

Hydrolysis of PtdIns4,5P$_2$

The key event of the inositol lipid pathway is the hydrolysis of PtdIns4,5P$_2$ to give Ins1,4,5P$_3$ and DG (Fig. 2). This transduction event is carried out by a PtdIns4,5P$_2$ phosphodiesterase, a membrane-bound enzyme which cleaves the lipid molecule at the point where the phosphate on the 1-position of the inositol ring is attached to carbon 3 of the glycerol backbone. As this is a hydrolytic cleavage, the enzyme adds a hydroxyl ion to break the phospho-diester bond to liberate Ins1,4,5P$_3$. In addition to using a free hydroxyl, the enzyme can use a resident hydroxyl (specifically that located at the 2-position of the inositol ring), whereupon it liberates inositol 1:2 cyclic 4,5-trisphosphate (cyclic Ins1,4,5P$_3$) (Wilson et al. 1985). As yet, formation of cyclic Ins1,4,5P$_3$ has been described only in in vitro experiments and it is not known whether this form is produced in vivo. Further discussion will thus concentrate on Ins1,4,5P$_3$, mindful of the fact that the cell may produce the cyclic form or a mixture of the two.

Oncogenes and Inositol Lipid Metabolism

It has been suggested that *src* (Sugimoto et al. 1984) and *ros* (Macara et al. 1984) might encode the inositol lipid kinases which phosphorylate PtdIns to PtdIns4,5P$_2$. However, subsequent studies showed that when these gene products are purified they retain their tyrosine kinase activity, but lose all inositol lipid kinase activity (Sugano and Hanafusa 1985; Sugimoto and Erikson 1985).

While *src* does not appear to play a direct role in providing substrate, there is much indirect evidence which suggests that this oncogene has some effect on inositol lipid metabolism. Cells transformed with *src* have an enhanced turnover of inositol lipids and are capable of growing in a calcium-deficient medium (Durkin et al. 1981), which would be consistent with an enhanced input from the Ins1,4,5P$_3$/Ca^{2+} limb of the signal pathway. An increase in intracellular calcium or an enhanced sensitivity to calcium would also account for the decrease in junctional permeability observed in *src*-transformed cells (Chang et al. 1985). Indeed, Chang et al. (1985) have suggested that the *src* gene may increase intracellular levels of Ins1,4,5P$_3$, thereby stimulating the mobilization of calcium.

At what point in the inositol lipid pathway might *src* act? One possibility, which would account for all the known actions of *src*, is that it functions to enhance the activity of PtdIns4,5P$_2$ phosphodiesterase (Sugano and Hanafusa 1985), leading to an increase in the formation of Ins1,4,5P$_3$ and DG. The latter would then activate protein kinase C which operates a feedback loop to enhance the phosphorylation of PtdIns to PtdIns4,5P$_2$, thus accounting for the increased labeling of inositol lipids seen in *src*-transformed cells. Alternatively, *src* may activate turnover by phosphorylating the inositol lipid kinases directly. The location of the *src* gene product in the plasma membrane is consistent with the notion that it participates in signal transduction. There is an urgent need to explore further its possible role in promoting the hydrolysis of PtdIns4,5P$_2$.

Information from receptors is transmitted to the PtdIns4,5P$_2$ phosphodiesterases by means of a G-protein (Berridge and Irvine 1984), a class of protein which shares many properties with the products of the *ras* oncogenes. One interesting possibility, therefore, is that *ras* proteins might function as the G-protein in the inositol lipid signal pathway (Berridge and Irvine 1984 and Masters and Bourne, this Vol.). Some support for this possibility comes from the observation that *ras*-transformed cells produce more DG and inositol phosphates than do normal cells (Chiarugi et al. 1985; Fleischman et al. 1986). However, other studies argue that *ras* does not operate in the same signal pathway used by *sis, myc,* and the tumour promotors which all have been linked with inositol lipid hydrolysis; rather, *myc* and *ras* act cooperatively with each other in cell transformation and have therefore been placed in separate complementation groups (Land et al. 1983 and by Land, this Vol.). This separation is supported by the observation that tumour promotors strongly enhance the growth of *ras*- but nor *myc*-transformed cells (Dotto et al. 1985). Studies with revertants have also placed *ras* and *sis* into separate groups and furthermore, placed *ras* and *src* in the same group (Noda et al. 1983), which argues against a direct involvement of both oncogenes in the signal pathway used by *sis* (the inositol lipid signal pathway). The function of *src* and *ras* in signal transduction is therefore still very much an open question.

Second Messengers

The crucial event of the inositol lipid pathway is the hydrolysis of PtdIns4,5P$_2$ to give the second messengers Ins1,4,5P$_3$ and DG, which then mediate the mitogenic action of certain growth factors. This bifurcating signal pathway is complicated by the fact that these two initial second messengers may generate additional signal molecules. DG can either be phosphorylated to phosphatidic acid as the first step in the resynthesis of PtdIns or it can act as the substrate for a DG lipase that releases arachidonic acid, the precursor of various metabolites such as PGE$_1$ or PGF$_{2\alpha}$, which are potent activators of cell growth.

An alternative mechanism for metabolizing Ins1,4,5P$_3$ has recently been uncovered. The previously established pathway was a stepwise dephosphorylation to free inositol. The newly described pathway converts Ins1,4,5P$_3$ to inositol 1,3,4,5-tetrakisphosphate (Ins1,3,4,5P$_4$) and Ins1,3,4P$_3$ (Fig. 2). The latter has been detected in a number of cell types where its onset of formation in response to agonists was found to lag behind that of $Ins1,4,5P_3$ (Irvine et al. 1985). The $Ins1,3,4P_3$ is now known to come from $Ins1,3,4,5P_4$ (Batty et al. 1985), which is formed by the phosphorylation of $Ins1,4,5P_3$ (Irvine et al. 1986). It remains to be seen whether these new inositol phosphates have any second-messenger function.

With regard to the control of cell proliferation, an important function of the two primary second messengers is to alter intracellular levels of calcium and protons (see Moolenaar, this Vol.). The DG/protein kinase C pathway brings about an increase in intracellular pH by activating a Na^+/H^+ exchanger. This alkalinization of the cytosol is essential for the onset of DNA synthesis. Mutant cells which lack the exchanger are unable to grow unless the pH is raised above 7.0 (Pouysségur et al. 1984). The other major ionic event is controlled by $Ins1,4,5P_3$, which triggers the release of calcium from the endoplasmic reticulum (Berridge and Irvine 1984). The importance of these two signal pathways for the initiation of cell growth is apparent from experiments on lymphocytes which demonstrate that DNA synthesis can be triggered by combining the action of a calcium ionophore (which mimics the effect of $Ins1,4,5P_3$) with that of phorbol ester (mimics the stimulatory effect of DG on protein kinase C) (Guy et al. 1985).

Nuclear Events

A major problem in trying to understand how growth factors control cell proliferation is that there is a long period between the arrival of the stimulus and the onset of DNA synthesis. What is happening inside the nucleus and how are these events regulated by some of the second messengers associated with the different signal pathways? Within minutes of adding fresh serum or growth factors there is a burst of RNA synthesis, resulting in the appearance of at least 100 new mRNA transcripts, including transcripts of the *myc* and *fos* oncogenes (see Bravo and Müller, this Vol.). This activation of *myc* and *fos* transcription can be stimulated by phorbol esters or calcium ionophores, thus establishing a link with the ionic events associated with the inositol lipid pathway described in the preceding section. While we are still a long way from a full understanding of these events, the answers which are beginning to emerge should deepen our insight into the role of the various signal pathways in cellular growth control.

References

Batty IR, Nahorski SR, Irvine RF (1985) Rapid formation of inositol (1,3,4,5) tetrakisphosphate following muscarinic stimulation of rat cerebral cortical slices. Biochem J 232:211–215
Berridge MJ, Irvine RF (1984) Inositol trisphosphate, a novel second messenger in cellular signal transduction. Nature 312:315–321
Chang C-C, Trosko JE, Kung H-J, Bombick D, Matsumura F (1985) Potential role of the *src* gene product in inhibition of gap-junctional communication in NIH/3T3 cells. Proc Natl Acad Sci USA 82:5360–5364

Chiarugi V, Porciatti F, Pasquali F, Bruni P (1985) Transformation of BALB/3T3 cells with EJ/T24/H-RAS oncogene inhibits adenylate cyclase response to α-adrenergic agonist while increases muscarinic receptor dependent hydrolysis of inositol lipids. Biochem Biophys Res Commun 132:900 – 907

Cuttitta F, Carney DN, Mulshine J, Moody TW, Fedorko J, Fischler A, Minna JD (1985) Bombesin-like peptides can function as autocrine growth factors in human small-cell lung cancer. Nature 316:823 – 826

Dotto GP, Parada LF, Weinberg RA (1985) Specific growth response of *ras*-transformed embryo fibroblasts to tumour promotors. Nature 318:472 – 475

Downward J, Yarden Y, Mayes E, Scrace E, Totty N, Stockwell P, Ullrich A, Schlessinger J, Waterfield MD (1984) Close similarity of epidermal growth factor receptor and v-*erbB* oncogenes protein sequences. Nature 307:521 – 527

Durkin JP, Boynton AL, Whitfield JF (1981) The *src* gene product (pp60^{src}) of avian sarcoma virus rapidly induces DNA synthesis and proliferation of calcium-deprived rat cells. Biochem Biophys Res Commun 103:233 – 239

Fleischman LF, Chahwala SB, Cantley L (1986) *Ras*-transformed cells: altered levels of phosphatidylinositol-4,5-bisphosphate and catabolites. Science 231:407 – 410

Guy GR, Gordon J, Michell RH, Brown G (1985) Synergism between diacylglycerols and calcium ionophore in the induction of human B cell proliferation mimics the inositol lipid polyphosphate breakdown signals induced by crosslinking surface immunoglobulin. Biochem Biophys Res Commun 131:484 – 491

Irvine RF, Änggård EA, Letcher AJ, Downes CP (1985) Metabolism of inositol (1,4,5) trisphosphate and inositol (1,3,4) trisphosphate in rat parotid glands. Biochem J 229:505 – 511

Irvine RF, Letcher AJ, Heslop JP, Berridge MJ (1986) The inositol tris/tetrakisphosphate pathway-demonstration of Ins(1,4,5)P$_3$ 3-kinase activity in animal tissue. Nature 320:631 – 634

Land H, Parada LF, Weinberg RA (1983) Tumorigenic conversion of primary embryo fibroblasts requires at least two cooperating oncogenes. Nature 304:596 – 602

Macara IG, Marinetti GV, Balduzzi PC (1984) Transforming protein of avain sarcoma virus UR2 is associated with phosphatidylinositol kinase activity: Possible role in tumorigenesis. Proc Natl Acad Sci USA 81:2728 – 2732

Michell RH (1982) Inositol lipid metabolism in dividing and differentiating cells. Cell Calcium 3:429 – 440

Nishizuka Y (1984) The role of protein kinase C in cell surface signal transduction and tumor promotion. Nature 308:693 – 697

Noda M, Selinger Z, Scolnick EM, Bassin RH (1983) Flat revertants isolated from Kirsten sarcoma virus-transformed cells are resistant to the action of specific oncogenes. Proc Natl Acad Sci USA 80:5602 – 5606

Pouysségur J, Sardet C, Franchi A, L'Allemain G, Paris S (1984) A specific mutation abolishing Na$^+$/H$^+$ antiport activity in hamster fibroblasts precludes growth at neutral and acidic pH. Proc Natl Acad Sci USA 81:4833 – 4837

Sherr CJ, Rettenmier CW, Sacca R, Roussel ML, Look AJ, Stanley ER (1985) The c-*fms* proto-oncogene product is related to the receptor for the mononuclear phagocyte growth factor, CSF-1. Cell 41:665 – 676

Sugano S, Hanafusa H (1985) Phosphatidylinositol kinase activity in virus-transformed and non-transformed cells. Mol Cell Biol 5:2399 – 2404

Sugimoto Y, Erikson RL (1985) Phosphatidylinositol kinase activities in normal and Rous sarcoma virus-transformed cells. Mol Cell Biol 5:3194 – 3198

Sugimoto Y, Whitman M, Cantley LC, Erikson RL (1984) Evidence that the Rous sarcoma virus-transforming gene product phosphorylated phosphatidylinositol and diacylgylcerol. Proc Natl Acad Sci USA 81:2117 – 2121

Wilson DB, Connolly TM, Bross TE, Majerus PW, Sherman WR, Tyler AN, Rubin LJ, Brown JE (1985) Isolation and characterization of the inositol cyclic phosphate products of polyphosphoinositide cleavage by phospholipase C. J Biol Chem 260:13496 – 13501

Protein Kinase C

Peter J. Parker and Axel Ullrich

Protein phosphorylation plays a central role in the regulation of cellular functions. Since the identification of the cAMP-dependent protein kinase much effort has been directed towards the discovery of other multifunctional protein kinases which are involved in transduction of intracellular signals from external stimuli. This review focuses on the recently identified protein kinase C (PKC), a multifunctional protein kinase, that appears to play such a role.

Characteristics of PKC

Nishizuka and colleagues first identified PKC as a proenzyme that requires mM concentrations of calcium for activity (Takai et al. 1977). Following activation there is an apparent shift in molecular mass from about 80 kDa to 50 kDa; this activation was shown to be irreversible and due to proteolytic cleavage by calpain, a protease present in crude preparations (Inoue et al. 1977). In further characterizing the enzyme, Nishizuka demonstrated that reversible activation could be achieved with ~10 μm calcium and acidic phospholipid and that the activation constants for both these effectors could be reduced by the neutral lipid diacylglycerol (DG, see Nishizuka 1984). The effect of DG is thus to allow activation of PKC at physiological calcium levels. While many of these studies were carried out on partially purified PKC, further studies have confirmed that the purified enzyme retains a calcium, phospholipid and DG dependency and it is these characteristics that provide an operational definition of the protein.

PKC seems to be ubiquitous and has been purified to apparent homogeneity from a number of mammalian species (Kikkawa et al. 1982; Wise et al. 1982; Parker et al. 1984). It is acidic, with a pI between pH 5.0 and 5.5, and is composed of a single polypeptide chain with an apparent molecular weight of 79–83 kDa in SDS-polyacrylamide gels. The specific activity of the purified enzyme from bovine brain is about 4750 nmol min^{-1} mg^{-1} using histone H-I as substrate. This yields a turnover number of about 380 min^{-1}, which compares favourably with other protein kinases.

Various reports have examined the dependence of PKC activity on long chain fatty acids and on phospholipids acting either independently or synergistically. It remains to be determined whether those agents found to be activators provide a strictly reversible response and whether this can be effectively

Oncogenes and Growth Control
Edited by P. Kahn and T. Graf
© Springer-Verlag Berlin Heidelberg 1986

reproduced with purified protein. In the case of phosphatidyl-serine (PS), it appears that the interaction with PKC is not on a 1:1 basis, but of the order of between 4:1 and 10:1 (Hannun et al. 1985). This suggests that it is the micellar structure and presumably surface charge distribution that governs this interaction.

In analyzing a series of permeable diacylglycerols, Czech and colleagues found that among the saturated forms those containing C_8 or C_{10} acyl groups were the most potent activators of PKC in vitro (Davis et al. 1985). They also showed that the chloro-, deoxy- and thiol-analogues of 1,2 di-octanoyl-glycerol are inactive, indicating a stringent requirement for the 3-OH group (Davis et al. 1985); 1,3 diacylglycerol and triglycerides are also inactive (Mori et al. 1982).

While a number of proteins have been shown to be substrates for PKC in vitro, the sites phosphorylated have been identified in only a few cases. Interestingly, in two apparently physiological membrane-associated substrates (EGF receptor and c-*src*) the targets of phosphorylation lie within basic stretches expected to be close to the phospholipid bilayer (Cochet et al. 1984; Davis and Czech 1985; Downward et al. 1985; Gould et al. 1985).

Recent studies with synthetic peptide substrates emphasize the importance of basic residues on both the amino- and carboxyl-sides of the target serine or threonine residue (Ferrari et al. 1985; Turner et al. 1985). However, there appears to be no consensus sequence within these peptide substrates, and consequently it may prove difficult to predict phosphorylation sites for PKC.

PKC in Signal Transduction

Recent studies employing defined lipid/detergent and mixed lipid vesicles have provided evidence for an absolute dependence upon DG in the activation of PKC (Boni and Rando 1985; Hannun et al. 1985), further emphasizing a role for this molecule as a second messenger. These in vitro observations suggest that the activation of PKC by DG might also be of crucial importance in vivo.

A number of agonists acting at the cell surface stimulate the production of DG as a consequence of phosphatidylinositol breakdown (see Berridge, this Vol.); it has been proposed that the DG produced activates PKC and that this effect is crucial in the overall response to the agonist. Evidence that this interpretation is correct has come from in vivo studies with synthetic DG forms and phorbol esters where PKC activation can be obtained independently of phosphatidylinositol breakdown.

Castagna and co-workers were the first to show that biologically active phorbol esters could mimic DG in the activation of PKC (Castagna et al. 1982). The relative potencies of these agents as activators of PKC were found to be similar to their relative tumour-promoting abilities; furthermore the activation constants resembled the binding constants determined in intact cells. Subsequent investigations defined PKC as the major receptor for phorbol esters (reviewed by Nishizuka 1984). The implication of these observations is that the physiological responses elicited by phorbol esters through their high affinity binding site are effected through the activation of PKC.

Besides phorbol esters, a number of structurally related and unrelated tumour promoters such as mezerin (Miyake et al. 1984), teleocidin and debromoaplysiatoxin (Fujiki et al. 1984), can mimic the effect of DG in the activation of PKC. These observations reinforce the idea that the activation of PKC may be a crucial event in tumour promotion.

Studies on the phorbol ester binding sites in cultured cells revealed that, as with a number of hormone/growth factor receptors, the phorbol ester receptors are downregulated in response to ligand exposure in a time-dependent manner. This "downregulation" is paralleled by a loss of PKC activity and a severe decrease in the steady-state levels of the 80 kDa PKC protein (S. Stabel, M. Rodriguez-Pena, S. Young, E. Rozengurt and P. J. Parker, submitted).

In response to agonists that produce DG, cytosolic PKC has been shown to shift to an EDTA-sensitive membrane-associated form (reviewed in Anderson et al. 1985). An even more pronounced redistribution is observed after phorbol ester treatment (see Anderson et al. 1985). It is not yet established whether this is a biologically relevant phenomenon. It has been suggested that a redistribution of PKC is necessary for the interaction with membrane substrates. However, in our laboratory we found that the cytosolic 50 kDa fragment of PKC can readily phosphorylate the EGF receptor, an intrinsic membrane protein, suggesting that an association with the membrane is not a prerequisite for interaction with target substrates.

It is attractive to suppose that the membrane-associated form of PKC reflects the presence of an activator, and is causally related to the proteolytic activation and subsequent degradation of the kinase. In vitro, PKC can be cleaved into a catalytic unit of PKC that is constitutively active even in the absence of Ca^{2+} and lipid. This in vitro cleavage is greatly accelerated by Ca^{2+}/PS/diolein (Kajikawa et al. 1983). In vivo, the catalytic fragment also appears to be generated following agonist exposure (Tapley and Murray 1984). This might explain the ability of PKC to come into physical contact with non-membrane target proteins; it could also explain the downregulation of PKC by phorbol ester exposure discussed above; this is a relatively slow phenomenon and we would suggest represents a gradual change in steady state levels of the 80 kDa PKC molecule as a consequence of its continual cleavage to a catalytic fragment, which is eventually degraded (Fig. 1). The steady-state levels of the 80 kDa form would become vanishingly low if $K_2 \gg K_0$. The steady-state levels of the 50 kDa fragment are dependent upon the relationship between K_3 and K_2, and since a significant level of 50 kDa does not appear to be present following long-term downregulation of 80 kDa in mouse fibroblasts (unpublished observations), we would expect that $K_3 > K_2$. Unlike the situation of chronic phorbol ester exposure, the relatively transient nature of diacylglycer-

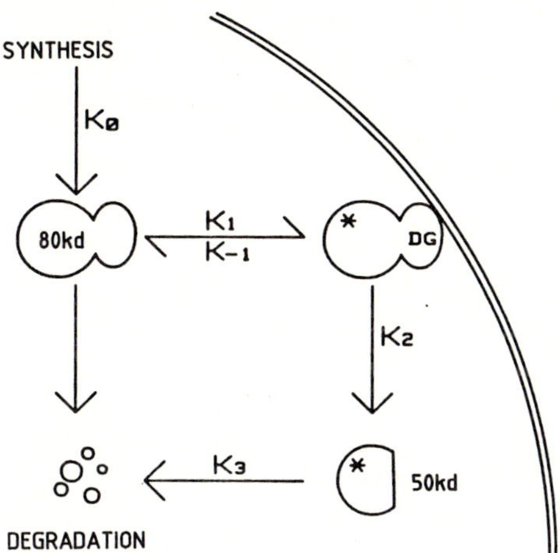

Fig. 1. PKC – schematic distribution and proteolysis. * Denotes active forms of PKC

ol production in vivo would be expected to generate only short-term changes in 80 kDa and 50 kDa levels.

The scheme shown in Fig. 1 is a simplified model and should represent the lowest level of complexity on which other regulatory mechanisms are superimposed. Thus the cleavage of the 80 kDa protein would be controlled by the presence of, for example, calpstatin (a specific inhibitor of calpain) and may also be controlled by the phosphorylation state of the 80 kDa form itself. In addition, this model raises the possibility that agents known not to function through phosphatidylinositol turnover may nevertheless influence the activity of PKC through production of the 50 kDa species.

There is circumstantial evidence to support a role for PKC in the control of many biological responses, including endocrine and exocrine secretion, neurotransmitter release, platelet granule release, neutrophil activation, Na^+/H^+ exchange, cell-cell interaction, cell surface expression of receptors and gene expression (reviewed by Nishizuka 1984). This somewhat abbreviated list of involvements reflects both the ubiquitous nature of PKC and the fact that specific cell lineages respond differently to a given stimulus.

There are many reports which indicate that phorbol esters are mitogenic or co-mitogenic for cells in culture; this in turn suggests that PKC is involved in stimulating cell division. It has been shown that the dose-response curves for different phorbol esters acting as co-mitogens in Swiss 3T3 cells follows their high affinity binding curves (Collins and Rozengurt 1982). Furthermore this mitogenic activity is mimicked by oleylacetylglycerol (Rozengurt et al. 1984). These findings are consistent with the interpretation that these proliferative responses are evoked through PKC. The observation that the activation of the PKC pathway is often not a sufficient stimulus for cell division is not surprising in view of the complex growth factor requirements for most primary cells. Nevertheless, certain growth factors do promote phosphatidylinositol turnover and are likely to function in part through this pathway (see Berridge, this Vol.). The involvement of PKC in normal growth control may also have parallels in the maintenance of transformed cells. For example, it has been documented that cells transformed by v-*src* (Diringer and Friis 1977) or by v-*abl* (Fry et al. 1985) show a more rapid rate of phosphatidylinositol turnover, although there is no direct evidence for elevated DG levels or an activated PKC in these cells.

There has been a mountain of literature generated over the past few years dealing with potential physiological substrates for PKC. The brevity of this review precludes a comprehensive discussion. However, it is instructive to consider one target for which there is a relatively detailed understanding. Treatment of cells with certain agents such as phorbol esters induces the loss of high affinity binding sites for EGF and their apparent conversion to low affinity sites (e.g. Collins et al. 1983). This effect may be caused by the phosphorylation of the EGF receptor at threonine 654 (Cochet et al. 1984; Davis and Czech 1985). It has been shown that following in vitro phosphorylation by PKC of threonine 654 there is a reduction in affinity for EGF (Downward et al. 1985).

These studies fulfill the criteria necessary for defining the EGF receptor as a target for PKC in vivo, namely (1) the protein is phosphorylated by PKC at a specific site in vitro and this site is found to become phosphorylated in vivo when cells are exposed to agents that activate PKC (2) the consequence of this phosphorylation event in vivo can be observed in a suitably reconstituted system in vitro. It should be noted that this does not exclude the possibility that other protein kinases can evoke the same phosphorylations and responses via different pathways.

PKC – A Family of Kinases

The complete predicted amino acid sequence of bovine brain PKC has been obtained from the nucleotide sequence of a series of overlapping cDNA clones (P. J. Parker, L. Coussens, N. Totty, L. Rhee, S. Young, E. Chen, S. Stabel, M. D. Waterfield, A. Ullrich, in press). The PKC structure can be divided into three domains. The first domain contains a cysteine-rich unit (Fig. 2a) that is repeated, giving a precise duplication of the cysteine spacing. Similar cysteine-rich regions have been noted elsewhere, including the external domains of a number of transmembrane proteins. In relation to the functional properties of PKC, it is of interest that phospholipase A_2 enzymes are cysteine-rich (see Slotboom et al. 1982). Thus this cysteine-rich domain in PKC may be involved in the binding of DG, presumably in conjunction with phospholipid (see above). The second domain contains a potential calcium binding site (Fig. 2b). This consists of an "E-F hand-like" structure that shows some structural homology to calmodulin and related calcium binding proteins. Like calmodulin, PKC is inhibited by phenothiazines (Mori et al. 1980; Schatzman et al. 1981), which is consistent with some structural homology. However,

A.

32   96
```
 HEVKNHRFIARFFKQPTFCSHCTDFIWGFGKQGFQCQVCCFVVHKRCHEFVTFSCPGADKGPDTD
 DPRSKHKFKIHTYGSPTFCDHCGSLLYGLIHQGMKCDTCDMNVHKQCVINVPSLCGMDHTEDRGR
```
97   161

B.
277   317
```
 LLNQEEGEYYNVPIPEGDEEGNVELRQKFEKAKLGPAGNKV
                   *  *  *  *  *  *
```

C.
```
1 MVLGKGSFGKVMLSERKGTDEL----YAVKILKKDVV------IQDDDVEC-TMVEKRVL
2 MVLGKGSFGKVMLADRKGTEEL----YAIKILKKDVV------IQDDDVEC-TMVEKRVL
3 IKLGTGSFGRVMLVKHMETGNH----YAMKILDKQKV------VKLKQIEH-TLNEKRIL
4 EILGRGVSSVVRRCIHKEPTCKE---YAVKIIDVTGGGSFSAEEVQELREA-TLKEVDIL
5 AKLGQGCFGEVWMGTWNDTTR-----VAIKTLKPGTM---SPEA--FLQEAYVM--KK-L
6 KVLGSGAFGTVYKGLWIPEGEKVTIPVAIKELREATSPKANKEI---LDEAYVMASVD--

1 ALPGKPPFL-TQLHSCFQTMDRLYFVMEYVNGGDLMYHIQQVGRFKE--PHAVFYAAEIA
2 ALLDKPPFL-TQLHSCFQTVDRLYFVMEYVNGGDLMYHIQQVGKFKE--PQAVFYAAEIS
3 QAVN-FPFL-VKLEFSFKDNSNLYMVMEYVPGGEMFSHLRRIGRFSE--PHARFYAAQIV
4 RKVSGHPNI-IQLKDTYETNTFFFLVFDLMKKGELFDYLTEKVTLSE--KETRKIMRALL
5 RHE-K---I-VQLYAVV-SEEPIYIVIEYMSKGSLLDFLKGEMGKYLRLPQLVDMAAQIA
6 -NPHVCRLLGICLTSTVQ-----LITQLMPFGCLLDYVRHKDNIGSQY--LLNWCVQIA

1 IGLFFIQ-SKGIIYRDLKLDNVMLDSEGHIKIADFGMCKENI-WD-GVTTKTF--CGT--
2 IGLFFLH-KRGIIYRDLKLDNVMLDSEGHIKIADFGMCKEHM-MD-GVTTRTF--CGT--
3 LTFEYLH-SLDLIYRDLKPENLLIDQQGYIQVTDFGFAKR------GVKGRTWTLCGT--
4 EVICALH-KLNIVHRDLKPENILLDDDMNIKLTDFGFSCQ---LDPGEKLKLREVCGT--
5 SGMAYVE-RMNYVHRDLRAANILVGENLVCKVADFGLARL-I-EDNEYTAR----QGAKF
6 KGMNYLEDRR-LVHRDLAARNVLVKTPQHVKITDFGLAKLLGAEEKEYHAE-----GGKV

1 PD-YIAPEIIAYQ------PYGKSVDWWAFGVLLYEMLA-GQAPFEGEDEDELFQSIMEH
2 PD-YIAPEIIAYQ------PYGKSVDWWAYGVLLYEMLA-GQPPFDGEDEDELFQSIMEH
3 PE-YLAPEIILSK------GYHKAVDWWALGVLIYEMAA-GYPPFADQPIQIYEKIVSG
4 PS-YLAPEIIECSMNDNHPGYGKEVDMWSTGVIMYTLLA-GSPPFWHRKQMLMLRMIMSG
5 PIKWTAPEAALY------GRFTIKSDVWSFGILLTELVTKGRVPYPGMVNREVLEQVERG
6 PIKWWALESILH------RIYTHQSDVWSYGVTVWELMTFGSKPYDGIPASEISSILEKG

1 NVAY---PKSMSKEAVAI-CKGLITKHPGKRLGCGP
2 NVSY---PKSLSKEAVSI-CKGLMTKHPGKRLGCGP
3 KVRF--P-SHFS---SD-LKDLLRNLLQVDLTKRF
4 NYQFGSP---EWDDYSDTVKDLVSRFLVVQPQKRY
5 YRMPC-PP---ECPES--LHDLMCQCWRKDPEERP
6 ERLPQ-PP---ICTI-D-VYMIMVKCWMIDADSRP
```

Fig. 2A–C. PKC – structural features. **A** Repeat unit within PKC. **B** Potential Ca^{2+}-binding domain. * Denotes residues involved in co-ordination of Ca^{2+}. **C** Homology of kinase domains between PKCs and other kinases. *1* PKCα; *2* PKCβ; *3* cAMP-dependent protein kinase; *4* γ subunit of phosphorylase kinase; *5* v-*src* tyrosine kinase; *6* EGF receptor tyrosine kinase

there are other potential calcium binding sites present in this domain and the assignment of site(s) requires direct experimentation. The third, catalytic domain shows close homology with other members of the protein kinase family and includes a putative nucleotide binding site (Fig. 2c). Outside the "PKC family" (see below) the closest relationship is seen with the cAMP-dependent protein kinase, which is 44% homologous in this region.

The proteolytic fragmentation of PKC into a constitutively active unit of about 50 kDa would fit with the initial cleavage between the putative Ca^{2+} binding domain and the kinase domain. This region contains a number of basic residues that are potential targets for calpain (or trypsin, which also readily activates PKC in vitro). The constitutively active nature of this 50 kDa fragment suggests that in the intact molecule the N-terminal "regulatory" domain maintains the catalytic domain in an inactive state. Presumably this inhibition is relieved by the binding of activators; similarly, selective denaturation of this regulatory domain may relieve inhibition.

In determining the sequence of bovine brain PKC through cDNA cloning it has become apparent that there is a family of PKC-related molecules (A. Ullrich, L. Coussens, L. Rhee, E. Chen and P. J. Parker, in press). In the case of two members of this family, designated PKC-α and PKC-β, there is an overall identity of 80%. Further work has so far uncovered one other member of this family that is as closely related (manuscript in preparation). The nucleotide sequence of PKC-α corresponds precisely to the protein sequence determined from bovine brain (P. J. Parker, L. Coussens, N. Totty, L. Rhee, S. Young, E. Chen, S. Stabel, M. D. Waterfield and A. Ullrich, in press). Using a combination of α- and β-specific oligonucleotide probes and α- and β-specific antisera it is evident that there is differential expression of mRNAs and proteins in a variety of cells and tissues. The extent to which this complexity of expression may govern tissue-specific responsiveness is clearly of great interest.

Summary and Perspectives

PKC plays a role in signal transduction from a number of cellular agonists, including certain mitogens such as growth factors; it also acts as one of the major transducing agents of phorbol ester action. An understanding of the means by which the cellular responses are evoked through PKC requires a detailed description of the target proteins and their physiological significance. The conditions under which PKC is active on different target proteins need to be assessed particularly with respect to their accessibility. Furthermore, the possibility has to be investigated that the generation of diacylglycerol is not the only mechanism responsible for the control of PKC.

The existence of a family of PKC proteins requires careful studies of the potentially different roles played by each of these proteins. For example, it will be important to establish whether some aspects of cell-specific responses are attributable in part to cell-specific expression of different PKC proteins. In conjunction with a detailed analysis of structure and function of these proteins, such studies will provide new and fascinating insights into the control of cellular growth and differentiation.

References

Anderson WB, Estival A, Tapiovaara H, Gopalakrishna R (1985) Altered subcellular distribution of protein kinase C (a phorbol ester receptor). Possible role in tumour promotion and the regulation of cell growth: relationship to changes in adenylate cyclase activity. In: Cooper DMF, Seamon KB (eds) Advances in cyclic nucleotide and protein phosphorylation research, vol 19. Raven, New York, pp 287–306

Boni LT, Rando RR (1985) The nature of protein kinase C activation by physically defined phospholipid vesicles and diacylglycerols. J Biol Chem 260:10819–10825

Castagna M, Takai Y, Kaibuchi K, Sano K, Kikkawa U, Nishizuka Y (1982) Direct activation of calcium-activated phospholipid-dependent protein kinase by tumor-promoting phorbol esters. J Biol Chem 257:7847–7851

Cochet C, Gill GN, Meisenhelder J, Cooper JA, Hunter T (1984) C-kinase phosphorylates the epidermal growth factor receptor and reduces its epidermal growth factor-stimulated tyrosine protein kinase activity. J Biol Chem 259:2553–2558

Collins MKL, Rozengurt E (1982) Binding of phorbol esters to high affinity sites on murine fibroblastic cells elicits a mitogenic response. J Cell Physiol 112:42–50

Collins MKL, Sinnett-Smith JW, Rozengurt E (1983) Platelet-derived growth factor treatment decreases the affinity of the epidermal growth factor receptor of Swiss 3T3 cells. J Biol Chem 258:11689–11693

Davis RJ, Czech MP (1985) Tumour-promoting phorbol diesters cause the phosphorylation of epidermal growth factor receptors in normal human fibroblasts at threonine 654. Proc Natl Acad Sci USA 82:1974–1978

Davis RJ, Ganong BR, Bell RM, Czech MP (1985) Structural requirements for diacylglycerols to mimic tumour promoting phorbol diester action on epidermal growth factor receptor. J Biol Chem 260:5315–5322

Diringer H, Friis R (1977) Changes in phosphatidylinositol metabolism correlated to growth state of normal and Rous sarcoma virus-transformed Japanese quail cells. Cancer Res 37: 2979–2984

Downward J, Waterfield MD, Parker PJ (1985) Autophosphorylation and protein kinase C phosphorylation of the epidermal growth factor receptor. J Biol Chem 260:14538–14546

Ferrari S, Marchiori F, Borin G, Pinna LA (1985) Distinct structural requirements of Ca^{2+}/phospholipid-dependent protein kinase and cAMP-dependent protein kinase as evidenced by synthetic peptide substrates. FEBS Lett 181:72–77

Fry MJ, Gebhardt A, Parker PJ, Foulkes JG (1985) Phosphatidylinositol turnover and transformation of cells by Abelson murine leukaemia virus. EMBO J 4:3173–3178

Fujiki H, Tanaka Y, Miyake R, Kikkawa U, Nishizuka Y, Sugimura T (1984) Activation of calcium-activated phospholipid-dependent protein kinase by new classes of tumour promoters — teleocidin and debromoaplysiatoxin. Biochem Biophys Res Commun 120:339–343

Gould KL, Woodgett JR, Cooper JA, Buss JE, Shalloway D, Hunter T (1985) Protein kinase C phosphorylates pp60^{src} at a novel site. Cell 42:849–857

Hannun YA, Loomis CR, Bell RM (1985) Activation of protein kinase C by Triton X-100 mixed micelles containing diacylglycerol and phosphatidylserine. J Biol Chem 260:10039–10043

Inoue M, Kishimoto A, Takai Y, Nishizuka Y (1977) Studies on a cyclic nucleotide independent protein kinase and its proenzyme in mammalian tissues. J Biol Chem 252:7610–7616

Kajikawa N, Kishimoto A, Shiota M, Nishizuka Y (1983) Ca^{2+}-dependent neutral protease and proteolytic activation of Ca^{2+}-activated phospholipid-dependent protein kinase. Meth Enzymol 102:279–289

Kikkawa U, Takai Y, Minakuchi R, Inohara S, Nishizuka Y (1982) Calcium-activated, phospholipid-dependent protein kinase from rat brain. J Biol Chem 257:13341–13348

Miyake R, Tanaka Y, Tsuda Y, Kaibuchi K, Kikkawa U, Nishizuka Y (1984) Activation of protein kinase C by non-phorbol tumor promoter, mezerin. Biochem Biophys Res Commun 121:649–656

Mori T, Takai Y, Minakuchi R, Yu B, Nishizuka Y (1980) Inhibitory action of chlorpromazine, dibucaine and other phospholipid-interacting drugs on calcium-activated, phospholipid-dependent protein kinase. J Biol Chem 255:8378–8380

Mori T, Takai Y, Yu B, Takahashi J, Nishizuka Y, Fujikura T (1982) Specificity of the fatty acyl moieties of diacylglycerol for the activation of calcium-activated phospholipid-dependent protein kinase. J Biochem 91:427–431

Nishizuka Y (1984) The role of protein kinase-C in cell-surface signal transduction and tumour promotion. Nature 308:693–697

Parker PJ, Stabel S, Waterfield MD (1984) Purification to homogeneity of protein kinase C from bovine brain-identity with the phorbol ester receptor. EMBO J 3:953–959

Rozengurt E, Rodriguez-Pena A, Coombs M, Sinnett-Smith J (1984) Diacylglycerol stimulates DNA synthesis and cell division in mouse 3T3 cells: role of Ca^{2+}-sensitive phospholipid-dependent protein kinase. Proc Natl Acad Sci USA 81:5748–5752

Schatzman RC, Wise BC, Kuo JF (1981) Phospholipid-sensitive calcium-dependent protein kinase: Inhibition by anti-psychotic drugs. Biochem Biophys Res Commun 98:669–676

Slotboom AJ, Verheij HM, de Haas GH (1982) On the mechanism of phospholipase A_2. In: Hawthorne JN, Ansell GB (eds) Phospholipids. Elsevier, Oxford, p 359

Takai Y, Kishimoto A, Inoue M, Nishizuka Y (1977) Studies on a cyclic nucleotide independent protein kinase and its proenzyme in mammalian tissues. J Biol Chem 252:7603–7609

Tapley PM, Murray AW (1984) Modulation of Ca^{2+}-activated phospholipid-dependent protein kinase in platelets treated with a tumour-promoting phorbol ester. Biochem Biophys Res Commun 122:158–164

Turner RS, Kemp BE, Su H-de, Kuo JF (1985) Substrate specificity of phospholipid/Ca^{2+}-dependent protein kinase as probed with synthetic fragments of the myelin basic protein. J Biol Chem 260:11503–11507

Wise BC, Raynor RL, Kuo JK (1982) Phospholipid-sensitive Ca^{2+}-dependent protein kinase from heart. J Biol Chem 257:8481–8488

The Relevance of Protein Kinase C Activation, Glucose Transport and ATP Generation in the Response of Haemopoietic Cells to Growth Factors

T. Michael Dexter, Anthony D. Whetton, and Clare M. Hayworth

Haemopoietic cells differ from most other cell types in that their survival in vitro absolutely requires the presence of specific growth factors. For example, the proliferation and development of multipotent haemopoietic stem cells in vitro requires the presence of a growth factor called interleukin-3 (IL-3); in its absence, the cells die. Similarly, the committed progenitor cells which arise as a consequence of differentiation of the multipotent stem cells also require factors for their growth and development in vitro. These growth factors include IL-3 (which, in addition to acting on stem cells, also acts on the more mature differentiating cells) and a variety of lineage-restricted molecules such as granulocyte/macrophage colony-stimulating factor (GM-CSF) (Burgess and Nicola 1983; Gough et al. 1984; Metcalf 1985), erythropoietin (Eaves and Eaves 1984), macrophage colony-stimulating factor (M-CSF or CSF-1) (Stanley and Jubinsky 1984; Burgess and Nicola 1983; Metcalf 1985) to name just a few (see Gough, this Vol.). In the absence of these growth factors, the committed progenitor cells die (Metcalf 1977). In their presence, the primitive stem cells and progenitor cells undergo growth and differentiation to produce the mature functional end cells, erythrocytes, neutrophils, macrophages and so on (Metcalf 1977). It is also clear that the growth factors are required continuously throughout the developmental programme: removal of the growth factor at any stage during the development into the mature cells leads to a cessation of growth and to death of the developing clone. Presumably, this "programmed death" has some physiological significance and is almost certainly relevant to the process of leukaemogenesis.

To approach the question of the mechanisms underlying the death of cells in the absence of growth factors, we have exploited two recent advances in haematology. First has been the isolation, characterization and molecular cloning of many of the growth factors which are known to be active on haemopoietic cells (see Metcalf 1985 and Gough, this Vol.). Obviously, this means that large amounts of purified material are now available. Second has been the observation that in some circumstances IL-3 will promote the continuous proliferation in vitro of multipotent stem cells, granulocyte precursor cells and mast cells (Bazill et al. 1983; Schrader 1983). The latter is particularly important since it provides us with "pure" cell populations available in essentially unlimited numbers. These IL-3-dependent cell lines can be readily cloned in soft-gel media, have a diploid karyotype, are non-leukaemic, and absolutely require the presence of IL-3 for their continued survival and prolif-

Oncogenes and Growth Control
Edited by P. Kahn and T. Graf
© Springer-Verlag Berlin Heidelberg 1986

eration in vitro (Dexter et al. 1980; Bazill et al. 1983). In the absence of IL-3 the cells die and in this respect such lines are similar to freshly isolated marrow stem cells. Furthermore, the "normality" of some IL-3-dependent cell lines is emphasized by our recent finding that marrow stromal cells can replace the requirement for IL-3. In other words, when cultured on a marrow stromal cell layer, at least some cell lines behave like normal stem cells: they undergo self-renewal, differentiation and development to produce mature myeloid cells of several lineages (E. Spooncer and T. M. Dexter, in press). Therefore, it is likely that the effects of IL-3 on these cell lines reflect those exerted by the marrow stromal cells, and that we are dealing with a biologically relevant response analogous to the growth of normal stem cells in vivo. With this important proviso in mind, we have investigated the mode of action of IL-3.

IL-3 Affects the Primary Metabolism of Stem Cells

To determine the specificity of the IL-3 response, IL-3-dependent stem cell lines were washed free of growth factor and their response to other growth stimuli was measured. These stimuli included epidermal growth factor (EGF), fibroblast growth factor (FGF), steroid hormones and a variety of other agents which are known to modulate growth in other cell types. The effect was clear: none of the agents tested could replace IL-3 in maintaining viability of the cells. As part of these studies, however, we examined the effect of exogenously added ATP or ATP regenerating systems and found that ATP could maintain the viability of some of the cell lines for several days (Whetton and Dexter 1983). Contrary to established dogma, it appeared that the external ATP entered the cells and that this played a major role in maintaining viability. Indeed, further studies showed that a drop in ATP levels was one of the earliest changes elicited when cells were deprived of IL-3 and that addition of ATP to cells in the absence of IL-3 was sufficient to maintain intracellular ATP levels, at least in the short term.

Obviously, all cells require energy to survive and proliferate. In mammalian cells, ATP is derived primarily from the breakdown of sugars (normally glucose) by glycolysis and mitochondrial respiration. Most cells in vitro, however, preferentially use anaerobic glycolysis for ATP generation. This complex process begins with hexose transport across the cell membrane and proceeds through a series of enzyme reactions, culminating in a net gain of ATP and the formation of lactate. If IL-3 is withdrawn from the stem cells, the consequent fall in ATP levels indicates a breakdown in the normal intracellular generating system. Where could the IL-3 be operating? When we examined the bulk levels of activities of the rate-limiting enzymes involved in glycolysis, no changes were seen in the critical first few hours following removal of IL-3 from the stem cells. However, a major change was seen in the level of glucose transport across the cell membrane. Within a few minutes, hexose uptake was markedly decreased and continued to fall over the next few hours (Whetton et al. 1984). When IL-3 was added back to these cells, there was a rapid and IL-3 concentration-dependent increase in the rate of glucose uptake and generation of ATP. This stimulation of glucose uptake appeared to represent an effect of IL-3 on the glucose transport protein since (1) the effect was specific for the transport of D-glucose; L-glucose which is not "recognized" by the glucose transport protein, was not taken up by the cells; and (2) the transport of D-glucose could be blocked by the addition to cytochalasin B, a specific inhibitor of the glucose transport protein.

These data led us to propose a model for haemopoietic stem cell survival and proliferation based upon stimulation of the primary metabolism of the cells by growth factors such as IL-3. As pointed out previously, one of the fundamental findings regarding haemopoiesis in vitro is that growth factors are essential for the *survival* of the cells: in the absence of growth factors, the cells die. That a similar situation also exists in vivo is indicated by our work with long-term marrow cultures, where haemopoiesis can be maintained in vitro for many months (Dexter et al. 1977). The important feature of long-term marrow cultures is that haemopoietic stem cell proliferation and differentiation is maintained *in the absence of* added growth factors, provided that the stem cells are supplied with a marrow stromal cell environment. These stromal cells are presumably supplying the regulatory molecules which are essential for haemopoiesis. Whether or not these are the same molecules as IL-3, GM-CSF and so on still remains to be unequivocally demonstrated. However, a clear point which emerges from studies of these long-term cultures is that intimate cell-cell contact is required, suggesting that the growth factors are in some way bound to the surface of the stromal cells and exert their effect only when the stroma and the stem cells are in close contact. When the stem cells are separated from this stromal cell environment, they also die. Because the long-term marrow cultures provide a situation closely analogous to haemopoiesis in vivo, it is reasonable to suggest a similar mechanism underlying the response of cells to IL-3 or to stroma. The net result is the same: a "programmed" cell death in the absence of regulatory molecules. Such a mechanism may be important as a means of protecting the host animal from cells with a vast proliferative potential (like stem cells) which might otherwise grow out of control when they escape from the bone marrow (Dexter et al. 1985). Certainly, stem cells *are* found in the circulation, but they do not proliferate or develop there; they are dying cells. A simple mechanism to control their survival is to couple energy generation to growth-factor binding. Growth factor-dependent activation of glucose transport and primary metabolism would have this effect. The finding that detectable levels of IL-3 cannot be found in the serum or tissues of mice (Garland et al. 1983; Lord et al. 1986) supports this model.

Second Messenger Systems Generated as a Cellular Response to IL-3

The stimulation of glucose transport, described above, presumably occurs as a consequence of an enzyme cascade set in motion by the binding of IL-3 to its receptor. To determine which second messenger system is involved in this process, we have looked at the roles of cAMP and protein kinase C.

The activation of cAMP occurs in many cell types following stimulation by biological peptides, growth factors and chemicals, and a major role for cAMP has been proposed in cell proliferation (Ralph 1983). However, *removal* of IL-3 from IL-3-dependent stem cells had no major effects upon cAMP levels in the first few hours. Similarly, addition to IL-3 growth-factor deprived cells caused a *decrease* in cAMP (possibly due to stimulation of phosphodiesterases) followed by a slow return to "normal" levels. Thus, these studies gave no indications that cAMP has a major role to play in IL-3-mediated stem cell survival and proliferation (C. Heyworth, A. D. Whetton, T. M. Dexter, in press).

However, when protein kinase C levels were investigated, a marked response to IL-3 was seen (A. D. Whetton, C. Heyworth, T. M. Dexter, in press). In these experiments, the stem cells were deprived of IL-3 for a short period. Following re-addition of IL-3, there was a rapid (10′) translocation of protein kinase C from the cytosol to the membrane fraction of the cells, confirming the earlier work reported by Farrar et al. (1985) on the response of cells both to IL-2 and to IL-3. This result was intriguing in view of the recent report that protein kinase C can phosphorylate the glucose transport protein (Witters et al. 1985) since it raised the possibility that there is an association between the ability of IL-3 to stimulate glucose transport and its ability to "activate" protein kinase C.

Although speculative, this model makes some predictions. For example, phorbol esters (such as TPA) have a variety of effects on cells, most of which are probably associated with the activation of protein kinase C (Nishizuka 1984 and Parker and Ullrich, this Vol.). Therefore, if IL-3 is exerting its effects via activation of this kinase, it may be expected that some of these effects should be mimicked by treatment of the stem cells with TPA. This was found to be the case. When the IL-3-dependent stem cells were incubated with TPA in the absence of IL-3, there was a marked increase in survival and ^{3}HTdR incorporation over the first 24 h. In other words, TPA maintained a certain level of cell viability in the absence of growth factor. This effect was enhanced if the cells were also supplied with Ca^{2+} ionophore and in optimal conditions (100 ng ml^{-1} TPA and 100 ng ml^{-1} Ca^{2+} ionophore) the levels of survival and ^{3}HTdR incorporation seen in the cells were equivalent to $20-50$ units ml^{-1} of IL-3.

Is this effect associated with activation of protein kinase C? When this was measured, we found that TPA was as efficient as IL-3 in promoting the translocation (and presumably activation) of protein kinase C from the cytosol to the membrane fraction of the stem cells. Furthermore, this effect was also associated with an activation of glucose transport in the cells by TPA. In these experiments, the IL-3-dependent stem cells were washed free of growth factors and then incubated for 4 h with various additives (control medium or optimum stimulatory levels of TPA or IL-3). Both IL-3 and TPA stimulated glucose transport to the same extent; as discussed previously, this represented an effect upon the glucose transport protein, since the uptake could be blocked by addition of cytochalasin B. Furthermore, if cells were pre-incubated with optimal concentrations of TPA, no further stimulation of hexose uptake was seen on addition of IL-3. Conversely, if cells were pre-incubated with IL-3, no further stimulation of glucose transport was observed upon addition of TPA. This strongly suggests that both TPA and IL-3 are activating glucose transport through the same mechanism, leading us to propose the model shown in Fig. 1.

IL-3 binds to a cell surface receptor of 50 – 70 kDa and induces a transient decrease in the levels of cAMP as well as a translocation of protein kinase C from the cytosol to the cell membrane. Whether or not this activation of protein kinase C is mediated though a stimulation of phosphatidylinositol turnover is not clear at present. However, work presented by C. P. Downes (Smith, Kline & French Laboratories, England) at the European Artery Club meeting, Germany, 1986, strongly suggests that protein kinase C activation is occurring through a mechanism which does *not* involve increased PI turnover. How this is achieved is unknown. What is known is that the activation of protein kinase C can elicit a variety of phosphorylation reactions including the phosphorylation of the glucose transport protein (Nishizuka 1984; Farrar

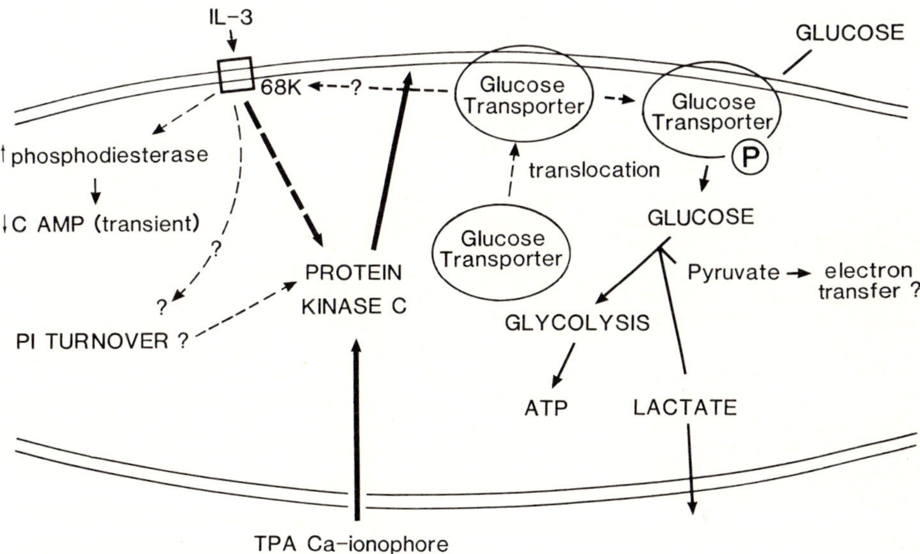

Fig. 1. Possible signal pathway of IL-3 in haematopoietic stem cells

et al. 1985). Perhaps this phosphorylation is associated with translocation of the glucose transporter itself to the cell membrane (Watanabe et al. 1984) or with modulation in the activity of the glucose transport protein. These possibilities are presently being explored using monoclonal antibodies directed against the glucose transporter. The end result, however, is that glucose flows into the cell, primary metabolism is activated, ATP is generated, and the cells survive. The validity of the model is emphasized by the work with TPA plus Ca^{2+} ionophore which in the short term elicits many of the events seen with IL-3, including the translocation of protein kinase C and the activation of glucose transport.

Acknowledgments. The authors are supported by the Cancer Research Campaign (TMD, CH) and the Leukaemia Research Fund (ADW).

References

Bazill GW, Haynes M, Garland J, Dexter TM (1983) Characterization and partial purification of a haemopoietic cell growth factor in WEHI-3 cell conditioned medium. Biochem J 210:747 – 759

Burgess AW, Nicola NA (1983) Growth factors and stem cells. Academic Press, Sydney, pp 93 – 124

Dexter TM, Allen TD, Lajtha LF (1977) Conditions controlling the proliferation of haemopoietic stem cells in vitro. J Cell Physiol 91:335 – 344

Dexter TM, Garland J, Scott D, Scolnick E, Metcalf D (1980) Growth of factor-dependent hemopoietic precursor cell lines. J Exp Med 152:1036 – 1047

Dexter TM, Heyworth C, Whetton AD (1985) The role of growth factors in haemopoiesis. Bio Essays 2:154 – 158

Eaves AC, Eaves CJ (1984) Erythropoiesis in culture. Clin Haematol 13:371 – 391

Farrar WL, Thomas TP, Anderson WB (1985) Altered cytosol/membrane enzyme redistribution on Interleukin-3 activation of protein kinase C. Nature 315:235 – 237

Garland JM, Aldridge A, Wagstaffe J, Dexter TM (1983) Studies on the in vivo production of a lymphokine activity, interleukin-3 (IL-3) elaborated by lymphocytes and a myeloid leukaemic line in vitro and the fate of IL-3 dependent cell lines. Br J Cancer 48:247 – 259

Gough NM, Gough J, Metcalf D, Kelso A, Grail D, Nicola NA, Burgess AW, Dunn AR (1984) Molecular cloning of cDNA encoding a murine haematopoietic growth regulator, granulocyte-macrophage colony-stimulating factor. Nature 309:763 – 767

Lord BI, Molineux J, Testa NG, Kelly M, Spooncer E, Dexter JM (1986) The kinetic response of haemopoietic precursor cells in vivo to highly purified recombinant Interleukin-3. Lymphokine Res (in press)

Metcalf D (1977) Haemopoietic colonies. In vitro cloning of normal and leukaemia cells. Springer, Berlin Heidelberg New York

Metcalf D (1985) The granulocyte-macrophage colony-stimulating factors. Science 229:16 – 29

Nishizuka J (1984) The role of protein kinase C in cell surface signal transduction and tumour promotion. Nature 308:693 – 698

Ralph RK (1983) Cyclic AMP, calcium and control of cell growth. FEBS Lett 161:1 – 8

Schrader JW (1983) Bone marrow differentiation in vitro. Crit Rev Immunol 4:197 – 277

Stanley ER, Jubinsky PT (1984) Factors affecting the growth and differentiation of haemopoietic cells in culture. Clin Haematol 13:329 – 348

Watanabe T, Smith MM, Robinson FW, Kono T (1984) Insulin action on glucose transport in cardiac muscle. J Biol Chem 259:13117 – 13122

Whetton AD, Dexter TM (1983) Effect of haematopoietic cell growth factor on intracellular ATP levels. Nature 303:629–631

Whetton AD, Bazill GW, Dexter TM (1984) Haematopoietic cell growth factor mediates cell survival via its action on glucose transport. EMBO J 3:409–413

Witters LA, Vater CA, Lienhard GE (1985) Phosphorylation of the glucose transporter in vivo and in vitro by protein kinase C. Nature 315:777–778

Cytoplasmic pH and Free Ca^{2+} in the Action of Growth Factors

Wouter H. Moolenaar

Among the immediate consequences of growth factor-receptor interaction are the activation of various ion transport systems in the plasma membrane and changes in intracellular ionic composition. Mitogen-induced alterations in the levels of cytoplasmic free Ca^{2+} ($[Ca^{2+}]_i$) and H^+ are of special interest, because these ions are thought to serve as second messengers that trigger and regulate cell proliferation. During the past few years our understanding of the ionic changes in growth factor-stimulated cells has increased dramatically. In particular, the molecular mechanisms responsible for a transient rise in $[Ca^{2+}]_i$ and for a sustained increase in cytoplasmic pH (pH_i) have been largely unraveled, and much effort is now being devoted to determining how these ionic signals might contribute to a proliferative response. This review describes the Ca^{2+} and pH signals generated by growth factor receptors, with emphasis on the activation of Na^+/H^+ exchange and the resultant rise in pH_i, which has a permissive effect on the initiation of DNA synthesis in mitogen-stimulated cells.

Rise in $[Ca^{2+}]_i$

Addition of serum, platelet-derived growth factor (PDGF), epidermal growth factor (EGF) or other mitogenic peptides to responsive cells evokes a rapid, but transient, severalfold increase in $[Ca^{2+}]_i$ (Moolenaar et al. 1984a, b, 1986a, b; Hesketh et al. 1985). Subsequent to growth factor binding, the $[Ca^{2+}]_i$ rise is initiated without a detectable lag period (<1 s). It usually peaks within $30-60$ s and then returns to its resting level during the next $5-10$ min. In general, the rapid $[Ca^{2+}]_i$ signals in response to extracellular stimuli are mediated by the second messenger inositol 1,4,5-trisphosphate (IP_3), which triggers the release of Ca^{2+} from the endoplasmic reticulum (see Berridge, this Vol.). Mitogens like PDGF, serum and thrombin indeed appear to mobilize Ca^{2+} from internal stores, as demonstrated by studies with the fluorescent Ca^{2+} indicator quin-2 and by IP_3 measurements. In contrast, the Ca^{2+} signal evoked by EGF in quin-2 loaded A-431 and 3T3 cells does not seem to originate from internal stores and has been attributed to the activation of a Ca^{2+} influx pathway ("channel" or carrier) in the plasma membrane (Moolenaar et al. 1986a; Hesketh et al. 1985). This interpretation is based mainly on the finding that the transient increase in $[Ca^{2+}]_i$ in response to EGF does not oc-

Oncogenes and Growth Control
Edited by P. Kahn and T. Graf
© Springer-Verlag Berlin Heidelberg 1986

cur in Ca^{2+}-free media and is abolished by Ca^{2+} entry blockers such as La^{3+} and Mn^{2+}. These results are intriguing because they suggest that there is a fundamental difference between the receptors for EGF and PDGF in terms of their [Ca^{2+}]$_i$-raising mechanisms; however, some caution is needed in interpreting the disappearance of the Ca^{2+}-quin-2 response to EGF when external Ca^{2+} is removed, since it is conceivable that intracellular quin-2 (a Ca^{2+} chelator) somehow interferes with the proper functioning of the EGF receptor, particularly in the absence of extracellular Ca^{2+}.

Regardless of the distinct mechanisms by which growth factors raise [Ca^{2+}]$_i$ (influx or intracellular release), it is generally accepted that Ca^{2+} may play an important role as a second messenger which regulates numerous cellular activities. Since the increase in [Ca^{2+}]$_i$ is short-lived, lasting only for 5 – 10 min after receptor stimulation, it obviously cannot directly mediate such late events as the initiation of protein synthesis and DNA synthesis, which begin only after many hours. Instead, the transient increase in [Ca^{2+}]$_i$ is more likely to trigger a sequence of early cellular changes occurring within minutes of growth factor binding. In this context, it is noteworthy that artificial elevation of [Ca^{2+}]$_i$ by means of an ionophore mimics EGF and PDGF in rapidly inducing the expression of the c-*fos* and c-*myc* proto-oncogenes (Bravo et al. 1985; Tsuda et al. 1985). Thus, Ca^{2+} may have a key role in mediating, either directly or indirectly, the early transcriptional effects of growth factors. Furthermore, one should consider the possibility that the rapid increase in [Ca^{2+}]$_i$ serves to trigger some of the early nonmitogenic responses to growth factors, such as cytoskeletal reorganizations, fluid endocytosis or chemotaxis.

Regulation of pH$_i$ by Na$^+$/H$^+$ Exchange

In addition to raising [Ca^{2+}]$_i$, mitogens rapidly induce a sustained increase in pHi of ~0.15 – 0.3 unit. This cytoplasmic alkalinization is due to the activation of an otherwise quiescent Na$^+$/H$^+$ exchange mechanism in the plasma membrane and persists for as long as the growth factor receptor remains occupied (reviewed by Moolenaar 1986a). Figure 1 schematically illustrates the time courses of both the transient increase in [Ca^{2+}]$_i$ and the shift in pH$_i$ induced by EGF in responsive cells.

Most cells maintain their pH$_i$ at 7.0 – 7.3, which is well above the electrochemical equilibrium value of 6.0 – 6.3 predicted by the Nernst equation from a transmembrane potential of about -60 mV. In vertebrate cells, the specific H$^+$-extruding mechanism which raises pH$_i$ appears to be Na$^+$/H$^+$ exchange in the plasma membrane (Roos and Boron 1981; Moolenaar 1986b). This Na$^+$/H$^+$ exchanger, whose molecular identity is not yet known, tightly regulates pH$_i$ by virtue of its sensitivity to cytoplasmic H$^+$.

The functioning of the plasma membrane Na$^+$/H$^+$ exchanger and its normal housekeeping role in pH$_i$ homeostasis are usually assessed by sudden

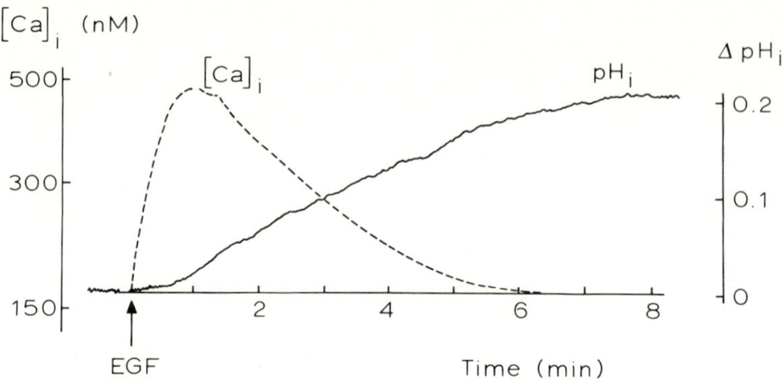

Fig. 1. Changes in $[Ca^{2+}]_i$ and pH_i following the addition of EGF to human A431 cells

acidifying of the cytoplasm, for example by an NH_4^+ prepulse, and monitoring the ensuing recovery of pH_i to its resting level. While the basal Na^+/H^+ exchange activity is normally very low and exactly balances the acidifying effects of H^+ influx and metabolic acid production, Na^+/H^+ exchange is dramatically accelerated as soon as an acute cytoplasmic acid load is applied. The excess cytoplasmic H^+ is then rapidly extruded from the cell, causing pH_i to return to its basal value within a few minutes. Na^+/H^+ exchange, and hence pH_i regulation, are inhibited by the diuretic amiloride and is driven by the steep transmembrane Na^+ gradient, which, in turn, is generated by the Na^+, K^+-ATPase.

It should be emphasized, however, that the value of the resting pH_i is not determined simply by the magnitude of the Na^+ gradient. Under normal con-

Table 1. Stimuli that raise pH_i by activating Na^+/H^+ exchange in their target cells[a]

Stimulus	Cell type
Serum	Fibroblasts, A431 cells
EGF	Fibroblasts, A431 cells
PDGF	Fibroblasts
Thrombin	Fibroblasts
Insulin	[b]
Vasopressin	3T3 cells
Vanadate	A431 cells
Lectins	T lymphocytes
Lipopolysaccharide	B-lymphoid cells (Rosoff et al. 1984)
Interleukin-2	T lymphocytes (Mills et al. 1985)
Phorbol ester/diacylglycerol	Various cell types
Hypertonicity (cell shrinking)	Lymphocytes, A431 cells

[a] Reported alkalinizations range from ~0.1 − 0.3 pH unit. For references see Moolenaar (1986a), unless indicated otherwise.

[b] Insulin alone fails to raise pH_i in most cell types, but it often potentiates the pH_i response to other mitogens.

ditions, the transmembrane Na$^+$ gradient could theoretically raise pH$_i$ by about one unit more alkaline. This raises the important question of how the activity of the Na$^+$/H$^+$ exchanger is regulated. It appears that the major determinant of the Na$^+$/H$^+$ exchange rate is cytoplasmic [H$^+$]: when pH$_i$ falls below a certain "threshold" (its resting value), the activity of the exchanger is increasingly stimulated. Aronson et al. (1982) have shown that cytoplasmic H$^+$ acts as an allosteric activator of the Na$^+$/H$^+$ exchanger by binding to an inward-facing regulatory site. According to this concept the protonation of the regulatory H$^+$ binding site sets the exchanger in motion, while a functionally distinct H$^+$ transport site mediates H$^+$ extrusion once the exchanger is activated. In principle, by changing the H$^+$ affinity of the regulatory site, extracellular stimuli could control the physiological state of the Na$^+$/H$^+$ exchanger and thereby affect the value of pH$_i$. Indeed, growth factors exploit this simple and elegant control mechanism to shift the steady-state pH$_i$ in the alkaline direction.

Mechanisms of Mitogen-Induced Rise in pH$_i$

A large number of mitogens and other extracellular stimuli have been reported to raise pH$_i$ in their target cells (summarized in Table 1). In general, the increase in pH$_i$ is detectable within $10-20$ s and is virtually complete within 10 min. The observed alkalinizations range from $0.1-0.3$ pH unit and are due to H$^+$ extrusion through the Na$^+$/H$^+$-exchanger.

The kinetic mechanism by which mitogens activate the exchanger is now fairly well understood. By comparing the pH$_i$-dependence of Na$^+$/H$^+$ exchange in quiescent and stimulated cells, it appears that growth factors and other stimuli act by increasing the sensitivity of the Na$^+$/H$^+$ exchanger for cytoplasmic H$^+$. This alkaline shift in pH$_i$ sensitivity of $0.1-0.3$ units most likely reflects an increase in the apparent affinity of the regulatory site for H$^+$ (roughly by a factor of 2). One could imagine that some ionizable group at the regulatory site acquires a greater pK$_a$ because its immediate environment becomes more negatively charged (for example by phosphorylation). It is important to note that, consistent with the above model, the Na$^+$/H$^+$ exchange rate is only transiently increased by external stimuli: the H$^+$-extruding activity returns to its prestimulation level once pH$_i$ has reached its new, more alkaline steady-state value.

By what biochemical steps do growth factors modify the pH$_i$-sensitivity of the Na$^+$/H$^+$ exchanger? Current evidence indicates that protein kinase C can somehow activate the exchanger. Tumor-promoting phorbol esters and synthetic diacylglycerols, which bind to and directly activate kinase C, mimic growth factors in raising pH$_i$ without producing a rise in [Ca$^+$]$_i$ (Moolenaar et al. 1984b). Thus, a rapid increase in [Ca^{2+}]$_i$ is not essential for activation of the exchanger. Of the physiological stimuli listed in Table 1, many, but not all, are known stimulators of inositol lipid breakdown and, hence, of kinase C activity in their specific target cells.

Recent studies seem to indicate that there are additional pathways, not involving kinase C, by which the Na^+/H^+ exchanger can be activated. Chronic treatment of cells with phorbol esters leads to the disappearance of functional protein kinase C. Yet, kinase C-depleted 3T3 cells can still raise their pH_i in response to EGF (Vara and Rozengurt 1985). Similarly, osmotic activation of the Na^+/H^+ exchanger in lymphocytes does not seem to require stimulation of protein kinase C (Grinstein et al. 1986). Furthermore, osmotic cell shrinking neither stimulates phospholipase C activity nor evokes a rise in $[Ca^+]_i$. Thus, the Na^+/H^+ exchanger can be activated by at least two separate pathways, one involving the phospholipase C-protein kinase C system and the other one(s) unknown.

It seems plausible to hypothesize that EGF might activate the Na^+/H^+ exchanger through the intrinsic tyrosine-specific protein kinase of the EGF receptor. However, evidence arguing against this possibility comes from experiments using anti-EGF receptor monoclonal antibodies (Defize et al. 1986). Some of these antibodies have the interesting property that they act as partial agonists which activate the tyrosine-specific protein kinase both in vitro and in vivo without inducing any change in $[Ca^{2+}]_i$ and pH_i. This result strongly suggests that the EGF receptor tyrosine kinase is not capable of activating the Na^+/H^+ exchanger to raise pH_i. Furthermore, these antibodies fail to induce DNA synthesis when added to quiescent fibroblasts, suggesting that the EGF receptor-linked pathway responsible for the Ca^{2+} and pH_i signals is required for the stimulation of cell proliferation (Defize et al. 1986).

Further exploration of the molecular mechanisms which activate the Na^+/H^+ exchanger is hampered by the fact that the molecular identity of the exchanger has not yet been determined; however, genetic approaches may soon lead to the molecular identification of the exchanger (Pouysségur 1985).

Biological Significance of a Rise in pH_i

A mitogen-induced shift in pH_i of about 0.2 units would be expected to have considerable effects on a host of pH-sensitive processes in the cell. Growing consensus among workers in the field holds that an alkaline pH_i shift has a *permissive* rather than a strictly triggering role in the response of cells to mitogens. In this sense pH_i differs from the more classical second messengers such as Ca^{2+} and cAMP. Of critical importance is the question of whether cytoplasmic alkalinization, mediated by Na^+/H^+ exchange, is essential for the initiation of DNA synthesis and cell division in response to growth stimuli. Perhaps the most convincing demonstration of a signalling role for Na^+/H^+ exchange and pH_i in the initiation of a mitogenic response has been made with fertilized sea urchin eggs, in which pH_i must rise by at least 0.2 units to permit

DNA synthesis to begin (reviewed by Whitaker and Steinhardt 1982). Using mutant fibroblasts which lack a functional Na$^+$/H$^+$ exchanger, Pouysségur et al. (1985a, b) elegantly showed that below a certain threshold value (around 7.2) pH$_i$ becomes limiting for cell proliferation and furthermore, that one of the critical pH$_i$-dependent steps in activated fibroblasts appears to be the stimulation of protein synthesis. Studies by others seem to confirm that a mitogen-induced rise in pH$_i$ is a permissive event which is necessary but not sufficient for progression through the S-phase, at least in fibroblasts (Bravo and McDonald-Bravo 1986; Moolenaar et al. 1986b). However, cytoplasmic alkalinization may have a less critical role in the mitogenic response of lymphocytes to interleukin-2 than in stimulated fibroblasts (Mills et al. 1985).

Shifts in pH$_i$ during the cell cycle have been observed in lower eukaryotes such as protozoa, slime molds and yeast (reviewed by Busa and Nuccitelli 1984). For example, during the *Dictyostelium* cell cycle, pH$_i$ oscillates (by an unknown mechanism) with the same period as the DNA replication cycle, with alkalinization occurring during DNA synthesis. This pH$_i$ oscillator may have an on-off triggering function rather than a permissive role in the timing and regulation of protein and DNA synthesis (Aerts et al. 1985). It will be of considerable interest to monitor pH$_i$ and the state of the Na$^+$/H$^+$ exchanger during the cell cycle of higher eukaryotes. Another challenge for future studies is to examine whether the ionic signals generated by growth factor receptors have their correlates in the action of certain oncogene products. One widely held idea is that oncogenes induce malignant growth at least partially via the constitutive activation of signal pathways normally involved in growth factor action. This might lead not only to permanently altered intracellular Ca^{2+} compartments but also to an uncontrolled activation of the Na$^+$/H$^+$ exchanger and, hence, to an elevated pH$_i$.

Acknowledgment. Research related to this review was supported by the Netherlands Cancer Foundation (Koningin Wilhelmina Fonds).

References

Aerts RJ, Durston AJ, Moolenaar WH (1985) Cytoplasmic pH and the regulation of the *Dictyostelium* cell cycle. Cell 43:653–657

Aronson PS, Nee J, Suhm MA (1982) Modifier role of internal H$^+$ in activating the Na-H exchanger in renal microvillus membrane vesicles. Nature 299:161–163

Bravo R, MacDonald-Bravo H (1986) Effect of pH on the induction of competence and progression to the S-phase in mouse fibroblasts. FEBS Lett 195:309–312

Bravo R, Burckhardt J, Curran T, Müller R (1985) Stimulation and inhibition of growth by EGF in different A431 cell clones is accompanied by the rapid induction of c-*fos* and c-*myc* proto-oncogenes. EMBO J 4:1193–1198

Busa WB, Nuccitelli R (1984) Metabolic regulation via intracellular pH. Am J Physiol 246:R409–R438

Defize LHK, Moolenaar WH, van der Saag PT, de Laat SW (1986) Dissociation of cellular responses to epidermal growth factor using anti-receptor monoclonal antibodies. EMBO J 5:1187–1192

Grinstein S, Mack E, Mills GB (1986) Osmotic activation of the Na$^+$/H$^+$ antiport in protein kinase C-depleted lymphocytes. Biochem Biophys Res Commun 134:8–13

Hesketh TR, Moore JP, Morris JDH, Taylor MV, Rogers J, Smith GA, Metcalfe JC (1985) A common sequence of calcium and pH signals in the mitogenic stimulation of eukaryotic cells. Nature 313:481–484

Mills GB, Cragoe EJ, Gelfand EW, Grinstein S (1985) Interleukin 2 induces a rapid increase in intracellular pH through activation of a Na$^+$/H$^+$ antiport. J Biol Chem 260:12500–12507

Moolenaar WH (1986a) Effects of growth factors on intracellular pH regulation. Annu Rev Physiol 48:363–376

Moolenaar WH (1986b) Regulation of cytoplasmic pH by Na$^+$/H$^+$ exchange. Trends Biochem Sci 11:141–143

Moolenaar WH, Aerts RJ, Tertoolen LGJ, de Laat SW (1986a) The epidermal growth factor-induced calcium signal in A431 cells. J Biol Chem 261:279–284

Moolenaar WH, Defize LHK, de Laat SW (1986b) Ionic signalling by growth factor receptors. J Exp Biol (in press)

Moolenaar WH, Tertoolen LGJ, de Laat SW (1984a) Growth factors immediately raise cytoplasmic free Ca^{2+} in human fibroblasts. J Biol Chem 259:8066–8069

Moolenaar WH, Tertoolen LGJ, de Laat SW (1984b) Phorbol ester and diacylglycerol mimic growth factors in raising cytoplasmic pH. Nature 312:371–374

Pouysségur J (1985) The growth factor-activatable Na$^+$/H$^+$ exchange system: a genetic approach. Trends Biochem Sci 10:453–455

Pouysségur J, Chambard J-C, Franchi A, L'Allemain G, Paris S, van Obberghen-Schilling E (1985a) Growth-factor activation of the Na$^+$/H$^+$ antiporter controls growth of fibroblasts by regulating intracellular pH. In: Feramisco J, Ozanne B, Stiles C (eds) Cancer cells, vol 3. Cold Spring Harbor Lab, Cold Spring Harbor, NY, pp 409–416

Pouysségur J, Franchi A, L'Allemain G, Paris S (1985b) Cytoplasmic pH, a key determinant of growth factor-induced DNA synthesis in quiescent fibroblasts. FEBS Lett 190:115–118

Roos A, Boron W (1981) Intracellular pH. Physiol Rev 61:296–434

Rosoff PM, Stein LF, Cantley LC (1984) Phorbol esters induce differentiation in a pre-B-lymphocyte cell line by enhancing Na$^+$/H$^+$ exchange. J Biol Chem 259:7056–7060

Tsuda T, Kaibuchi K, West B, Takai Y (1985) Involvement of Ca^{2+} in platelet-derived growth factor-induced expression of c-*myc* oncogene in Swiss 3T3 fibroblasts. FEBS Lett 187:43–46

Vara F, Rozengurt E (1985) Stimulation of Na$^+$/H$^+$ antiport activity by EGF and insulin occurs without activation of protein kinase C. Biochem Biophys Res Commun 130:646–653

Whitaker MJ, Steinhardt RA (1982) Ionic regulation of egg activation. Quart Rev Biophys 15:593–666

Epidermal Growth-Factor Mediation of S6 Phosphorylation During the Mitogenic Response: A Novel S6 Kinase

George Thomas

The 40S ribosomal protein S6 undergoes extensive phosphorylation during the processes of tissue regeneration, development, cell growth and transformation. Much attention has recently been focused on the role of this event in protein translation and the mechanisms by which it is regulated (Gressner and Wool 1974; Decker 1981; Nielsen et al. 1982; Thomas et al. 1982). This chapter reviews briefly what is known about S6 function in translation and then discusses in greater detail how its multiple phosphorylation is regulated by EGF and, to a lesser extent, by other mitogens during the mitogenic response.

S6

Ribosomal protein S6 is an integral ribosomal protein with an approximate molecular weight of 33,000 and which is present in one copy per 40S ribosomal subunit (Collatz et al. 1976). By immune electron microscopy and chemical crosslinking either to other ribosomal proteins or to ribosomal RNA, the protein has been mapped to the small head region of the 40S subunit in a position juxtaposed to the larger 60S subunit, an area of the ribosome which is involved in the binding of tRNA and mRNA. These findings are consistent with the fact that poly (U), a synthetic mRNA, can be crosslinked by UV irradiation to one set of 40S ribosomal proteins and can also protect a second set from chemical modification. S6 is the only protein common to both sets (reviewed by Martin-Pérez et al. 1984).

As many as 5 mol of phosphate can be incorporated into S6. Isolation and tryptic peptide mapping of each of the increasingly phosphorylated forms of S6 has revealed that these 5 mol of phosphate are added in a specific order (Martin-Pérez and Thomas 1983). It is thought, though not yet proven, that all the sites of phosphorylation may be clustered near the C-terminus of the protein (Wettenhall et al. 1983). Ribosomal RNA melting experiments argue that the phosphorylation of S6 induces a conformational change in the ribosome (Hallberg et al. 1981). A similar conclusion has been drawn from chemical modification studies correlating the accessibility of specific ribosomal proteins to the phosphorylation state of S6 (Kisilevsky et al. 1984). Finally, more recent studies in vitro demonstrate that phosphorylated ribosomes bind poly (U) and utilize it to synthesize polyphenylalanine more efficiently than nonphosphorylated ribosomes (Burkhard and Traugh 1983). These results

Oncogenes and Growth Control
Edited by P. Kahn and T. Graf
© Springer-Verlag Berlin Heidelberg 1986

have led to the hypothesis that the phosphorylation of S6 alters the affinity of the 40S ribosome for messenger RNA. Such an hypothesis is consistent with the fact that in many systems there are a number of alterations in the pattern of protein synthesis that occur during the time that S6 is becomming phosphylated, some of which are controlled at the translational level. What is needed now is direct evidence, either in vivo or in vitro, that the phosphorylation of S6 plays a role in inducing these changes in translation.

Activation of Quiescent Cells

When quiescent cells in culture are stimulated to proliferate by serum or specific growth factors such as epidermal growth factor (EGF), they exhibit a two- to three-fold increase in the rate of protein synthesis within 60 min (Thomas et al. 1982). This increase is controlled at the level of initiation and is essential for the activation of cell growth. In part, this large and rapid change in the initiation rate can be accounted for by the movement of both newly transcribed mRNA and a large pool of stored mRNA into actively translating polysomes. The latter accounts for approximately 80% of the mRNA in polysomes during the initial 6 h of the mitogenic response (Rudland et al. 1975). This alteration in mRNA expression, as described above, leads to at least 20 specific qualitative and quantitative changes in the pattern of translation which are controlled at both the pretranslational and translation levels (Thomas et al. 1981).

The increase in protein synthesis and the changes in the translation pattern are preceded by the multiple phosphorylation of S6. Pretreatment of cells with cycloheximide, which completely blocks the changes in protein synthesis, has no effect on the rate or extent to which S6 is phosphorylated. In contrast, methylxanthines, which inhibit S6 phosphorylation, inhibit both processes in a parallel dose-responsive manner (Thomas et al. 1980). Furthermore, those ribosomes containing the most highly phosphorylated forms of S6 have a selective advantage in entering polysomes (Thomas et al. 1982; Duncan and McConkey 1982). These findings support the notion that S6 phosphorylation is a prerequisite for the activation of protein synthesis, and thus cell growth, during the early stages of the mitogenic response.

EGF-Activated S6 Phosphorylation

The addition of 10^{-10} M – 10^{-9} M EGF to quiescent cells in culture leads to a sharp burst in S6 phosphorylation; this response reaches a maximal level between 10^{-9} M and 10^{-8} M EGF. The addition of other mitogens, such as insulin or prostaglandin $F_{2\alpha}$ ($PGF_{2\alpha}$) together with EGF, induces a synergistic response in S6 phosphorylation. EGF alone or in combination with the other mitogens also induces a parallel dose response in the initiation of protein and

DNA synthesis (Thomas et al. 1982; Nielsen-Hamilton et al. 1982; Chambard et al. 1983). At saturating concentrations of EGF, the increase in S6 phosphorylation can be detected within minutes, reaching a maximum by 60 min. At this point most of S6 is fully phosphorylated, such that on two-dimensional polyacrylamide gels all the protein has shifted to dervatives S6d and e, containing 4 to 5 mol of phosphate per mol of S6, respectively (Novak-Hofer and Thomas 1985). The phosphates are added to S6 in a specific order and recent evidence in vitro suggests that it is the interaction of the substrate and a single kinase (see below), rather than the sequential activation of a number of enzymes, which dictates the order of phosphorylation (Martin-Pérez et al. 1986). Between 1 and 2 h post-induction, S6 phosphorylation begins to slowly decrease, and by 3 to 4 h basal levels are restored. The activation of protein synthesis, though kinetically delayed, parallels the increase in S6 phosphorylation but remains persistently activated (Novak-Hofer and Thomas 1985). Recent evidence suggests that it is not only the extent of S6 phosphorylation which determines the eventual level to which protein synthesis rises, but also the length of time for which S6 remains phosphorylated. Thus, S6 phosphorylation seems to be important for triggering some early response in the activation of protein synthesis, but does not appear to be required for its maintenance.

EGF-Activated S6 Kinase

High speed extracts prepared from EGF-stimulated 3T3 cells are tenfold more efficient in phosphorylating S6 in vitro than comparable extracts prepared from quiescent cells. Extracts from cells treated with increasing concentrations of EGF closely mimicked the dose-response curve for S6 phosphorylation in the intact cell (Novak-Hofer and Thomas 1984). Comparison of the tryptic phosphopeptide pattern derived in vitro with the in vivo maps of S6 show that they are equivalent. Within 5 min of EGF treatment the S6 kinase is 80% activated, reaching a maximal activity between 15 and 30 min. The activity then slowly declines in a manner similar to that observed for S6 phosphorylation, returning to basal levels by 2 h post-induction. The kinetics for S6 kinase inactivation and for S6 dephosphorylation are very similar to those reported for the loss of EGF binding (Carpenter 1981; Novak-Hofer and Thomas 1985). Indeed, removal of EGF leads to an immediate loss of kinase activity and to the dephosphorylation of S6. Thus, the level to which the S6 kinase and S6 phosphorylation are activated appears to be closely controlled by the number of ligand-occupied EGF receptors on the cell surface.

Phosphatase Inhibitors

In searching for an increased S6 kinase activity, it was discovered that the presence of β-glycerol phosphate, a phosphatase inhibitor, was required during the preparation of the extract in order to recover full S6 kinase activity (Novak-Hofer and Thomas 1984). More recently, a number of other laboratories have also employed β-glycerol phosphate in their extraction procedure and have described the presence of a similar activity (Blenis et al. 1985; Erikson and Maller 1985; Tabarini et al. 1985). The idea that β-glycerol phosphate is acting as a phosphatase inhibitor was first suggested by the fact that it could be replaced by a number of other phosphatase inhibitors. The most efficient of these was phosphotyrosine followed by p-nitrophenolphosphate, β-glycerol phosphate, phosphoserine and finally sodium orthovanadate. Sodium fluoride was ineffective (Novak-Hofer and Thomas 1985). Secondly, whole cell extracts or high speed supernatants, when incubated at 30°C, showed a time-dependent loss of S6 kinase activity which was blocked at any point by the addition of the phosphatase inhibitor (Novak-Hofer and Thomas 1985; Ballou and Thomas, unpublished). Taken together, these results suggest that the S6 kinase or some regulatory component of the kinase is controlled by phosphorylation. In this regard, it is tempting to speculate that the inactivation of the S6 kinase is due to a specific phosphatase whose activity may also be regulated (Fig. 1). The role of phosphatases in controlling S6 kinase activity should be tested directly by employing purified phosphotyrosyl- and phosphoseryl-threonyl-protein phosphatases.

Regulation of S6 Phosphorylation

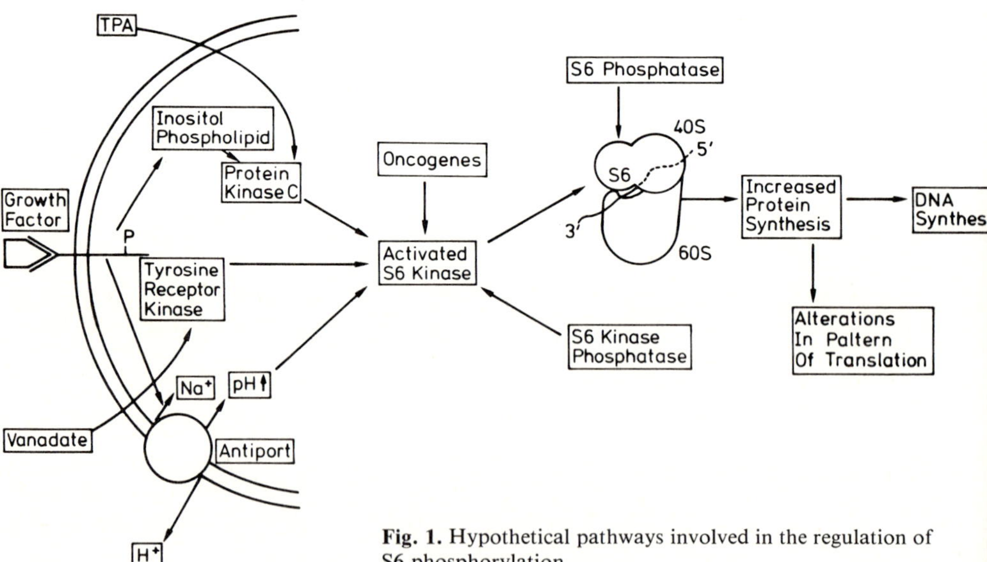

Fig. 1. Hypothetical pathways involved in the regulation of S6 phosphorylation

Activation of S6 Kinase by Sodium Orthovanadate

Sodium orthovanadate (referred to as vanadate), a phosphotyrosyl-protein phosphatase inhibitor, is a potent mitogen either alone or in the presence of other growth factors (Carpenter 1981; Smith 1983; Klarlund 1985). In the presence of EGF it has no effect on EGF binding or downregulation of EGF receptors. In searching for agents which could activate the S6 kinase at a post-receptor level, it was found that vanadate strongly activated the enzyme at concentrations as low as 10 μM, with maximal activation at 4 mM (Novak-Hofer and Thomas 1985). In contrast to EGF, the onset of activation is much slower, reaching a maximum between 30 and 60 min and remaining persistently activated. In the presence of both EGF and vanadate, under conditions where EGF receptors downregulate, the kinase also remains persistently activated. These findings imply that the kinase can be activated under conditions where the EGF receptor has downregulated. Indeed, addition of vanadate to cells which have been pretreated with EGF for 4 h and which do not respond to the further addition of EGF leads to the immediate reactivation of the S6 kinase and S6 phosphorylation (Novak-Hofer and Thomas 1985). Thus, vanadate in some way circumvents the downregulation of the EGF receptor and activates the S6 kinase. How this process might take place is discussed below.

Characteristics of the S6 Kinase

The results discussed above suggest that EGF and vanadate activate a single common enzyme. Analysis of EGF- and vanadate-stimulated cell extracts by anion exchange chromatography shows that both agents activate an enzyme which elutes at 0.34 M salt. The increase within 30 min is about tenfold relative to quiescent cell extracts. The activity elutes in a distinct position from that of the catalytic subunit of the cyclic AMP-dependent protein kinase and the phospholipid activated protein kinase C. Furthermore, addition of either the cyclic AMP-dependent protein kinase inhibitor PKI or phospholipid to the assay has no effect on the S6 kinase. The protein has an apparent molecular weight of 75,000 and a pI of 5.5. Incubation of the enzyme under optimal conditions leads to full phosphorylation of S6. In addition to EGF and vanadate, the tumor promoter TPA, insulin, platelet-derived growth factor, PGF-2α, serum and the oncogenes *ras* and *src* all activate the same enzyme, although the extent and kinetics of activation are quite different (Novak-Hofer, Luther, Siegmann, Friis and Thomas, unpublished). Many of these mitogenic agents are thought to use distinct signaling systems for stimulating growth and should therefore serve as useful probes in elucidating the mechanisms involved in the activation of the S6 kinase and S6 phosphorylation (see below).

Mechanisms Involved in Regulating S6 Phosphorylation

All the agents listed above lead to increased S6 phosphorylation and to the activation of the S6 kinase. A number of signaling systems (see other articles in this chapter) have been implicated in this process, including changes in pH, inositol phospholipid breakdown, and phosphorylation cascades initiated by the growth-factor receptor (see Fig. 1). For example, EGF might exert its effect through the receptor phosphorylation cascade, possibly activating a serine kinase such as the S6 kinase by tyrosine phosphorylation. EGF apparently does not work through protein kinase C, since it does not appear to induce the breakdown of inositol phospholipids. In contrast, PDGF may employ both the protein kinase C pathway (which would predict that this enzyme can activate the S6 kinase) and the pathway used by EGF. In this regard it is interesting that TPA, which presumably acts directly on protein kinase C, activates the S6 kinase to the same extent as EGF, whereas PDGF, which can use both pathways, induces twice the level of enzyme activity than either of the two agents alone. This raises the possibility that the S6 kinase may, under certain conditions, become multiply phosphorylated. Antibodies against the protein would greatly facilitate studies of this question.

From the data obtained with EGF and vanadate it is also clear that in addition to the S6 kinase and the different pathways involved in its activation, there are at least two phosphatases involved in controlling the extent of S6 phosphorylation, one of which operates on the kinase and the other on S6 itself (Fig. 1). These two activities appear to be independent. Certain conditions are known to induce a rapid shut-off of the kinase and dephosphorylation of S6; however, it is not clear whether this is due to increased activity of either of these phosphatases. It has been proposed that vanadate might activate cell growth through inhibiting phosphotyrosyl-protein phosphatases (Smith 1983); certainly the kinetics of vanadate-induced S6 kinase activation are consistent with such a model. To progress in our understanding of the mechanisms involved in controlling S6 phosphorylation it will be essential to purify the kinases and phosphatases involved in controlling this event.

References

Blenis J, Sugimoto Y, Biemann HP, Erikson RL (1985) Analysis of S6 phosphorylation in quiescent cells stimulated with serum growth factors, a tumor promoter, or by expression of the *Rous* sarcoma virus-transforming gene product. In: Feramisco J, Ozanne B, Stiles C (eds) Cancer cells: growth factors and transformation. Cold Spring Harbor Press, Cold Spring Harbor, NY, pp 381–388

Burkhard SJ, Traugh JA (1983) Changes in ribosome function by cAMP-dependent and cAMP-independent phosphorylation of ribosomal protein S6. J Biol Chem 258:14003–14008

Carpenter G (1981) Vanadate, epidermal growth factor and the stimulation of DNA synthesis. Biochem Biophys Res Commun 102:1115–1121

Chambard JC, Franchi A, LeCam A, Pouysségur J (1983) Growth factor-stimulated protein phosphorylation in G_0/G_1-arrested fibroblasts. J Biol Chem 258:1706–1713

Collatz E, Wool IG, Lin A, Stöffler G (1976) The isolation of eukaryotic ribosomal proteins. J Biol Chem 251:4666–4672

Decker S (1981) Phosphorylation of ribosomal protein S6 in avian sarcoma virus-transformed chicken embryo fibroblasts. Proc Natl Acad Sci USA 78:4112–4115

Duncan R, McConkey E (1982) Preferential utilization of phosphorylated 40S ribosomal subunits during initiation complex formation. Eur J Biochem 123:535–538

Erikson E, Maller JL (1985) A protein kinase from *Xenopus* eggs specific for ribosomal protein S6. Proc Natl Acad Sci USA 82:742–746

Gressner AM, Wool IG (1974) The phosphorylation of liver ribosomal proteins in vivo. J Biol Chem 249:6917–6925

Hallberg R, Wilson P, Sutton C (1981) Regulation of ribosome phosphorylation and antiobiotic sensitivity in tetrahymena thermophilia: a correlation. Cell 26:47–56

Kisilevsky R, Treloar MA, Weiler L (1984) Ribosome conformational changes associated with protein S6 phosphorylation. J Biol Chem 259:1351–1356

Klarlund JK (1985) Transformation of cells by an inhibitor of phosphatases acting on phosphotyrosine in proteins. Cell 41:707–717

Martin-Pérez J, Thomas G (1983) Ordered phosphorylation of 40S ribosomal protein S6 after serum stimulation of quiescent 3T3 cells. Proc Natl Acad Sci USA 80:926–930

Martin-Pérez J, Siegmann M, Thomas G (1984) EGF, $PGF_{2\alpha}$ and insulin induce the phosphorylation of identical S6 peptides in Swiss mouse 3T3 cells: effect of cAMP on early sites of phosphorylation. Cell 36:387–394

Martin-Pérez J, Rudkin BB, Siegmann M, Thomas G (1986) Activation of ribosomal protein S6 phosphorylation during meiotic maturation of *Xenopus laevis* oocytes: In vitro ordered appearance of S6 phosphopeptides. EMBO J 5:725–731

Nielsen PJ, Thomas G, Maller JL (1982) Increased phosphorylation of ribosomal protein S6 during meiotic maturation of *Xenopus* oocytes. Proc Natl Acad Sci USA 79:2937–2941

Nielsen-Hamilton M, Hamilton RT, Allen WR, Polter-Perigo S (1982) Synergistic stimulation of S6 ribosomal protein phosphorylation and DNA synthesis by epidermal growth factor and insulin in quiescent 3T3 cells. Cell 31:237–242

Novak-Hofer I, Thomas G (1984) An activated S6 kinase in extracts from serum- and epidermal growth factor-stimulated Swiss 3T3 cells. J Biol Chem 259:5995–6000

Novak-Hofer I, Thomas G (1985) Epidermal growth factor-mediated activation of an S6 kinase in Swiss mouse 3T3 cells. J Biol Chem 260:10314–10319

Rudland PS, Weil S, Hunter AR (1975) Changes in RNA metabolism and accumulation of presumptive messenger RNA during transition from the growing to the quiescent state of cultured mouse fibroblasts. J Mol Biol 96:745–766

Smith JB (1983) Vanadium ions stimulate DNA synthesis in Swiss mouse 3T3 and 3T6 cells. Proc Natl Acad Sci USA 80:6162–6166

Tabarini D, Heinrich J, Rosen O (1985) Activation of S6 kinase activity in 3T3-L1 cells by insulin and phorbol ester. Proc Natl Acad Sci USA 82:4369–4373

Thomas G, Siegmann M, Kubler AM, Gordon J, Jimenez de Asua L (1980) Regulation of 40S ribosomal protein S6 phosphorylation in Swiss mouse 3T3 cells. Cell 19:1015–1022

Thomas G, Thomas G, Luther H (1981) Transcriptional and translational control of cytoplasmic proteins after serum stimulation of quiescent Swiss 3T3 cells. Proc Natl Acad Sci USA 78:5712–5716

Thomas G, Martin-Pérez J, Siegmann M, Otto AM (1982) The effect of serum, EGF, $PGF_{2\alpha}$ and insulin on S6 phosphorylation and the initiation of protein and DNA synthesis. Cell 30:235–242

Wettenhall REH, Chesterman CN, Walker T, Morgan FJ (1983) Phosphorylation sites for ribosomal S6 protein kinases in mouse 3T3 fibroblasts stimulated with platelet-derived growth factor. FEBS Lett 162:171–176

Role of G Proteins in Transmembrane Signaling: Possible Functional Homology with the *ras* Proteins

Susan B. Masters and Henry R. Bourne

The G proteins comprise a family of membrane-associated proteins that transduce extracellular signals such as hormones or photons into a diverse array of cellular responses. All these proteins bind and hydrolyze GTP, and GTP in turn regulates their interactions with signal detectors (cell surface receptors) and effectors (membrane-associated enzymes and ion channels). Each G protein utilizes a common GTP-dependent mechanism to transduce signals between unique sets of detector and effector elements.

The best-characterized members of the G protein family include transducin-1 (T1), G_s, and G_i. T1, the G protein of the rod cell visual transduction system, couples photorhodopsin to a cyclic GMP phosphodiesterase (PDE) (Stryer 1986). G_s and G_i stimulate and inhibit, respectively, adenylate cyclase (AC) in response to activation of different sets of hormone receptors (Gilman 1984). G_i or a G_i-like protein regulates the activity of phospholipase C (PLC) in neutrophils and mast cells, an inwardly rectifying potassium channel in heart cells and voltage-dependent calcium channels in dorsal root ganglion neurons (reviewed by Stevens 1986). G_o, a GTP-binding protein abundant in brain, also belongs to the G protein family. Although G_o structurally resembles G_i and T1 (Hurley et al. 1984b), its function is unknown. Transducin-2 (T2), a second retinal G protein, mediates phototransduction in cone cells (Grunwald et al. 1986).

Several lines of evidence suggest that the 21 kDa protein products of the *ras* proto-oncogenes ($p21^{ras}$) are related to the signal transducing family of G proteins. Our knowledge of the structure and function of the G proteins prompts speculative comparisons of their biochemical roles with those of the cellular $p21^{ras}$ proteins and their transforming counterparts.

Structure of the G Proteins

All the G proteins are heterotrimers. The α-subunits bind and hydrolyze GTP and interact with effector elements; they are distinguished by their molecular weights and their differing susceptibilities to ADP-ribosylation by bacterial exotoxins, cholera toxin (CT) and pertussis toxin (PT) (Table 1). The primary sequences of the α-subunits of T1, T2 (for references, see Sugimoto et al. 1985), G_s and G_i (unpublished results) have been deduced from the corresponding cDNA sequences. Sequence comparison reveals strong overall

Oncogenes and Growth Control
Edited by P. Kahn and T. Graf
© Springer-Verlag Berlin Heidelberg 1986

Table 1. Biochemical characteristics of the G proteins

G protein	Subunit molecular weights	Susceptibility to ADP-ribosylation by		Detector	Effector
		CT	PT		
T1	α 39 β 36 γ 8	Yes	Yes	Photorhodopsin	$\uparrow$PDE
G_s	α 45, 53 β 35, 36 γ 6−10	Yes	No	Hormone receptors	$\uparrow$AC
G_i	α 41 β 35, 36 γ 6−10	No	Yes	Hormone receptors chemotactic receptors	$\downarrow$AC $\uparrow$PLC ? K^+ channel ? Ca^{2+} channel
G_0	α 39 β 35, 36 γ 6−10	No	Yes	Hormone receptors	?$\downarrow$AC ?$\uparrow$PLC ? K^+ channel ? Ca^{2+} channel

homology, especially in regions believed to participate in GTP binding and hydrolysis.

The G protein β- and γ-subunits, inseparable in the native state, function biochemically as a $\beta\gamma$-subunit which anchors the more hydrophilic α-subunits to the plasma membrane (Sternweis 1986). The β-subunits of G_s, G_i and G_o migrate on SDS acrylamide gels as a $35-36$ kDa doublet, while β_t migrates as a single band at 36 kDa (Roof et al. 1985). Peptide maps, immunoreactivity, and cDNA sequences indicate that the 36 kDa β-subunits of G_s, G_i, and T are identical (for references see Roof et al. 1985; Sugimoto et al. 1985). Similar evidence indicates that the 74-residue γ_t-chain, expressed only in the retina, differs from the $6-10$ kDa γ-chains of G_s, G_i and G_o (for references see Hildebrandt et al. 1985). In spite of γ-chain differences, $\beta\gamma_i$ and $\beta\gamma_t$ function interchangeably in supporting photorhodopsin-stimulated hydrolysis of GTP by α_{t1} in reconstitution experiments (Kanaho et al. 1984). Because $\beta\gamma_i$ and $\beta\gamma_o$ are more hydrophobic than $\beta\gamma_t$ (Sternweis 1986), it is likely that the γ subunits are probably largely responsible for anchoring G proteins to the cell membrane.

G Protein Function

Investigations of light activation of rod cell PDE and hormonal regulation of adenylate cyclase provide a general scheme for the function of G proteins in signal transduction (Fig. 1). Activation of the detector element (by light or

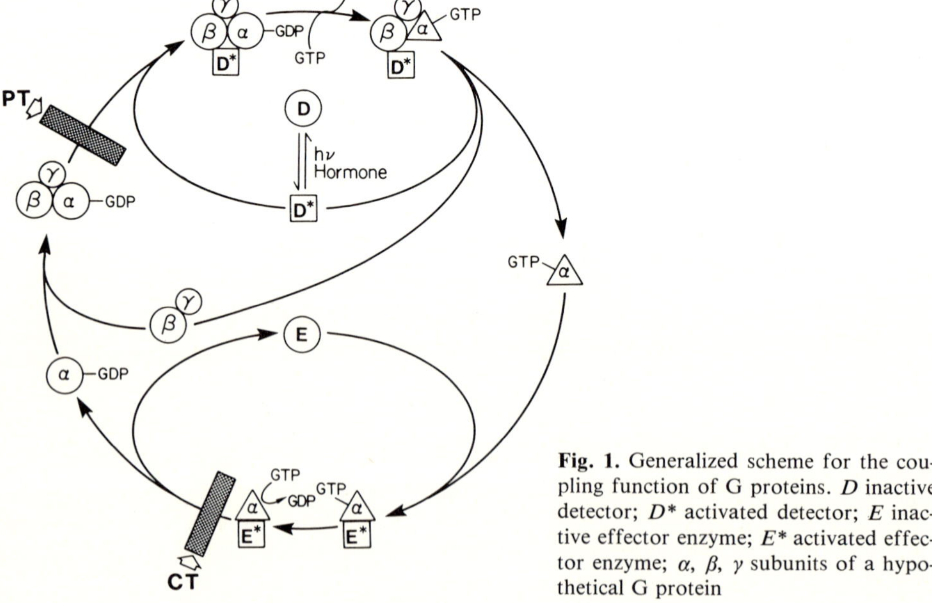

Fig. 1. Generalized scheme for the coupling function of G proteins. D inactive detector; D^* activated detector; E inactive effector enzyme; E^* activated effector enzyme; α, β, γ subunits of a hypothetical G protein

hormone) promotes binding of detector to $\alpha\beta\gamma$ and markedly accelerates the rate of exchange of GTP for bound GDP. Binding of GTP causes dissociation of α, $\beta\gamma$, and detector. Free α-GTP then binds to and activates the effector enzyme. Effector activation terminates when α hydrolyzes the GTP to GDP. The cycle continues until the activated detector becomes inactivated.

Toxin-catalyzed ADP ribosylation of the α-subunits of the G proteins specifically alters their function. CT preferentially ADP-ribosylates α_s and α_{t1} in the GTP-bound state; once modified, the α-subunits are stabilized in the GTP-bound conformation (Kahn and Gilman 1984). CT treatment of both α_s and α_{t1} leads to persistent effector activation. In contrast, PT preferentially ADP-ribosylates the GDP-bound, $\alpha\beta\gamma$ form of G_i and T1. PT treatment interrupts signal transduction by preventing the heterotrimer from binding activated detector (Van Dop et al. 1984).

Comparison of p21^{ras} Proteins and the G Proteins

Mammalian and viral p21^{ras} proteins resemble the G proteins in several ways, although detector and effector elements for p21^{ras} proteins have not been identified. Like the G proteins, the *ras* gene products are located on the cytoplasmic face of the plasma membrane. They also bind and hydrolyze GTP at rates comparable to those of the G proteins (Gibbs et al. 1984; Sweet et al. 1984). In addition, like the G proteins, p21^{ras} is active in its GTP-bound form. Several spontaneous or engineered mutations in the cellular *ras* genes increase the capacity of the mutant protein to promote malignant transformation and reduce the rate of GTP hydrolysis without decreasing GTP binding (see Marshall and Fasano, both this Vol.). Furthermore, antibodies directed against the putative GTP-binding domain of p21^{ras} block GTP binding and also prevent malignant transformation when they are injected into cells transformed with a *ras* oncogene (Clark et al. 1985). As with the G proteins, stabilization of p21^{ras} in the GTP-bound state enhances activity, while prevention of GDP-GTP exchange appears to decrease activity.

The primary structures of the p21^{ras} proteins, the G proteins and bacterial elongation factor Tu (EF-Tu) exhibit four short regions of homology (Table 2). X-ray crystallographic studies of GDP-bound EF-Tu demonstrate that these four regions, all flexible loops between β strands and α-helices, form the GDP binding site (Jurnak 1985). In EF-Tu, region 1 serves as a small binding pocket for the two phosphoryls of GDP. Lys24, conserved in all these polypeptides, interacts with the β-phosphoryl. Region 1 also contains a glycine residue that corresponds to gly12 of p21^{ras}. When this glycine is changed to a valine, the mutant p21 exhibits decreased GTPase activity and enhanced transforming capability. Interestingly, the corresponding residue is glycine in all the G proteins, but is valine in EF-Tu. Region 1 probably plays an important role in GTP hydrolysis.

Table 2. Conserved regions of G protein α-chains, c-Ha-*ras* and EF-Tu[a]

Region 1:

												1										
Gs	37:	A	T	H	R	L	L	L	L	G	A	G	E	S	G	K	S	T	I	:54
Gi	32:	R	E	V	K	L	L	L	L	G	A	G	E	S	G	K	S	T	I	:49
T1	29:	–	T	V	K	L	L	L	L	G	A	G	E	S	G	K	S	T	I	:45
T2	32:	K	T	V	K	L	L	L	L	G	A	G	E	S	G	K	S	T	I	:49
cHras	2:	T	E	Y	K	L	V	V	V	G	A	G	G	V	G	K	S	A	L	:19
EF-Tu	10:	P	H	V	N	V	G	T	I	G	H	V	D	H	G	K	T	T	L	:27

Region 2:

					2				
Gs	205:	F	D	V	G	G	Q	R	:211
Gi	200:	F	D	V	G	G	Q	R	:206
T1	195:	F	D	V	G	G	Q	R	:201
T2	199:	F	D	V	G	G	Q	R	:205
cHras	56:	L	D	T	A	G	Q	E	: 62
EF-Tu	79:	V	D	C	P	G	H	A	: 85

Region 3:

						3				
Gs	227:	I	I	F	V	V	A	S	S	S	:235
Gi	222:	I	I	F	C	V	A	L	S	A	:230
T1	217:	I	I	F	I	A	A	L	S	A	:225
T2	221:	I	I	F	C	A	A	L	S	A	:229
cHras	78:	F	L	C	V	F	A	I	N	N	: 86
EF-Tu	101:	A	I	L	V	V	A	A	T	–	:108

Region 4:

						4					
Gs	270:	V	I	L	F	L	N	K	Q	D	L	:279
Gi	265:	I	I	L	F	L	N	K	K	D	L	:274
T1	260:	I	V	L	F	L	N	K	K	D	V	:269
T2	264:	I	V	L	F	L	N	K	K	D	L	:273
cHras	112:	V	L	V	–	G	N	K	C	D	L	:120
EF-Tu	130:	I	I	V	F	L	N	K	C	D	M	:139

[a] Boxes include identical or conserved residues. Sequences are those of the α-subunits of G_s, G_i, T1, T2, EF-Tu, and c-Ha-*ras*. Primary sequences of c-Ki-*ras* and N-*ras* are identical to c-Ha-*ras* in these four regions. The single letter code for amino acids is: A, alanine; C, cysteine; D, aspartate; E, glutamate; F, phenylalanine; G, glycine; H, histidine; I, isoleucine; K, lysine; L, leucine; M, methionine; N, asparagine; P, proline; Q, glutamine; R, arginine; S, serine; T, threonine; V, valine; W, tryptophan; Y, tyrosine.

Region 2 contains a conserved aspartate (asp80 in EF-Tu) whose side chain forms a salt bridge with a Mg^{2+} ion in EF-Tu (Jurnak 1985). This region may be involved in GTPase activity; substitution of threonine for ala59 or leucine for glu61 in $p21^{ras}$ reduces the rate of GTP hydrolysis by these activated $p21^{ras}$ proteins (Gibbs et al. 1984; Temeles et al. 1985). Region 3 is a strikingly hydrophobic region that forms part of the binding pocket for GDP in EF-Tu (Jurnak 1985). The number of residues between regions 2 and 3 is constant for all these proteins. Region 4 contains an invariant leucine and asparagine that form part of the binding pocket for the guanine ring in EF-Tu (Jurnak 1985). Aspartate138 in EF-Tu, another invariant residue, interacts with the amino constituent on the guanine ring of GDP (Jurnak 1985).

In addition, the carboxy termini of the p21^{ras} proteins exhibit homology with the γ subunit of T1 (Hurley et al. 1984a). The fourth residue from the C-terminus in all these polypeptides is a cysteine. In p21^{ras} this cysteine is a site for fatty acylation, a covalent modification that enables p21^{ras} to attach to the plasma membrane (Willumsen et al. 1984). By analogy, acylation of the cysteine residue in γ subunits may contribute to the anchoring function of the G protein $\beta\gamma$ complex.

Possible Roles of p21^{ras} Proteins in Signal Transduction

Saccharomyces cerevisiae contains two RAS proteins that show extensive homology in their amino termini with mammalian *ras* proteins (see Fasano, this Vol.). The yeast RAS proteins regulate yeast AC in a GTP-dependent manner. p21^{ras} does not stimulate or couple to mammalian AC (for references, see Bourne 1985), although it can support GTP-dependent AC activity when expressed in RAS-deficient yeast cells. The latter finding suggests that the binding site of yeast AC for activation by p21^{ras} and yeast RAS proteins resembles the corresponding binding site of the putative mammalian effector(s) of p21^{ras}.

Each of the three distinct p21^{ras} proteins may couple to a different effector. For two reasons, hormonally regulated PLC is a potential effector candidate. Firstly, hormonal activation of PLC by a G_i-like receptor-effector coupling protein produces two intracellular messengers thought to be involved in regulation of cell proliferation by peptide growth factors (Berridge and Irvine 1984). The two messengers are inositol trisphosphate, which mobilizes intracellular calcium, and diacylglycerol, which stimulates protein kinase C (see Berridge and Parker and Ullrich, both this Vol.). Two reports (Chiarugi et al. 1985; Fleischman et al. 1986) show that transformation of several cell lines with viral *ras* genes enhances the turnover of phosphoinositides. Confirmation of a role of p21^{ras} in hormonal stimulation of PLC will require demonstration of a direct effect of p21^{ras} in a reconstituted system. By analogy to the G proteins, other potential effectors for the *ras* proteins include ion channels and enzymes that regulate cyclic nucleotide metabolism.

Evidence for a $\beta\gamma$ component in the *ras* system is lacking. Because detergent is required to solubilize p21^{ras} proteins, and because *ras* proteins show homology with the γ subunit of T, p21^{ras} may perform functions assigned to the $\beta\gamma$ subunits of the G proteins.

What type of signals or detectors could activate p21^{ras}? The ability of mutationally activated and viral p21^{ras} to induce neoplastic transformation suggests that the cellular homologs of the viral p21^{ras} proteins may transduce signals which promote cell proliferation. If so, we might expect the *ras* proteins to mediate signals received from peptide growth factors such as epidermal growth factor or platelet-derived growth factor. Until an effector response is identified, it will be extremely difficult to determine which, if any, cell surface receptors couple to the *ras* proteins.

How Useful Is the G Protein Analogy?

The evidence summarized here suggests that *ras* oncogene products and their normal counterparts may act as GTP-dependent signal transducers analogous to the G proteins. Lessons from the G protein field should help direct investigations of the role of mammalian p21^{ras}. For example, if the model is correct, a detector element that couples to p21^{ras} should stimulate GTP binding and GTPase activity of *ras* proteins.

It is possible, however, that p21^{ras} does not transduce extracellular signals at all. Instead, its cellular role may more closely resemble those of GTP-binding proteins, including EF-Tu, that function in ribosomal protein synthesis. These elongation and initiation factors do not seem to transduce regulatory signals. Instead, they act to promote association of other molecules (Kaziro 1978). For example, the GTP-bound form of EF-Tu is required for association of aminoacyl-tRNA and mRNA at the acceptor site on the ribosome; following GTP hydrolysis, EF-Tu dissociates from the aminoacyl-tRNA-mRNA-ribosome complex and a new peptide bond forms. It is possible that p21^{ras} similarly promotes assembly of a macromolecular complex in cell membranes or juxtaposes cytoplasmic or cytoskeletal proteins with membrane proteins. These intriguing possibilities remain to be investigated.

The G protein, p21^{ras} protein and elongation factor families probably evolved from an ancestral GTP binding protein. The common theme of harnessing the binding and hydrolysis of GTP to repetitive cycles of association and dissociation of protein molecules features in widely divergent processes, including ribosomal protein synthesis, signal transduction, and regulation of cell growth. Although the G and *ras* proteins are close relatives, their functional homologies are not yet defined.

Acknowledgements. Work from this laboratory discussed in this review was supported by grants from the National Institutes of Health and the March of Dimes.

References

Berridge MJ, Irvine RF (1984) Inositol trisphosphate, a novel second messenger in signal transduction. Nature 312:315−321
Bourne HR (1985) Transducing proteins. Yeast RAS and Tweedledee's logic. Nature 317:16−17
Chiarugi V, Porciatti F, Pasquali F, Bruni P (1985) Transformation of Balb/3T3 cells with EJ/T24/H-*ras* oncogene inhibits adenylate cyclase response to β-adrenergic agonist while increases muscarinic receptor dependent hydrolysis of inositol lipids. Biochem Biophys Res Commun 132:900−907

Clark R, Wong G, Arnheim N, Nitecki D, McCormick F (1985) Antibodies specific for amino acid 12 of the *ras* oncogene product inhibit GTP binding. Proc Natl Acad Sci USA 82:5280–5284

Fleischman LF, Chahwala SB, Cantley L (1986) *Ras*-transformed cells: Altered levels of phosphatidylinositol-4,5-bisphosphate and catabolites. Science 231:407–410

Gibbs JB, Sigal IS, Poe M, Scolnick EM (1984) Intrinsic GTPase activity distinguishes normal and oncogenic *ras* p21 molecules. Proc Natl Acad Sci USA 81:5704–5708

Gilman AG (1984) G proteins and dual control of adenylate cyclase. Cell 36:577–579

Grunwald GB, Gierschik P, Nirenberg M, Spiegel A (1986) Detection of α-transducin in retinal rods but not cones. Science 231:856–860

Hildebrandt JD, Codina J, Rosenthal W, Birnbaumer L, Neer EJ, Yamazaki A, Bitensky MW (1985) Characterization by two-dimensional peptide mapping of the γ subunits of N_s and N_i, the regulatory proteins of adenylyl cyclase, and of transducin, the guanine nucleotide-binding protein of rod outer segments of the eye. J Biol Chem 260:14867–14872

Hurley JB, Fong HKW, Teplow DB, Dreyer WJ, Simon MI (1984a) Isolation and characterization of a cDNA clone for the γ subunit of bovine retinal transducin. Proc Natl Acad Sci USA 81:6948–6952

Hurley JB, Simon MI, Teplow DB, Robishaw JD, Gilman AG (1984b) Homologies between signal transducing G proteins and *ras* gene products. Science 226:860–862

Jurnak F (1985) Structure of the GDP domain of EF-Tu and location of the amino acids homologous to *ras* oncogene proteins. Science 230:32–36

Kahn RA, Gilman AG (1984) ADP-ribosylation of G_s promotes the dissociation of its α and β subunits. J Biol Chem 259:6235–6240

Kanaho Y, Tsai SC, Adamik R, Hewlett EL, Moss J, Vaughan M (1984) Rhodopsin-enhanced GTPase activity of the inhibitory GTP-binding protein of adenylate cyclase. J Biol Chem 259:7378–7381

Kaziro Y (1978) The role of guanosine 5′-triphosphate in polypeptide chain elongation. Biochim Biophys Acta 505:95–127

Roof DS, Applebury ML, Sternweis PC (1985) Relationships within the family of GTP-binding proteins isolated from bovine central nervous system. J Biol Chem 260:16242–16249

Sternweis PC (1986) The purified α subunits of G_0 and G_i from bovine brain require $\beta\gamma$ for association with phospholipid vesicles. J Biol Chem 261:631–637

Stevens CF (1986) Neurotransmitters. Modifying channel function. Nature 319:622

Stryer L (1986) Cyclic GMP cascade of vision. Annu Rev Neurosci 9:87–119

Sugimoto K, Nukada T, Tanabe T, Takahashi H, Noda M, Minamino N, Kangawa K, Matsuo H, Hirose T, Inayama S, Numa S (1985) Primary structure of the β-subunit of bovine transducin deduced from the cDNA sequence. FEBS Lett 191:235–240

Sweet RW, Yokoyama S, Kamata T, Feramisco JR, Rosenberg M, Gross M (1984) The product of *ras* is a GTPase and the T24 oncogenic mutant is deficient in this activity. Nature 311:273–275

Temeles GL, Gibbs JB, D'Alonzo JS, Signal IS, Scolnick EM (1985) Yeast and mammalian *ras* proteins have conserved biochemical properties. Nature 313:700–703

Van Dop C, Yamanaka G, Steinberg F, Sekura RD, Manclark CR, Stryer L, Bourne HR (1984) ADP-ribosylation of transducin by pertussis toxin blocks the light-stimulated hydrolysis of GTP and cGMP in retinal photoreceptors. J Biol Chem 259:23–26

Willumsen BM, Norris K, Papageorge AG, Hubbert NL, Lowy DR (1984) Harvey murine sarcoma virus p21 *ras* protein: biological and biochemical significance of the cysteine nearest the carboxy terminus. EMBO J 3:2581–2585

The *ras* Gene Family

Christopher J. Marshall

The *ras* gene family in mammalian cells consists of five closely related members: the Harvey and Kirsten *ras* genes (c-Ha-*ras1* and c-Ki-*ras2*), an inactive pseudogene of each (c-Ha-*ras2* and c-Ki-*ras1*; see Hall 1984) and the N-*ras* gene. There are also several distinctly related genes designated *rho* (ras-homologous; Madaule and Axel 1985). In addition to their sequence homology, the genes of the *ras* family are related by the fact that they encode protein products of approximately 21 kDa (p21^{ras}) which share the properties of GTP/GDP binding, weak GTPase activity and the presence of C-terminal cysteine residues (potential palmitation sites for membrane localization). In the past several years, a growing body of evidence has implicated the *ras* genes as playing a role in a wide variety of human cancers. The evidence comes largely from studies indicating that approximately 10–20% of most types of human malignancies contain an "activated" *ras* oncogene which is capable of inducing transformed foci following transfer into recipient cells such as mouse NIH-3T3. This chapter discusses three aspects of the *ras* genes: gene structure and expression, mechanism of activation in human tumours, and biochemistry of the p21^{ras} protein. Additional aspects are covered in the chapters by Fasano and by Masters and Bourne in this Volume.

Structure and Expression of *ras* Family Genes

The c-Ki-*ras2*, c-Ha-*ras1* and N-*ras* genes encode polypeptides containing 188 or 189 amino acids (aa) derived from four exons (see Fig. 1). The c-Ki-*ras2* gene contains two fourth exons; alternative splicing yields messages containing one or the other exon (McGrath et al. 1983). The three introns between the coding exons have very different sizes in the different *ras* genes, but interrupt the coding sequences at the same points. Due to the presence of additional introns in the untranslated portions of the c-*ras* genes and to variations in intron length, these genes have very different sizes (approximately 40 kb, 6.6 kb and 14 kb, respectively).

Further analysis of the c-Ha-*ras1* and N-*ras* genes has revealed some features of their transcription. Both genes have multiple transcription initiation sites which lack upstream TATA and CAAT boxes (Hall and Brown 1985; Ishi et al. 1985). However, the promoter region contains multiple copies of the GC box (GGGCGG) and its inverted repeat. These boxes appear to be charac-

Oncogenes and Growth Control
Edited by P. Kahn and T. Graf
© Springer-Verlag Berlin Heidelberg 1986

```
N-ras       MTEYKLVVVGAGGVGKSALTIQLIQNHFVDEYDPTIEDSYRKQVVIDGET      50
c-Ha-ras1   ..................................................
c-Ki-ras2   ..................................................
v-Ha-ras    ...............R..................................
v-Ki-ras    ...........S........................Q.............

N-ras       CLLDILDTAGQEEYSAMRDQYMRTGEGFLCVFAINNSKSFADINLYREQI     100
c-Ha-ras1   ........................................T...E..HQ.....
c-Ki-ras2   ........................................T...E..HH.....
v-Ha-ras    ........T...............................T...E..HQ.....
v-Ki-ras    ........T...............................T...E..HH....L

N-ras       KRVKDSDDVPMVLVGNKCDLPTRTVDTKQAHELAKSYGIPFIQTSAKTRQ     150
c-Ha-ras1   ......................AA...ESR..QD..R.....T.........
c-Ki-ras2   ......E...............S........QD..R...............
v-Ha-ras    ......................AG...ESR..QD..R.....T.........
v-Ki-ras    ......E...............S........Q...R...............

N-ras       GVQDAFYTLVREIRQYRMKKLNSSDDGTQGCMGLPCVVMT
c-Ha-ras1   ................HKLR...PP.ESGP...SCK..LS.
c-Ki-ras2   ..D............KHKE.MSKDGKKKKKKSKT.K..I..
v-Ha-ras    ................HKLR...PP.ESGP...SCK..LS.
v-Ki-ras    R...............L..ISKQQKTPGCVKIKK..IM.
```

Fig. 1. Amino and sequences of human *ras* genes and of v-Ki-*ras* and v-Ha-*ras* genes. Dots indicate identity of amino acids to the N-*ras* sequence

teristic of genes lacking TATA boxes and may function to bind proteins involved in transcription. Whether the c-Ki-*ras2* promoter has similar features remains to be resolved. Each of the *ras* genes appears to produce multiple transcripts. The c-Ha-*ras1* gene has a major transcript of 1.2 kb as well as a 5.8 kb transcript (Shih and Weinberg 1982). The N-*ras* gene has two transcriptional starts 10 base pairs apart, and readthrough of a weak polyadenylation signal provides two routes for polyadenylation, generating 2 kb and 4 kb messages (Hall and Brown 1985). For the c-Ki-*ras2* gene, messages of 5.5, 3.8 and 1.2 kb have been detected (Capon et al. 1983; McCoy et al. 1984).

The pattern of transcription of *ras* genes in different tissues has not yet been analyzed but is of particular interest because various *ras* genes are activated at different frequencies in different tumour types. For example, the activated *ras* gene in bladder tumours is usually c-Ha-*ras1* (Fujita et al. 1985; I. C. Summerhayes, personal communication), while N-*ras* is the most frequently activated *ras* gene in acute myeloid leukemia (Bos et al. 1985; D. Toksoz, personal communication). Similar observations have been made with tumours induced experimentally by chemical carcinogens (see Balmain, this Vol.). Anal-

Table 1. Comparison of p21 amino acid sequences

	Pos. 5 – 17	Pos. 54 – 68	Pos. 114 – 121
Human *ras*	K L V V V G A G G V G K S	D I L D T A G Q E E Y S A M R	V G N K C D L
Drosoph. ras1	· · · · · · P · · · · · ·	· · · · · · · · · · · · · · ·	· · · · · · ·
Drosoph. ras2	V F C A S A · D · T I I L	· T A G H E E F S A M R E Q Y	· · · · · · ·
Dictyost. ras	· · · I · · G · · · · · ·	· · · · · · · · · · · · · · ·	· · · · A · ·
Yeast *RAS1*	· I · · · G · · · · · · ·	· · · · · · · · · · · · · · ·	· · · · L · ·
Yeast *RAS2*	· I · · · G · · · · · · ·	· · · · · · · · · · · · · · ·	· · · · S · ·
Transducin α	· · L L L · · · E S · · ·	V R A M · T L N I Q · G S · ·	P L · · K · V
Aplysia rho	· · · I · G D · A C · · T	A L W · · · · · · D · D R L ·	· · · · K · ·

Dots indicate identity of amino acids with the human N-*ras*, Ha-*ras*, and Ki-*ras* sequence in the first line

ysis of the tissue-specific pattern of *ras* gene expression might also partly explain why the human genome contains three genes which encode very similar proteins.

Comparison of *ras* genes from non-vertebrate species, including *Drosophila* (Neuman-Silberberg et al. 1984), *Dictyostelium* (Reymond et al. 1984) and yeast (Powers et al. 1984), of the *rho* genes of the sea snail *Aplysia* (Madaule and Axel 1985) and of transducin (Tanabe et al. 1985) has indicated which domains in their gene products are most highly conserved (see Table 1). Except for the yeast *RAS* genes, protein size is also conserved (approximately 21 kDa). The regions in which the human N, Ki- and Ha-*ras* genes diverge from each other (121 – 140 and 165 – 185, based on the amino acid numbering of human *ras*; see Fig. 1) are the same regions in which they diverge from the *rho* genes (Madaule and Axel 1985). The sites at which mutations activate *ras* to a transforming gene occur in two regions of homology (codons 12 – 13 and 61). These regions, together with residues 116 – 119 and 144 – 160, appear to form important components of the GTP binding site (McCormick et al. 1985; Tanabe et al. 1985).

Activation of *ras* Genes in Human Tumours

There are two ways in which *ras* proto-oncogenes can be activated: by mutations in the coding sequence, leading to an altered protein product, and by overexpression of the normal gene product. Although transformation by overexpression of the normal p21^{ras} has been clearly demonstrated in experimental situations (Chang et al. 1982), it appears to be relatively rare in human malignancies. A few human tumours have been shown to contain amplified copies of apparently unmutated *ras* genes (e.g. Fujita et al. 1985) and presumably contain elevated levels of p21. Elevated p21 expression in tumours has also been directly demonstrated at the RNA (Spandidos and Kerr 1984) and protein levels (Hand et al. 1984). However, the meaning of these results is unclear, since increased levels of p21^{ras} could merely be a consequence of higher proliferation rates in the tumour compared with normal tissue.

Activation of *ras* genes in most human tumours which have been examined appears to be due to mutations in the coding sequence. Approximately 40 of these mutant *ras* genes have been analyzed and in all cases the mutation responsible for transforming activity has been found to reside in codon 12, 13 or 61. While codon 12 and 61 mutations are readily detectable in the NIH-3T3 focus assay, mutations in codon 13 are poor at inducing foci and are more easily detected by assaying directly for tumorigenicity of the transfected NIH-3T3 cells (Bos et al. 1985; C. J. M., unpublished results). Each of the mutations is a single base change, and all possible base pair mutations have been detected. At codon 12 (GGN), all six possible amino acid replacements resulting from single base pair substitutions have been observed, while 2/6 possible replacements at codon 13 (GGT) and 4/6 at codon 61 (CAA/G) have been observed. Curiously, while nearly 50% of the mutations analyzed in c-Ki-*ras2* have been G→C transversions, this mutation has not yet been seen in N-*ras* or Ha-*ras*.

To examine the spectrum of amino acid replacements at codon 12 that can lead to transforming activity, Seeburg et al. (1984) constructed c-Ha-*ras1* genes bearing all possible mutations. Other than proline substitution, all amino acid replacements resulted in transforming activity. Similarly, 15 out of 17 amino acid substitutions at codon 61 lead to transforming activity (Der et al. 1986). Deletion of codon 12 or insertion of an extra amino acid at codon 12 also leads to transforming activity (Chipperfield et al. 1985). Taken together, these results suggest that activation is due to the loss of some regulatory function. Interestingly, the different mutations at codons 12 and 61 appear to differ in the "strength" of their transforming activity in the NIH-3T3 transfection assay (Seeburg et al. 1984; Der et al. 1986; C. Onorati-Steinman, personal communication).

Two additional sites at which mutations can lead to transforming activity have been identified by in vitro mutagenesis of the c-Ha-*ras1* proto-oncogene (Fasano et al. 1984). These sites (codons 59 and 63) are adjacent to codon 61, where naturally occurring activating mutations have been found. It is noteworthy that the activating mutation of ala(59) to thr(59) creates a phosphorylation site, one which is present in v-Ki-*ras* and v-Ha-*ras*.

Biochemistry of Normal and Mutant p21^{ras}

With our considerable knowledge of the mutations which activate *ras* genes in tumours, much interest has now shifted to studying the biochemistry of the mammalian p21^{ras} proteins. This was initially difficult because even v-*ras*-transformed cells produce only small amounts of *ras* protein ($10^4 - 10^5$ molecules/cell). This problem has recently been solved through the use of bacterial expression systems, which produce milligram amounts of protein but have the disadvantage that the protein is not acylated. However, the finding that bacterially produced *ras* protein is functional when injected into cells (e.g., Birchmeier et al. 1985) indicates that it is probably acylated after microinjection.

Earlier studies of the viral *ras* proteins showed that they are able to bind the guanine nucleotides GTP and GDP (Scolnick et al. 1979) and that they are localized at the inner surface of the plasma membrane (Willingham et al. 1980). These properties are now known to be shared by the normal and mutant cellular *ras* genes. However, the viral *ras* proteins are phosphorylated at position 59 (Shih et al. 1980), where an ala→thr substitution has occurred, while the normal and mutant c-*ras* proteins, which do not show the thr(59) substitution, are not phosphorylated. Localization at the inner surface of the cell membrane is essential to the function of both normal and transforming p21^{ras} proteins (Willumsen et al. 1984). Membrane localization requires acylation by a palmitic acid residue at cys(186), which is close to the C-terminus of the proteins. All *ras* and *ras*-related genes sequenced so far, including G protein genes, contain C-terminal cysteine residues which are potential acylation sites and occur in a motif of cys-X-X-terminus, where X is an aliphatic uncharged amino acid. The palmitic acid is added to p21^{ras} post-translationally and appears to turn over more rapidly than the half-life of the protein, which is about 20 h (T. McGhee, personal communication).

Both the normal and mutant forms of *E. coli*-produced p21^{ras} bind GTP and GDP with affinities of about 10^{-8} M. However, the normal p21$^{c-Ha-ras1}$ protein hydrolyses GTP at about 2 mmol/min/mol protein (Gibbs et al. 1984), whereas transforming *ras* proteins with val(12), arg(12), thr(59) or leu(61) show at least a tenfold reduction in GTPase activity (McGrath et al. 1984; Temeles et al. 1985). These results suggest that mutations at these sites generate a transforming protein because they lead to reduced GTPase activity. This idea is consistent with the observation that virually any amino acid substitution or deletion at codon 12 or 61 activates p21^{ras}. However, several findings raise questions which have not yet been answered about the significance of the reduction in GTPase. First, while p21^{N-ras} with val(12) or lys(61) mutations shows a seven to tenfold reduction in GTPase activity, an asp(12) mutant appears to have 60% of the wild-type activity (M. Trahey, R. Milley, F. McCormick, C. Marshall and A. Hall, unpublished results). Second, mutations at codon 61 do not show a correlation between reduced GTPase activity and potency of the transforming allele (Der et al. 1986). Third, the observation that overproduction of normal protein can lead to transformation is also not readily explainable by alterations in GTPase activity.

To understand how mutations in *ras* genes can lead to transforming activity, a knowledge of the three-dimensional structure of wild-type and mutant p21^{ras} proteins is required. Successful crystallization and X-ray crystallographic analysis of the proteins has not yet been reported; however, the strong sequence homology (40%) between p21^{ras} and the GTP binding domains of the *E. coli* protein synthesis elongation factor EF-Tu, whose crystal structure has been solved, has allowed the generation of a model of p21 structure

(McCormick et al. 1985 and chapter by Masters and Bourne). One important prediction of the model is that codons 12, 13 and 61, the sites for activating mutations, are involved in the guanine nucleotide binding region. This notion is supported by the finding that an antibody raised against a peptide spanning codon 12 abolishes GTP binding by p21^{ras} (Clark et al. 1985). It is unclear which residues are responsible for GTP hydrolysis, but mutations at codon 12 or 61 presumably lead to conformational changes which affect the hydrolysis of GTP. The idea that activating mutations produce conformational changes is supported by the fact that these mutant p21^{ras} gene products show altered electrophoretic mobility (Seeburg et al. 1984; Der et al. 1986). The model of McCormick et al. (1985) also predicts that binding of GTP to p21^{ras} creates a conformational change compared with the GDP form. It is therefore possible that the transforming mutations mimic the conformational change induced by GTP binding and thereby place the protein in a constitutively activated form. Such changes might also result in reduced GTPase activity, but this reduction would not be essential for transformation.

Physiological Functions of p21^{ras}

The sequence and functional homologies between p21^{ras} proteins and G proteins argue that p21^{ras} proteins mediate signals involved in the control of cell proliferation. One model is that binding of a particular ligand (such as a growth factor) to its receptor stimulates p21^{ras} to bind GTP and interact with an effector protein. This interaction would be terminated by hydrolysis of GTP to GDP. The transforming activity of mutants defective in GTPase could be due to a prolonged interaction with effector proteins. It is also possible that transforming p21^{ras} proteins bind GTP even in the absence of receptor binding and are therefore constitutively activated. Some evidence that p21^{ras} interacts with other proteins has been reported. First, p21^{Ha-ras} appears to interact with the EGF receptor (Kamata and Feramisco 1984). Second, genetic evidence argues that the yeast *ras* proteins interact with adenylate cyclase (see article by Fasano), although it is still not clear whether this interaction is indirect (Toda et al. 1985); moreover, it has not been seen in vertebrate systems (Beckner et al. 1985; Birchmeier et al. 1985).

Clearly, our understanding of the physiological role of the *ras* proteins is at a very preliminary stage. Further work will need to resolve questions such as which signal(s) the *ras* proteins mediate, how they interact with other molecules, and whether the different *ras* proteins have separate roles, perhaps interacting with different receptor or effector systems.

References

Beckner SK, Hatton S, Shih TY (1985) The *ras* oncogene product is not a regulatory component of adenylate cyclase. Nature 317:71–72

Birchmeier C, Broek D, Wigler M (1985) *RAS* proteins can induce meiosis in *Xenopus* oocytes. Cell 43:615–621

Bos JL, Toksoz D, Marshall CJ, Verlaan de Vries M, Veeneman GH, van der Eb AJ, Van Boom JH, Janssen JWG, Steenvoorden ACM (1985) Amino-acid substitutions at codon 13 of the N-*ras* oncogene in human acute myeloid leukaemia. Nature 315:726–730

Capon DJ, Seeburg PH, McGrath JP, Hayflick JS, Edman U, Levinson AD, Goeddel DV (1983) Activation of Ki-*ras*-2 gene in human colon and lung carcinomas by two different point mutations. Nature 304:507–513

Chang EH, Furth ME, Scolnick EM, Lowy DR (1982) Tumorigenic transformation of mammalian cells induced by a normal human gene homologous to the oncogene of Harvey murine sarcoma virus. Nature 297:479–483

Chipperfield RG, Jones SS, Lo K-M, Weinberg RA (1985) Activation of Ha-*ras* p21 by substitution, deletion and insertion mutations. Mol Cell Biol 5:1809–1813

Clark R, Wong G, Arnheim N, Nitecki D, McCormick F (1985) Antibodies specific for amino acid 12 of the *ras* oncogene product inhibit GTP binding. Proc Natl Acad Sci USA 82:5280–5284

Der CJ, Finkel T, Cooper GM (1986) Biological and biochemical properties of human ras^H genes mutated at codon 61. Cell 44:167–176

Fasano O, Aldrich T, Tamanoi F, Taparowsky E, Furth M, Wigler M (1984) Analysis of the transforming potential of human H-*ras* by random mutagenesis. Proc Natl Acad Sci USA 81:4008–4012

Fujita J, Srivastava SK, Kraus MH, Rhim JS, Tronick SR, Aarsonson SA (1985) Frequency of molecular alterations affecting *ras* proto-oncogenes in human urinary tract tumours. Proc Natl Acad Sci USA 82:3849–3853

Gibbs JB, Sigal IS, Poe M, Scolnick EM (1984) Intrinsic GTPase activity distinguishes normal and oncogenic *ras* p21 molecules. Proc Natl Acad Sci USA 81:5704–5708

Hall A (1984) The *ras* gene family. In: MacLean N (ed) Oxford surveys on eukaryotic genes, vol 1. Oxford Univ Press, Oxford, pp 111–144

Hall A, Brown R (1985) Human N-*ras*: cDNA cloning and gene structure. Nucl Acids Res 13:5255–5268

Hand PH, Thor A, Wunderlich D, Murano R, Caruso A, Schlom J (1984) Monoclonal antibodies of predefined specificity detect activated *ras* gene expression in human mammary and colon carcinomas. Proc Natl Acad Sci USA 81:5227–5231

Ishi S, Merlino GT, Pastan I (1985) Promoter region of the human Harvey *ras* proto-oncogene: Similarity to EGF receptor proto-oncogene promoter. Science 230:1378–1381

McCormick F, Clark BFC, LaCour TFM, Kjeldgaard M, Norskow-Lauritsen L, Nyborg J (1985) A model for the tertiary structure of p21, the product of the *ras* oncogene. Science 230:78–82

McCoy M, Bargmann CI, Weinberg RA (1984) Human colon carcinoma Ki-*ras*-2 oncogene and its corresponding proto-oncogene. Mol Cell Biol 4:1577–1582

McGrath JP, Capon DJ, Smith DH, Chen EY, Seeburg PH, Goeddel DV, Levinson AD (1983) Structure and organization of the human Ki-*ras* proto-oncogene and a related processed pseudogene. Nature 304:501–504

McGrath JP, Capon DJ, Goeddel DV, Levinson AD (1984) Comparative biochemical properties of normal and activated human p21 *ras* protein. Nature 310:644–649

Madaule P, Axel R (1985) A novel *ras*-related gene family. Cell 41:31–40

Neuman-Silberberg FS, Schejter E, Hoffmann FM, Shilo B-Z (1984) The Drosophila *ras* oncogenes: Structure and nucleotide sequence. Cell 37:1027–1033

Powers S, Kataoka T, Fasano O, Goldfarb M, Strathern J, Broach J, Wigler M (1984) Genes in S. cerevisiae encoding proteins with domains homologous to the mammalian *ras* proteins. Cell 36:607–612

Reymond CD, Gomer RH, Mehdy MC, Firtel RA (1984) Developmental regulation of a *Dictyostelium* gene encoding a protein homologous to mammalian *ras* proteins. Cell 39:141 – 148

Scolnick EM, Papageorge AG, Shih TY (1979) Guanine nucleotide-binding activity as an assay for src protein of rat-derived murine sarcoma virus. Proc Natl Acad Sci USA 76:5355 – 5359

Seeburg PH, Colby WW, Capon DJ, Goeddel DV, Levinson AD (1984) Biological properties of human c-Ha-*ras*-1 genes mutated at codon 12. Nature 312:71 – 75

Shih C, Weinberg RA (1982) Isolation of a transforming sequence from a human bladder carcinoma cell line. Cell 29:161 – 169

Shih TY, Papageorge AG, Stokes PE, Weeks MO, Scolnick EM (1980) Guanine nucleotide-binding and autophosphorylating activities associated with the p21^{src} protein of Harvey murine sarcoma virus. Nature 287:686 – 691

Spandidos DA, Kerr IB (1984) Elevated expression of the human *ras* oncogene family in premalignant and malignant tumours of the colorectum. Br J Cancer 49:681 – 688

Tanabe T, Nakada T, Nishikawa Y, Sagimoto K, Suzuki H, Takahashi H, Noda M, Haga T, Ichiyama A, Kangawa K, Minamino N, Matsuo H, Numa S (1985) Primary structure of the α-subunit of transducin and its relationship to *ras* proteins. Nature 315:242 – 245

Temeles GL, Gibbs JB, D'Alonzo JS, Sigal IS, Scolnick EM (1985) Yeast and mammalian *ras* proteins have conserved biochemical properties. Nature 313:700 – 703

Toda T, Uno I, Ishikawa T, Powers S, Kataoka T, Broek D, Broach J, Matsamoto K, Wigler M (1985) In yeast *RAS* proteins are controlling elements of the cyclic AMP pathway. Cell 40:27 – 36

Willingham MC, Pastan I, Shih TY, Scolnick EM (1980) Localization of the *src* gene product of the Harvey strain of MSV to the plasma membrane of transformed cells by electron microscopic immunocytochemistry. Cell 19:1005 – 1014

Willumsen BM, Christensen A, Hubbert NL, Papageorge AG, Lowy DR (1984) The p21 *ras* C-terminus is required for transformation and membrane association. Nature 310:583 – 586

RAS Genes and Growth Control in the Yeast *Saccharomyces cerevisiae*

OTTAVIO FASANO

Ras[1] genes activated by point mutations have dramatic effects on the proliferation of mammalian cells (see Marshall, this Vol. and Land, this Vol.). However, the physiological function of the proteins they encode is not known. The identification of *RAS* genes in yeast has made it possible to study their function using the sophisticated genetic and biochemical techniques available in this organism. This review describes data which indicate that *RAS* proteins are elements of the cAMP effector pathway in *S. cerevisiae*, and then discusses *RAS* function at the molecular level.

Altered Growth and Metabolism in Yeast Cell Following Genetic Manipulation of *RAS* Genes

S. cerevisiae contains two genes, designated as *RAS1* and *RAS2*, which are related to the human and viral *ras* genes by extensive DNA sequence homology (DeFeo-Jones et al. 1983; Powers et al. 1984). Yeast also contain a third *ras*-related gene, designated YPT, which is characterized by a lower degree of homology (Gallwitz et al. 1983). The *RAS1* and *RAS2* genes have open reading frames potentially encoding proteins of 309 and 322 amino acids, respectively (Powers et al. 1984; Dhar et al. 1984). The homology between yeast and human *ras* proteins is shown schematically in Fig. 1. Remarkably, perfect sequence conservation is seen in the amino acid residues which are critical for oncogenic activation of the mammalian *ras* gene products (corresponding to residues in position 12, 13, 59, 61 and 63; Fasano et al. 1984 and Marshall, this Vol.) and in the surrounding amino acids.

The function of the yeast *RAS* proteins has been extensively analyzed by site-directed mutagenesis combined with the technique of gene replacement (Rothstein 1983), in which the normal *RAS* genes in intact yeast cells are replaced with in vitro manipulated alleles. The analysis involved: (1) introducing point mutations at specific sites in the *RAS* genes; (2) completely inactivating the *RAS1* and/or *RAS2* genes by replacing them with in vitro engineered al-

---

[1] Consistent with standard nomenclature, yeast *RAS* genes and their products as well as dominant mutations are indicated by capital letters, while recessive mutations and human *ras* genes are indicated with small letters. Collectively, human and yeast *ras* genes are referred to by small letters.

Oncogenes and Growth Control
Edited by P. Kahn and T. Graf
© Springer-Verlag Berlin Heidelberg 1986

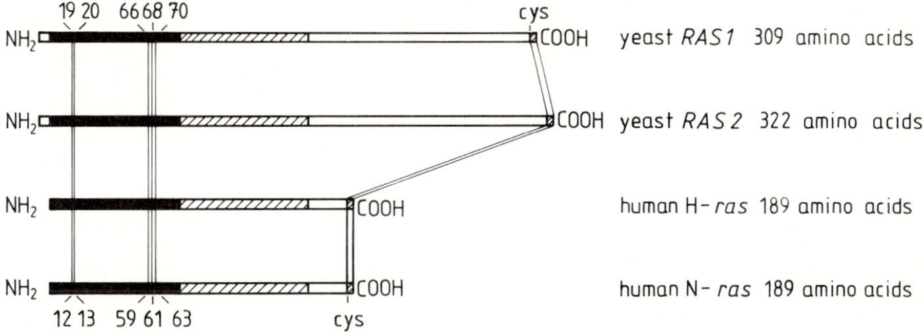

Fig. 1. Homology between yeast and human *ras* proteins. Regions in *black* indicate more than 90% homology of the amino acid level; *dashed areas* indicate lower but still significant homology. *White areas* show regions of no homology

leles in which the coding region is disrupted by the insertion of an unrelated DNA sequence; and (3) partially inactivating the function of the *RAS2* gene by replacing it with an in vitro mutagenized allele encoding a defective gene product. The results of these experiments, which are summarized in Table 1, indicated that deletion of both the *RAS1* and *RAS2* genes led to loss of spore viability, while functional inactivation of these genes led to growth arrest, predominantly in the G_1 phase of the cell cycle. Neither of the two *RAS* genes alone was essential for cell viability, suggesting that the *RAS1* and *RAS2* gene products performed complementary functions. This conclusion was further supported by the finding that the inability of cells with a *wtRAS1* and a disrupted *RAS2* allele (*RAS1 ras2*) to grow on nonfermentable carbon sources could be suppressed by increasing the gene dosage of *RAS1* (Tatchell et al. 1985). Interestingly, a gly(19) to val(19) substitution in the *RAS2* gene had opposite effects to those produced by inactivating the *RAS2* function, and furthermore, behaved as a dominant mutation (Kataoka et al. 1984), suggesting that this substitution led to a more active gene product.

Since mutations which activate the oncogenic potential of mammalian *ras* genes are dominant, these findings also suggested that the analogy between yeast and human *ras* proteins is both structural and functional. Further support for this hypothesis came from the demonstration that the yeast *RAS1* protein containing an internal deletion of amino acids 186 to 302 and activated by a single amino acid substitution in position 68 induced morphological transformation of NIH-3T3 mouse fibroblasts (DeFeo-Jones et al. 1985). Conversely, human or viral *ras* proteins expressed in yeast cells compensated the lethality of yeast mutants in which both *RAS* genes were inactivated (Kataoka et al. 1985b, DeFeo-Jones et al. 1985).

Table 1. Summary of phenotypes observed after replacement of the yeast *RAS1* and *RAS2* by genetically manipulated alleles

Genotypes		Phenotypes of haploid cells		Phenotypes of diploid cells	Germination of spores	References
RAS1 gene	*RAS2* gene	Glycogen and trehalose levels	Growth on glucose-based media	Frequency of sporulation		
Wild-type	Wild-type	Normal	Normal	Normal	Yes	
Disrupted	Wild-type	n.d.	Normal	Slightly increased	Yes	Fasano et al., unpublished
Wild-type	Disrupted	Increased	Normal	Increased	Yes	Kataoka et al. 1984, Fraenkel 1985
Disrupted	Disrupted	–	–	–	No	Tatchell et al. 1984, Kataoka et al. 1984
Disrupted	Replaced by a defective allele (*ras2^{ts1}*)	n.d.	90% of the cells arrest in G1 at the nonpermissive temperature	Increased	Yes	Fasano et al., unpublished
Wild-type	Replaced by *RAS2^{val19}*	Decreased	Reduced viability after nutrient deprivation, suggesting inability to undergo proper Go arrest	Decreased	Yes	Kataoka et al. 1984, Toda et al. 1985

Similarity Between Phenotypes of Yeast Cells with Altered *RAS* Genes and Mutants of the cAMP Effector Pathway

Some of the properties of yeast cells with impaired RAS genes, including both G_1 arrest and premature sporulation, had been observed previously in mutants in which the regulatory subunit of the cAMP-dependent protein kinase showed decreased sensitivity to cAMP, presumably leading to low activity of the kinase catalytic subunit (*CYR3* mutants; Matsumoto et al. 1983a; Uno et al. 1982). Mutants with a defective catalytic subunit of adenylate cyclase (*cyr1* mutants) had a similar phenotype, and furthermore, accumulated abnormally high levels of the storage carbohydrate trehalose (Matsumoto et al. 1982; Uno et al. 1983). Conversely, mutants with increased protein kinase activity (*bcy1* mutants) were unable to enter G_0 arrest following nutrient deprivation, showed increased degradation of reserve carbohydrates and failed to sporulate (Matsumoto et al. 1982, 1983a, 1983b; Uno et al. 1983). These phenotypes were similar to those induced by the activated $RAS2^{\text{val}(19)}$ allele (Table 1). Taken together, these results suggested that *RAS* proteins are elements of the cAMP effector pathway.

Genetic evidence for this hypothesis came from experiments which exploited the properties of *bcy1* mutants. Matsumoto et al. (1982) had shown that the *bcy1* mutation, which leads to decreased biosynthesis of the cAMP-dependent protein kinase regulatory subunit, resulted in permanent activation of the catalytic subunit and allowed yeast cells to grow in the absence of a functional adenylate cyclase and without the addition of cAMP. Indeed, the *bcy1* mutation was originally isolated and named on the basis of its ability to *b*ypass *cy*clase inactivation. This finding suggested that the inactivation of genes other than the catalytic subunit of adenylate cyclase which are involved in the regulation of cAMP levels should not be lethal in a *bcy1* background. Furthermore, the ability of the *bcy1* mutation to rescue lethal mutants might indicate that lethality was caused by a decreased intracellular concentration of cAMP. Toda et al. (1985) subsequently found that *bcy1* suppressed the lethal effect of the functional absence of *RAS* protein: *ras1 ras2 bcy1* cells were fully viable, whereas *ras1 ras2* spores with the *wtBCY1* allele failed to germinate. This result led to the hypothesis that *RAS* gene products were positive effectors of the intracellular cAMP levels and that at least part of their effects were mediated by the cAMP-dependent protein kinase.

Biochemical Analysis of the Function of *RAS* Proteins

One prediction of the above hypothesis is that yeast cells with impaired *RAS* function have decreased intracellular concentrations of cAMP, while cells with the activated $RAS2^{\text{val}(19)}$ gene should have increased cAMP levels. This prediction was confirmed by experiments which showed that *ras1 ras2 bcy1* strains have $<0.1\,\text{pmol cAMP mg}^{-1}$ of protein, while $RAS1\ RAS2^{\text{val}(19)}$

strains have up to 10 pmol cAMP mg^{-1} protein. *WtRAS1 RAS2* strains contain about 2 pmol cAMP mg^{-1} protein (Toda et al. 1985).

The hypothesis that *RAS* proteins play a role in the biosynthesis of cAMP was tested biochemically by examining the in vitro adenylate cyclase activity present in yeast membranes which lack *RAS* proteins. Since removal of *RAS* proteins from membrane preparations by biochemical techniques is not possible, Toda et al. (1985) instead used *ras1 ras2 bcy1* yeast cells, which contain *RAS* genes disrupted by marker insertion and are therefore a convenient source of *RAS* protein-depleted membranes. (The *bcy1* mutation was required to ensure viability in the absence of functional *RAS* genes.) Membranes purified from these cells showed no detectable Mg-dependent adenylate cyclase activity, either in the presence or absence of GTP. However, synthesis of cAMP in vitro was observed after the addition of purified *RAS* proteins (Broek et al. 1985). The *RAS2* protein was threefold more effective than *RAS1*. Furthermore, concomitant expression of the yeast genes encoding the adenylate cyclase catalytic subunit and the *RAS2* protein in mutant *E. coli* cells which lack a functional adenylate cyclase led to cAMP production (Uno et al. 1985). These results suggested that the *RAS2* protein modulates the activity of the adenylate cyclase catalytic subunit by a direct interaction, but did not exclude the possibility that additional yeast and/or *E. coli* cellular components also mediate this interaction. However, other authors (Kataoka et al. 1985a) reported that the addition of purified *RAS2* protein did not significantly stimulate the activity of the yeast adenylate cyclase catalytic subunit in *E. coli*. A more detailed investigation will be required to resolve this question.

Like their mammalian counterparts, yeast *RAS1* and *RAS2* proteins share limited but significant amino acid sequence homology with other GTP-binding proteins, such as the bacterial elongation factor Tu and the regulatory components of the adenylate cyclase system (G proteins) (see Masters and Bourne, this Vol.). They also share certain functional properties with these GTP-binding proteins, including the ability to bind GDP and GTP with nearly 1:1 stoichiometry (Tamanoi et al. 1984; Temeless et al. 1985; De Vendittis and Fasano, unpublished results) and to hydrolyze GTP. The finding that oncogenic activation of mammalian *ras* genes generally results in decreased GTPase activity (see Marshall, this Volume and references therein) has led to the suggestion that activation might be the consequence of stabilizing the active (GTP-bound) state by point-mutations, resulting in constitutive activity of the *ras* proteins. In the yeast system, the GDP-bound form of the purified *RAS2* protein is tenfold less effective than the GTP-bound form in stimulating the adenylate cyclase activity of membranes from *ras1 ras2 CRI4* yeast cells. Furthermore, GTP hydrolysis is not required for the synthesis of cAMP (Broek et al. 1985; De Vendittis and Fasano, unpublished results). These results suggest that the role of GTP hydrolysis might be limited to the conversion of *RAS* proteins from an active to a functionally inactive state. The finding that the GDP-bound form of a purified yeast *RAS2* protein with the gly(19) to val(19) substitution is only twofold less effective than the GTP-bound form in stimulating cAMP-synthesis in the in vitro adenylate cyclase assay (De Vendittis and Fasano, unpublished) might mean that the mutant $RAS2^{val(19)}$ protein does not respond to the conversion from the GTP- to the GDP-bound form by efficiently turning off the functionally stimulated state of the protein.

We have also shown that regeneration of the GTP-bound *RAS2* protein from the GDP-bound form occurs in vitro by displacement of the bound GDP by GTP. The rate of nucleotide exchange is very slow and the rate-limiting step appears to be the dissociation of GDP. ATP and inorganic polyphosphates at millimolar concentrations stimulate the rate of formation of the GTP-bound form of purified *RAS* proteins. Provided that the rate of nucleotide exchange is also limiting in vivo, these results suggest that the intracellular ATP and inorganic polyphosphate levels might directly regulate the function of the *RAS2* protein (De Vendittis and Fasano, unpublished). Since the *RAS2* gene product appears to mediate the cellular response to nutrient limitation (Toda et al. 1985), a variation of ATP and inorganic polyphosphate levels according to metabolic conditions might either inhibit or *RAS*-dependent pathways by modulating the rate of regeneration of the GTP-bound form.

Elements of the Pathway Controlling Cell Proliferation

The studies described above demonstrate that some effects of the *RAS* gene products are mediated by cAMP, but do not exclude the existence of either alternative or complementary pathways. The known elements of the intracellular pathways controling cell proliferation and their relationships to one another are shown schematically in Fig. 2. These elements include those which regulate the function of *RAS* proteins as well as those whose function is regulated by *RAS* proteins. The elements of the cAMP pathway mentioned previously fall within the latter class. Cells may also contain elements which do not act through the cAMP pathway, as suggested by the finding that injection of the human Ha-*ras* protein into *Xenopus* oocytes induces progression from prophase to metaphase without any significant change in cAMP levels (Birchmeier et al. 1985). It is also possible that the *ras* proteins in organisms other than yeast function (either partially or completely) by different mechanisms.

The identity of the elements regulating *RAS* function is unknown, even though the structural and functional similarity between *RAS* proteins and G proteins suggests that the former might couple the function of adenylate cyclase, and possibly other target system(s), to unidentified receptors. There is presently no evidence for macromolecular effectors which increase the displacement of *RAS* protein-bound GDP by GTP and thereby stimulate the activity of *RAS*-dependent pathways, as has been shown for mammalian G proteins (Cassel and Selinger 1978; Lad et al. 1980). However, the nucleotide exchange reaction provides both a potential target for regulation and a useful in

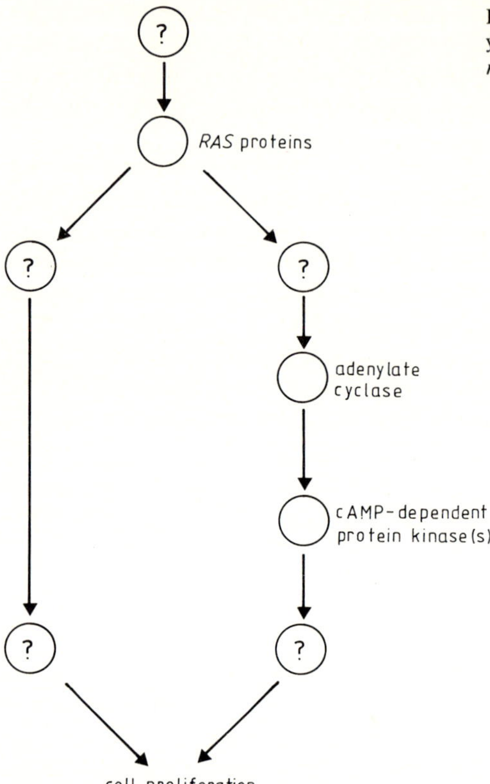

Fig. 2. Elements of the *RAS* pathway in the yeast *Saccharomyces cerevisiae. Question marks* indicate unknown components

vitro assay for detection of regulatory macromolecular elements in cell extracts.

The genetic approach is likely to be useful in identifying additional elements of the efferent pathway in yeast, as suggested by the previously mentioned finding that the *bcy1* mutation suppresses the lethality of *RAS* inactivation. More recently, deletion of the N-terminus of the adenylate cyclase gene was reported to result in increased enzyme activity and in the ability to suppress the lethality of *ras1 ras2* double mutants (Kataoka et al. 1985a). These examples lead to the general prediction that mutations which activate elements of the pathway that are downstream of *RAS* should suppress the lethality of *ras1 ras2* mutants. To facilitate the isolation of such mutants we have constructed a yeast strain dependent for growth on a temperature-sensitive *RAS2* gene and have found dominant mutations which suppress G_1 arrest at the nonpermissive temperature. Some of these mutations also suppress the lethality of *ras1 ras2* double mutants; these are currently being analyzed further (Fasano et al., unpublished results). Several groups have exploited the inability of *RAS1 ras2* cells to grow on nonfermentable carbon sources in order to identify recessive mutations able to bypass the requirement for *RAS* function (Fraenkel 1985; Tatchell et al. 1985).

Concluding Remarks

Studies with normal and mutant *RAS* genes in both yeast and mammalian cells indicate that deregulation of functions which are normally tightly controlled can occur via genetic lesions such as point mutations and deletions. Understanding the mechanisms involved in deregulation of the elements of the growth control pathway in yeast should therefore help to unravel the problem of oncogenesis at the molecular level.

Acknowledgments. I would like to thank Drs. G. Cesareni, P. Charnay, E. De Vendittis and A. Vitelli for critical reading of the manuscript and W. Moses for secretarial assistance.

References

Birchmeier C, Broek D, Wigler M (1985) *RAS* proteins can induce meiosis in *Xenopus* oocytes. Cell 43:615 – 621

Broek D, Samiy N, Fasano O, Fujiyama A, Tamanoi F, Northup J, Wigler M (1985) Differential activation of yeast adenylate cyclase by *wt* and mutant *RAS* proteins. Cell 41:763 – 769

Cassel D, Selinger Z (1978) Mechanism of adenylate cyclase activation through the β-adrenergic receptor: Catecholamine-induced displacement of bound GDP by GTP. Proc Natl Acad Sci USA 9:4155 – 4159

DeFeo-Jones D, Scolnick E, Koller R, Dhar R (1983) *ras*-related gene sequences identified and isolated from *Saccharomyces cerevisiae*. Nature 306:707 – 709

DeFeo-Jones D, Tatchell K, Robinson LC, Sigal IS, Vass WC, Lowy DR, Scolnick EM (1985) Mammalian and yeast *ras* gene products: Biological function in their heterologous systems. Science 228:179 – 184

Dhar R, Nieto A, Koller R, Defeo-Jones D, Scolnick E (1984) Nucleotide sequence of two *ras*[H]-related genes isolated from the yeast *Saccharomyces cerevisiae*. Nucl Acids Res 12:3611 – 3618

Fasano O, Aldrich T, Tamanoi F, Taparowsky E, Furth M, Wigler M (1984) Analysis of the transforming potential of the human H-*ras* gene by random mutagenesis. Proc Natl Acad Sci USA 81:4008 – 4012

Fraenkel DG (1985) On *ras* gene function in yeast. Proc Natl Acad Sci USA 82:4740 – 4744

Gallwitz D, Dorath C, Sander C (1983) A yeast gene encoding a protein homologous to the human C-has/bas proto-oncogene product. Nature 306:704 – 707

Kataoka T, Powers S, McGill C, Fasano O, Strathern J, Broach J, Wigler M (1984) Genetic analysis of yeast *RAS1* and *RAS2* genes. Cell 37:437 – 445

Kataoka T, Broek D, Wigler M (1985a) DNA sequence and characterization of the *S. cerevisiae* gene encoding adenylate cyclase. Cell 43:493 – 505

Kataoka T, Powers S, Cameron S, Fasano O, Goldfarb M, Broach J, Wigler M (1985b) Functional homology of mammalian and yeast *RAS* genes. Cell 40:19 – 26

Lad PM, Nielsen TB, Preston MS, Rodbell M (1980) The role of the guanine nucleotide exchange reaction in the regulation of the β-adrenergic receptor and in the actions of catecholamines and cholera toxin on adenylate cyclase in turkey erythrocyte membranes. J Biol Chem 255:988 – 995

Matsumoto K, Uno I, Oshima Y, Ishikawa T (1982) Isolation and characterization of yeast mutants deficient in adenylate cyclase and cyclic AMP-dependent protein kinase. Proc Natl Acad Sci USA 79:2355 – 2359

Matsumoto K, Uno I, Ishikawa T (1983a) Initiation of meiosis in yeast mutants defective in adenylate cyclase and cyclic AMP-dependent protein kinase. Cell 32:417 – 423

Matsumoto K, Uno I, Ishikawa T (1983b) Control of cell division in *Saccharomyces cerevisiae* mutants defective in adenylate cyclase and cAMP-dependent protein kinase. Exp Cell Res 146:151 – 161

Powers S, Kataoka T, Fasano O, Goldfarb M, Strathern J, Broach J, Wigler M (1984) Genes in *S. cerevisiae* encoding proteins with domains homologous to the mammalian *ras* proteins. Cell 36:607 – 612

Rothstein RJ (1983) One-step gene disruption in yeast. Methods Enzymol 101:202 – 211

Tamanoi F, Walsh M, Kataoka T, Wigler M (1984) A product of yeast *RAS2* gene is a guanine nucleotide binding protein. Proc Natl Acad Sci USA 81:6924 – 6928

Tatchell K, Chaleff D, Defeo-Jones D, Scolnick E (1984) Requirement of either of a pair of *ras*-related genes of *Saccharomyces cerevisiae* for spore viability. Nature 309:523 – 527

Tatchell K, Robinson LC, Breitenbach M (1985) *RAS2* of *Saccharomyces cerevisiae* is required for gluconeogenic growth and proper response to nutrient limitation. Proc Natl Acad Sci USA 82:3785 – 3789

Temeless GL, Gibbs JB, D'Alonzo JS, Sigal IS, Scolnick EM (1985) Yeast and mammalian *ras* proteins have conserved biochemical properties. Nature 313:700 – 703

Toda T, Uno I, Ishikawa T, Powers S, Kataoka T, Broek D, Cameron S, Broach J, Matsumoto K, Wigler M (1985) In yeast, *RAS* proteins are controlling elements of adenylate cyclase. Cell 40:27 – 36

Uno I, Matsumoto K, Ishikawa T (1982) Characterization of cyclic AMP-requiring yeast mutants altered in the regulatory subunit of protein kinase. J Biol Chem 257:14110 – 14115

Uno I, Matsumoto K, Aduchi K, Ishikawa T (1983) Genetic and biochemical evidence that trehalose is a substrate of cAMP-dependent protein kinase in yeast. J Biol Chem 258:10867 – 10872

Uno I, Mitsuzawa H, Matsumoto K, Tanaka K, Oshima T, Ishikawa T (1985) Reconstitution of the GTP-dependent adenylate cyclase from products of the yeast *CYR1* and *RAS2* genes in *Escherichia coli*. Proc Natl Acad Sci USA 82:7855 – 7859

IV Gene Expression and Nuclear Oncogenes

To trigger cell proliferation, the mitogenic signal initiated at the cell surface must eventually reach the nucleus and activate the expression of specific genes whose products are required for DNA synthesis and/or cell division. Elucidation of the mechanism(s) by which gene expression is controlled in the nucleus is a problem of central importance to many different areas of biological research. In particular, differentiation and development involve coordinate expression of genes in a tissue-specific manner, and much of this regulation seems to occur at the level of transcription. The mechanisms by which transcription is regulated can be examined from two viewpoints: that of the DNA sequences which encode regulatory information ("*cis*-acting elements"); and that of the molecules which participate in turning transcription on and off ("*trans*-acting factors"). This section examines some of the approaches which are currently being used to uncover these mechanisms.

The first several articles discuss the different *cis*- and *trans*-acting elements which are thought to play a role in the control of globin gene expression in erythroid cells, in the regulation of gene expression by steroid hormones and in immunoglobulin gene regulation in B-cells. A significant advance has been the discovery of so-called enhancer sequences, that is, sequences which can activate or suppress gene expression in an orientation-independent manner over large distances of the genome. Enhancers are thought to act through their capacity to bind specific factors, leading to the stimulation of neighboring promoter sequences. The mechanisms by which the activated enhancer conveys signals to the promoter are not yet understood. The discovery that some enhancers are tissue-specific in their function implies the existence of tissue specific *trans*-acting factors that are likely to play a key role in cell differentiation. DNA methylation is another mechanism which participates in controlling the transcriptional activity of genes, perhaps by inducing conformational changes in the chromatin and/or by affecting the interaction of *trans*-acting factors with DNA.

The roles of *cis*- and *trans*-acting elements are also discussed in the adenovirus and human T-cell leukemia/lymphotropic virus (HTLV) systems. During evolution, both DNA and RNA tumor viruses acquired powerful enhancers and promoters which utilize the transcriptional machinery of the host cell to express their genes. In addition, adenoviruses and HTLVs encode *trans*-acting elements which can positively or negatively regulate not only viral gene transcription but also that of certain host genes. Among the cellular genes which are trans-activated by HTLV I and II are those encoding the T-cell growth factor, interleukin-2, and its receptor. However, it is controversial whether expression of these genes leads to autocrine stimulation.

The remainder of the section discusses oncogenes which encode DNA binding proteins that are suspected of playing a role in regulating gene expression; these include c-*myc*, c-*fos*, c-*myb* and p53. The finding that the former three proto-oncogenes are induced upon growth factor stimulation has led to the speculation that their gene products participate in the mitogenic response, a conclusion which also seems to apply to p53. However, it is still an open question whether the expression of these genes is essential for triggering cell division and if so, whether additional gene activities are necessary.

Regulation of Human Globin Gene Expression

Patrick Charnay

At different stages of human development the various globin genes are sequentially activated and expressed in cells of the erythroid lineage. This precisely controlled process raises a number of important issues concerning the molecular basis of gene regulation. These include: (1) the specificity of globin gene expression in erythroid cells; (2) the sequential activation of the different genes within the α-like and β-like globin gene clusters during development, a phenomenon which has been referred to as hemoglobin switching; and (3) the balanced production of α-like and β-like globin polypeptide chains, implying a coordinate expression of genes located on different chromosomes.

This chapter will concentrate on the regulation of globin gene transcription, which appears to be a major level at which the expression of these genes is regulated (Groudine et al. 1981; Hofer et al. 1982; Chao et al. 1983; Wright et al. 1984). One model consistent with available data is that transcriptional activation of globin genes during erythroid cell differentiation involves the appearance or disappearance of regulatory molecules which interact with specific DNA sequences close to or within the gene. These putative regulatory molecules, which modulate initiation of transcription, are termed *trans*-acting factors, while the target DNA sequences are called *cis*-acting elements. Major efforts have recently gone into developing systems which make it possible to characterize the globin gene *cis*-acting regulatory elements and *trans*-acting factors. For a discussion of the organization, structure, and expression of human globin genes, which has been the subject of several recent reviews, the reader is referred to Karlsson and Nienhuis (1985) and to the references therein.

Promoter Elements

A prerequisite for understanding the mechanisms of globin gene regulation is the identification of the *cis*-acting DNA sequences required for accurate and efficient initiation of transcription. Detailed studies of these sequences in the rabbit β-globin gene and the mouse β-major globin gene, carried out by introducing mutations into cloned genes followed by the transfer of the mutant genes into cultured cells, has led to the identification of the β-globin promoter (Dierks et al. 1983; Charnay et al. 1985). Three distinct regions upstream from the mRNA cap site are required for efficient transcription: the ATA box lo-

Oncogenes and Growth Control
Edited by P. Kahn and T. Graf
© Springer-Verlag Berlin Heidelberg 1986

cated 30 bp from the mRNA cap site (-30), the CAAT box located around -75 and the distal sequence element CCACACCC (CACCC box) located around -90. The first two elements are conserved among most eukaryotic promoters and the CACCC box element is common to mammalian and avian β-like globin genes (Dierks et al. 1983). It is therefore likely that these elements are also required for efficient transcription of the human β-globin gene. Indeed, naturally occurring point mutations in the ATA box or in the distal sequence element result in β-thalassemia in humans and affect the expression of the cloned gene in cultured cells (reviewed by Orkin and Kazazian 1984). Recent data suggest that the CACCC box element and one of the two CAAT boxes 5' to the fetal β-like γ-globin genes are also required for efficient transcription of these genes (Anagnou et al. 1985). Finally, deletion of the sequences located between positions -87 and -55 upstream from the human α1-globin gene and including the CAAT box, severely decreases the transcription of the gene in monkey COS cells (Mellon et al. 1981).

These different transcriptional control elements have been defined by transfection experiments in nonerythroid cells. In some cases, they have been shown to be required for efficient expression of the gene in erythroid cells (Wright et al. 1984; Charnay et al. 1985). They might therefore constitute basic recognition signals for transcription initiation. However, the possibility that they are involved in tissue- or developmental-specific regulation of gene expression has not yet been investigated.

cis-Acting Regulatory Elements

Analysis of the *cis*-acting elements which regulate globin gene expression during development requires the introduction of the cloned genes into erythroid cells. Two different approaches have been taken: transfection of erythroid cell lines and introduction of the genes into the germ line of mice. Treatment of mouse erythroleukemia (MEL) cells (derived from transformed adult erythroid precursor cells) with one of several chemical inducers leads to the induction of erythroid differentiation. When cloned β-globin genes are introduced into MEL cells by transfection, they are coregulated with the endogenous mouse β-globin genes (Chao et al. 1983; Wright et al. 1983). These results indicate that the DNA fragments used in the transformation experiments contain the *cis*-acting regulatory sequences required for transcriptional activation of the genes during erythroid cell differentiation. Analysis of the expression of mutant and hybrid genes transfected into MEL cells suggests the existence of at least two distinct regulatory elements, located 5' and 3' to the mRNA cap site (Charnay et al. 1984; Wright et al. 1984; Charnay et al. 1985). In another cell line, K562 (human erythroid cells with an embryonic/fetal pattern of globin gene expression), cloned human ε-, γ- and β-globin genes also behave like the endogenous genes. Here, the embryonic ε- and the fetal γ-genes are expressed actively, while expression of the adult β-gene is not detected (Kioussis et al. 1985). These data suggest that the DNA fragments used in the experiment carry *cis*-acting sequences involved in the developmental regulation of β-like globin genes.

Such elements are also revealed by analyzing the expression of globin gene constructs in transgenic mice. Human β-globin gene expression in transgenic mice is restricted to the erythroid lineage and parallels the expression of the endogenous mouse β-globin genes (Magram et al. 1985; Townes et al. 1985). Furthermore, constructs with only 48 bp of 5′ flanking sequences are expressed appropriately, indicating that they contain the *cis*-acting sequences necessary for normal developmental regulation of the gene (Townes et al. 1985). This last result is consistent with the finding cited above that a regulatory sequence is located 3′ to the mRNA cap site. The analysis of human γ-globin gene expression in transgenic mice has led to a surprising result: the expression of the fetal γ-gene is restricted to the embryonic period (Chada et al. 1986). The study of hybrid γ/β-globin genes in transgenic animals should lead to identification of the sequences which determine the timing of globin gene expression during development.

Although cloned β-globin genes are correctly expressed after transfer into cultured cells or into transgenic mice, their level of expression is highly variable from one clone or line to another and is generally lower than that of the endogenous gene (Chao et al. 1983; Wright et al. 1983; Magram et al. 1985; Townes et al. 1985). These differences in levels of mRNA transcribed from the exogenous and endogenous genes have been interpreted either as an effect of chromosomal location or structure, of how much of the adjacent flanking sequences were present or of the pattern of DNA modification. These results differ from those obtained via P element-mediated transformation in *Drosophila*, which leads to similar expression levels of both endogenous and exogenous genes.

Tissue-specific enhancers appear to be essential elements in the regulation of the expression of some cellular genes (see Schlokat and Gruss, this Vol.). So far, no enhancer has been found in the vicinity of the human α- or β-globin genes. However, a possible enhancer element active in K562 cells has been localized recently 3′ to the fetal $^A\gamma$-globin gene (Bodine and Ley 1985). The involvement of this enhancer in β-like globin gene regulation has not yet been demonstrated.

Naturally occurring mutations might also lead to the discovery of regulatory sequences. Recently, several types of hereditary persistence of fetal hemoglobin, in which overexpression of one of the γ-globin genes occurs in the adult, have been attributed to specific point mutations in the 5′ flanking region of the γ-gene (reviewed by Karlsson and Nienhuis 1985). Although it has been proposed that these mutations might affect negative *cis*-acting regulatory elements, their actual effect on γ-globin gene transcription remains to be elucidated.

Trans-Acting Factors

Genetic evidence (Gianni et al. 1983) as well as the finding that expression of various cloned globin genes in erythroid cells or transgenic mice is regulated suggest the existence of cellular *trans*-acting factors. These factors have not yet been well characterized; however, recently developed in vitro assays promise to fill this gap. In the case of the chicken adult β^A-globin gene, several factors present specifically in adult chicken erythroid nuclei were found to bind to specific sites 5' to the gene (Emerson and Felsenfeld 1984). The binding of these factors allows the reconstitution of a nuclease hypersensitive region characteristic of erythrocytes. It is not known, however, whether these factors can modulate the level of expression of the gene. More recently, it was shown that a K562 nuclear extract specifically enhances the transcription of human ε-, γ- and β-globin genes in a HeLa cell-free transcription system (Bazett-Jones et al. 1985). This finding constitutes the first direct evidence for the existence of regulatory *trans*-acting factors.

Chromatin Structure and DNA Methylation

Activation of eukaryotic gene expression has been associated with modifications of the DNA methylation pattern and of the chromatin configuration in the vicinity of the gene, as detected by changes in nuclease sensitivity of the DNA (see Doerfler, this Vol.). In the case of the human β-like globin gene cluster, there is an inverse correlation between DNA methylation and the expression of the different genes, although the effect is confined to a subset of the methylation sites (Van der Ploeg and Flavell 1980; Mavilio et al. 1983). Furthermore, the activity of each gene correlates with the presence of DNAase I hypersensitive sites in its 5' flanking DNA region (Groudine et al. 1983; Tuan et al. 1985). Similarly, activation of mouse β-major globin gene transcription during MEL cell differentiation is associated with the appearance of new DNAase I hypersensitive sites in the 5' flanking DNA and within the second intervening sequence (Sheffery et al. 1982; Hofer et al. 1982). Although a causal relationship between the appearance of a DNAase I hypersensitive site and globin gene transcriptional activation has not been established, the chromatin changes have been shown either to coincide with gene activation or to precede it (Hofer et al. 1982; Groudine et al. 1983).

Further evidence that chromosomal structure might be important for globin gene activity have been provided recently by cell fusion experiments. Transfer of human chromosome 11 (which carries the β-like globin gene cluster) from either erythroid or nonerythroid adult cells into MEL cells leads to inducible expression of the adult β-globin gene in hybrid cells. In contrast, the fetal γ-globin genes are not expressed efficiently. However, treatment of the hybrid cells with 5-azacytidine before induction of differentiation causes hypomethylation of the globin gene region and selective activation of the human γ-globin genes (Ley et al. 1984). If the human chromosome 11 is derived from cells in which the γ-genes are efficiently expressed, such as early fetal erythroid cells or the erythroid cell line HEL, terminal differentiation of the hybrid cells also leads to activation of γ-globin gene expression (Papayannopoulou et al. 1985a, b). A likely interpretation of these results is that particular chromatin configurations participate in the modulation of γ-globin gene expression and can be maintained during chromosome transfer and cell division. An alternative interpretation is that the hybrid cells express the γ-globin genes because they inherited from the γ-globin expressing parental cell a *trans*-acting element able to activate the γ-globin genes.

A Model for Human Globin Gene Regulation

Human α-globin genes introduced into MEL cells by chromosome-mediated gene transfer are correctly regulated following induction of erythroid cell differentiation. In contrast, cloned human α-globin genes cotransfected into MEL cells are efficiently transcribed prior to induction but no further increase is observed when the cells differentiate (Charnay et al. 1984). The behavior of cloned α-globin genes is thus strikingly different from that of cloned β-globin genes since, as mentioned above, the β-gene is correctly regulated following DNA transfer into MEL cells. The behavior of the two genes is also different in nonerythroid cells: while the cloned human α-globin gene is readily transcribed after transfection into HeLa cells, efficient expression of the β-globin gene requires linkage to a transcriptional enhancer sequence (Mellon et al. 1981; Humphries et al. 1982; Treisman et al. 1983). To explain these observations, we proposed a model for the regulation of human α- and β-globin gene expression (Charnay et al. 1984). The model is based on the following hypothesis. Endogenous β-globin genes in erythroid cells are subjected to at least two levels of regulation: a negative control level, involving alterations in chromatin structure and/or changes in the state of DNA methylation in the vicinity of the gene, and a positive control level, involving interaction with an erythroid cell-specific *trans*-acting factor. Both regulatory events occur during MEL cell differentiation. In contrast, the endogenous α-globin genes are only subjected to the negative control. When α-or β-globin genes are introduced into MEL cells by DNA transfer, they assume a chromatin configuration that is required for efficient transcription and are not subjected to negative control. However, maximal expression of the β-globin gene but not the α-globin gene requires the erythroid cell-specific factor which is provided during MEL cell differentiation.

I would like to expand this model to take into account the behavior of the human γ-globin genes. It was recently observed that the $^A\gamma$-globin gene introduced MEL cells by DNA transfer is activated upon cell differentiation, although the endogenous embryonic and fetal mouse β-like globin genes are not expressed (Anagnou et al. 1986; Charnay and Henry 1986). I postulate that, like the *β*-globin gene, human *γ*-globin gene activation requires modifications in chromatin configuration as well as positive activation by a *trans*-acting factor. In addition, I propose that this *trans*-acting factor is either identical or very similar to the factor involved in *β*-globin gene activation, while the mechanisms involved in negative control of the two genes are significantly different. This model would explain the regulated expression of a cloned *γ*-globin gene in MEL cells, assuming that, like the *β*-globin gene, it is only subjected to the positive regulatory step and that the *β*-globin gene *trans*-activator can act on the *γ*-globin gene. It would also be consistent with the data obtained after chromosome-mediated *γ*-globin gene transfer into MEL cells: *γ*-genes originating from cells in which they are actively transcribed would be in the active chromatin configuration and would thus be expressed efficiently after MEL cell differentiation. Alternatively, an active chromatin configuration of the genes might be generated following 5-azacytidine treatment of the cells. The existence of one or more differences in the steps required to activate the *γ*- and *β*-globin genes would explain their differential expression during development in humans and in transgenic mice (Chada et al. 1986). Finally, the model is also consistent with the recent finding that nuclear extracts from K562 cells can enhance the in vitro transcription of the human *β*-, *ε*- and *γ*-globin genes, although the *β*-globin gene is not expressed in these cells (Bazett-Jones et al. 1985). This result suggests the existence of a *β*-globin gene family transcription factor which is necessary but not sufficient for efficient transcription in vivo.

Conclusions

The application of recombinant DNA techniques has led to remarkable progress in the past few years in the understanding of the structure and expression of human globin genes. Our knowledge of the molecular mechanisms responsible for the regulation of the genes is more limited and still very fragmentary. Nevertheless, it appears that the stage is now set for the definition of the *cis*-acting elements and the purification of some of the *trans*-acting factors involved in the regulation. Finally, recent data suggest the existence of multiple levels of transcriptional control. It is the combination of these multiple regulatory mechanisms which might be responsible for the complex pattern of globin gene expression during development.

Acknowledgments. I would like to thank M. Goossens, T. Maniatis and S. Ottolenghi for critical reading of the mansucript, R. Axel for discussion and W. Moses for typing the manuscript.

References

Anagnou NP, Moulton AD, Keller G, Karlsson S, Papayannopoulou T (1985) Cis-acting sequences that affect the expression of the human fetal γ-globin genes. In: Stamatoyannopoulos G, Nienhuis AW (eds) Experimental approaches for the study of hemoglobin switching. Liss, New York, pp 163–182

Anagnou NP, Karlsson S, Moylton AD, Keller G, Nienhuis AW (1986) Promoter sequences required for function of the human γ-globin gene in erythroid cells. EMBO J 5:121–126

Bazett-Jones DP, Yeckel M, Gottesfeld JM (1985) Nuclear extracts from globin-synthesizing cells enhance globin transcription in vitro. Nature 317:824–828

Bodine DM, Ley TJ (1985) A possible enhancer element lies 3′ to a human fetal [A]γ-globin gene. Blood 66:68a

Chada K, Magram J, Costantini F (1986) Expression of a human fetal globin gene in transgenic mice: implications for the evolution of hemoglobin switching. Nature 319:685–689

Chao MV, Mellon P, Charnay P, Maniatis T, Axel R (1983) The regulated expression of β-globin genes introduced into mouse erythroleukemia cells. Cell 32:483–493

Charnay P, Henry L (1986) Regulated expression of cloned human fetal [A]γ-globin genes introduced into marine erythroleukemia cells. Eur. J. of Biochem. (in press)

Charnay P, Treisman R, Mellon P, Chao M, Axel R, Maniatis T (1984) Differences in human α- and β-globin gene expression in mouse erythroleukemia cells: the role of intragenic sequences. Cell 38:251–263

Charnay P, Mellon P, Maniatis T (1985) Linker scanning mutagenesis of the 5′-flanking region of the mouse β-major-globin gene: sequence requirements for transcription in erythroid and non-erythroid cells. Mol. Cell. Biol. 5:1498–1511

Dierks P, von Ooyen A, Cochran MD, Dobkin C, Reiser J, Weissmann C (1983) Three regions upstream from the cap site are required for efficient and accurate transcription of the rabbit β-globin gene in mouse 3T6 cells. Cell 32:695–706

Emerson BM, Felsenfeld G (1984) Specific factor conferring nuclease hypersensitivity at the 5′ end of the chicken adult β-globin gene. Proc. Natl. Acad. Sci. USA 81:95–99

Gianni AM, Bregni M, Cappellini MD, Fiorelli G, Taramelli R, Giglioni B, Comi P, Ottolenghi S (1983) A gene controlling fetal hemoglobin expression in adults is not linked to the non-α globin cluster. EMBO J 6:921–925

Groudine M, Peretz M, Weintraub H (1981) Transcriptional regulation of hemoglobin switching in chicken embryos. Mol Cell Biol 1:281–288

Groudine M, Kohwi-Shigematsu T, Gelinas R, Stamatoyannopoulos G, Papayannopoulou T (1983) Human fetal to adult hemoglobin switching: changes in chromatin structure of the β-globin locus. Proc Natl Acad Sci USA 80:7551–7555

Hofer E, Hofer-Warbinek R, Darnell JE (1982) Globin RNA transcription: a possible termination site and demonstration of transcription control correlated with altered chromatin structure. Cell 29:887–893

Humphries RK, Ley T, Turner P, Moulton AD, Nienhuis AW (1982) Differences in human α-, β- and γ-globin gene expression in monkey kidney cells. Cell 30:173–183

Karlsson S, Nienhuis AW (1985) Developmental regulation of human globin genes. Annu Rev Biochem 54:1071–1108

Kioussis D, Wilson F, Khazaie K, Grosveld F (1985) Differential expression of human globin genes introduced in K562 cells. EMBO J 4:927–931

Ley TJ, Chiang YL, Haidaris D, Anagnou N, Wilson VL, Anderson WF (1984) DNA methylation and regulation of the human β-globin-like genes in mouse erythroleukemia cells containing human chromosome 11. Proc Natl Acad Sci USA 81:6618–6622

Magram J, Chada K, Costantini F (1985) Developmental regulation of a cloned adult β-globin gene in transgenic mice. Nature 315:338–340

Mavilio F, Giampaolo A, Care A, Migliaccio G, Calandrini M, Russo E, Pagliardi GL, Mastober-ardino G, Marinucci M, Peschle C (1983) Molecular mechanisms of human hemoglobin switching: selective undermethylation and expression of globin genes in embryonic, fetal, and adult erythroblasts. Proc Natl Acad Sci USA 80:6907–6911

Mellon P, Parker V, Gluzman Y, Maniatis T (1981) Identification of DNA sequences required for transcription of the human α1-globin gene in a new SV40 host-vector system. Cell 27:279–288

Orkin SH, Kazazian HH (1984) The mutation and polymorphism of the human β-globin gene and its surrounding DNA. Annu Rev Genet 18:131–171

Papayannopoulou T, Lindsley D, Kurachi S, Lewison K, Hemenway T, Melis M, Anagnou NP, Najfeld V (1985a) Adult and fetal human globin genes are expressed following chromosomal transfer into MEL cells. Proc Natl Acad Sci USA 82:780–784

Papayannopoulou T, Brice M, Kurachi S, Hemenway T, Lewison K, Stamatoyannopoulos G (1985b) A time dependent, chromosome-11 linked mechanism controls fetal to adult hemoglobin switching. Blood 66:133a

Sheffery M, Rifkind RA, Marks PA (1982) Murine erythroleukemia cell differentiation: DNase I hypersensitivity and DNA methylation near the globin genes. Proc Natl Acad Sci USA 79:1180–1184

Townes TM, Lingrel JB, Chen HY, Brinster RL, Palmiter RD (1985) Erythoid-specific expression of human β-globin genes in transgenic mice. EMBO J 4:1715–1723

Treisman R, Green MR, Maniatis T (1983) Cis- and trans-activation of globin gene transcription in transient assays. Proc Natl Acad Sci USA 80:7428–7432

Tuan D, Solomon W, Li, Qiliang, London IM (1985) The β-like globin gene domain in human erythroid cells. Proc Natl Acad Sci USA 82:6384–6388

Van der Ploeg LHT, Flavell RA (1980) DNA methylation in the human β-globin locus in erythroid and nonerythroid tissues. Cell 19:947–958

Wright S, de Boer E, Grosveld FG, Flavell RA (1983) Regulated expression of the human β-globin gene family in murine erythroleukemia cells. Nature 305:333–336

Wright S, Rosenthal A, Flavell R, Grosveld F (1984) DNA sequences required for regulated expression of β-globin genes in murine erythroleukemia cells. Cell 38:265–273

Regulation of Gene Expression by Steroid Hormones

MIGUEL BEATO

Regulation of gene expression in eukaryotes appears to utilize the same mechanisms that have been described in prokaryotic systems, namely the interaction of regulatory proteins with specific sequences near the regulated promoters. Examples of such regulatory proteins are the T-antigen of SV40 (see Schlokat and Gruss, this Vol.), the protein that regulates heat shock gene expression and the steroid hormone receptors. This chapter discusses DNA sequences that bind steroid hormone receptors, mechanisms of transcriptional activation by these receptors and the role of chromatin configuration in this process.

Glucocorticoid Regulatory Elements

Our understanding of the mechanism of hormonal regulation of gene expression received a great impulse after the discovery of specific DNA sequences in the mouse mammary tumor virus (MMTV) promoter that are recognized by the glucocorticoid receptor (GR) and that are involved in the hormonal stimulation of transcription (Chandler et al. 1983; Scheidereit et al. 1983). The receptor binding sequences of MMTV are able to confer dexamethasone inducibility to heterologous promoters when placed at different distances either upstream or downstream of the promoter, and therefore behave as hormone-dependent enhancers (Chandler et al. 1983; Ponta et al. 1985). This was the first demonstration of a glucocorticoid regulatory element (GRE), soon followed by the identification of similar elements in several other regulated genes including the human metallothionein II_A gene (Karin et al. 1984), the chicken lysozyme gene (Renkawitz et al. 1984), the rabbit uteroglobin gene (Cato et al. 1984), the human growth hormone gene (Slater et al. 1986), the Moloney murine sarcoma virus (DeFranco and Yamamoto 1986; Miksicek et al. 1986), and the mouse ribosomal promoter (C. Scheidereit et al. 1986).

A comparison of the receptor binding sites in the genes mentioned above allows the derivation of a consensus sequence, 5'-GGN ACANNNTGTYCT-3', that exhibits a certain degree of inverted symmetry centered around position 8. The hexanucleotide motif TGTYCT is strictly preserved in 22 sites that we have analyzed so far (Beato 1986). The left half of the consensus sequences is less well preserved although the position 5 is a cytosine in 80% of the

Oncogenes and Growth Control
Edited by P. Kahn and T. Graf
© Springer-Verlag Berlin Heidelberg 1986

sequences. These data, in conjunction with methylation protection studies, suggest that a dimer of the receptor contacts two subsequent turns of the DNA double helix through the major groove in a head to head orientation (Scheidereit and Beato et al. 1984). There is, however, no direct evidence that a dimer of the receptor molecules is bound to regulatory elements in vivo.

Regulatory Elements for Different Steroid Hormones

After the first identification of GRE the question arose as to whether similar elements exist for each individual steroid hormone. The progesterone receptor (PR) of rabbit uterus binds to sites in the chicken lysozyme gene and in the MMTV promoter region that coincide with the binding sites for the gluco-corticoid receptor (Ahe et al. 1985,1986). In the case of MMTV we know that the binding sites for GR and PR are both functional, since a heterologous promoter linked to these sites can be induced by either glucocorticoids or progestins (Cato et al. 1986). A similar situation seems to apply for the promoter distal regulatory element of the chicken lysozyme gene, that is also able to mediate induction by both glucocorticoid and progesterone (Renka-witz et al. 1984). Within this regulatory element, however, the receptors for glucocorticoids and progesterone recognize different features of the DNA sequences as determined by methylation protection studies (Ahe et al. 1986).

Recent experiments with cells containing the regulatory elements of MMTV have shown that they can mediate induction not only by glucocorti-coids and progesterone, but also by estrogens and androgens (A. C. B. Cato and H. Ponta, personal communication). Thus it seems that certain hormone regulatory elements (HRE) are able to mediate hormonal induction of adjacent promoters by all four classes of sex steroid hormones. How is then the hormone specificity of the response preserved in cells where receptors for other hormones are present? It is conceivable that the induction resulting from binding of a hormone receptor to the regulatory site of a gene requires an interaction with other protein factors that are specific for the individual hormone receptors. If this is the case, the presence within cells of such special factors will determine whether or not a particular hormone is able to activate a promoter.

Mechanism of Transcriptional Activation

How could binding of the hormone receptor to the HRE enhance the tran-scriptional efficiency of the adjacent promoter? As with other enhancer ele-ments (see Schlokat and Gruss, this Vol. and Charnay, this Vol.), proteins bound to the HRE must be able to act in both directions of the DNA over a relatively long distance and independently of the orientation of the HRE. The favored model for explaining these properties of enhancers postulates a protein-protein interaction between different enhancer-recognizing proteins (in our case receptors) and other transcriptional factors that interact with relevant elements of the promoter. In cases where the distance between the regulatory element and the promoter is long, this interaction could only occur by looping out of the DNA sequences located in between. Support for this type of model comes from studies with SV40, in which the effects observed after introducing DNA segments of defined length between the enhancer and the 21 base pair repeat depend on whether the length of DNA sequence is an odd or an even multiple of 5 base pairs. Odd multiples inhibit enhancer function whereas even multiples have little inhibitory effect, suggesting a requirement for stereospecific alignement of these two regulatory sequences on one side of the double helix (Takahashi et al. 1986).

An alternative explanation for the relative distance- and orientation-independence of the enhancer activity is that binding of the enhancer-activating proteins to the regulatory elements alters the structure of the DNA. This alteration can be propagated along the DNA double helix and leads to a change in structure at relevant regions of the promoter. Such changes could then be recognized by elements of the transcription machinery as signals for more efficient promoter utilization. Propagation of conformational changes along the DNA requires that the DNA molecule is restrained in its freedom of rotation, a requirement that could be accomplished in vivo through the generation of DNA domains anchored at the nuclear matrix. To test this possibility we have studied the influence of receptor binding to the HRE of MMTV on the topology of closed circular plasmids. We have found that binding of the 94 kDa intact glucocorticoid receptor results in the introduction of two positive supercoils turns into relaxed circular plasmid or, alternatively, in the relaxation of two negative superhelical turns in negatively supercoiled plasmids (M. Carballo and M. Beato, unpublished). This effect is dependent on specific binding of the receptor because deletion of part of the HRE diminishes the topological changes and no effect is seen with other DNAs such as that from the plasmid pBR322. In addition, a 40 kDa proteolytic product of the receptor, containing only the steroid and the DNA binding site, still shows specific binding to the HRE but does not change the topology of the corresponding circular plasmids.

To our knowledge, there is only one other purified protein that has been shown to generate this type of topological change on circular plasmids, namely the bacterial enzyme DNA gyrase (Liu and Wang 1978). This effect of gyrase has been attributed to the wrapping of the DNA around the surface of the enyzme, known to participate in the catalytic mechanism (Liu and Wang 1978; Kirchhausen et al. 1985). Whatever the mechanism, it is clear that the topological changes induced by receptor binding could generate sufficient energy to stabilize conformational changes of the DNA within a closed domain. It should now be possible to directly test this possibility by analyzing the influence of receptor binding on the secondary structure of the adjacent promoter region. This could be done with enzymes that distinguish alternative structures of the DNA (e.g., single-strand nucleases), or by means of chemical reagents (e.g., diethyl pyrocarbonate), that have been used to study the transition of the B- to the Z-form of DNA (Herr 1985; Johnston and Rich 1985). Once such a change in DNA structure has been detected, one could look for proteins that interact with this region of the DNA.

Negative Regulation of Transcription

An interesting aspect of hormonal regulation is the observation that the same hormones can act as positive or negative modulators of transcription, as known for bacterial DNA binding regulatory proteins. For instance, the cyclic AMP receptor protein of *E. coli* usually functions as a positive modulator of catabolite-activated genes, but represses the transcription of its own gene (Aiba 1983). Examples of negative transcriptional regulation by glucocorticoids are well established as, for instance, the repression of the proopiomelanocortin gene in pituitary cells (Birnberg et al. 1983; Eberwine and Roberts 1984). Recently, it has been shown that sequences upstream of the transcription initiation site of this gene are sufficient to confer glucocorticoid repression to a heterologous promoter (Israel and Cohen 1985). It will be interesting to test whether in this region of the DNA there is a binding site for GR, and how this binding results in repression of the adjacent promoter. One possibility would be that changes in topology induced by GR binding to such negative regulatory sequences will differ from those that we have described above. Alternatively, the receptor could interfere with binding of general or specific factors, thus leading to a negative regulatory response.

Chromatin Structure and Differentiation

How do the different cell types manage to direct hormonal regulation to a particular set of genes and to prevent other potentially regulatable genes from being influenced by the hormones? This decision is probably made during the course of cellular differentiation. One plausible model would be that during a critical mitosis the cells organize certain regions of the genome into a particular chromatin structure that makes them available for regulation by DNA regulatory proteins whereas other regions, are organized into a silent or inaccessible conformation. This organization of chromatin needs the potential to perpetuate itself through cell divisions. Certainly this complex process has to be a result of the interplay of several factors. One possible candidate for maintaining the repressed conformation of chromatin structure is histone H1 and different subtypes thereof (Weintraub 1985). This mechanism, however, does not account for the specificity of gene repression and of cell-type-dependent regulation. In addition, it cannot explain the fact that during mitosis specific

DNA regulatory proteins have to be present in excess over DNA in order to bind to the replicating DNA molecules and signal them to an open or a repressed chromatin structure (Brown 1984).

Once the cells are differentiated, those genes whose expression is regulated usually adopt a special chromatin structure as demonstrated by hypersensitivity to DNase I (Weintraub et al. 1981). In genes that are regulated by steroid hormones very often hormone-dependent DNase I hypersensitive sites are localized in the vicinity of receptor binding sites (Zaret and Yamamoto 1984). An interesting example is the uteroglobin gene in rabbit endometrium in which the DNA region that is bound by the progesterone and the glucocorticoid receptor is localized 2.6 kb upstream of the transcription initiation site (Cato et al. 1984). When the DNase I hypersensitive sites are mapped in the uteroglobin region, a series of hormone-dependent sites are localized at -2.5 kb, close to these receptor-binding sites (M. Jantzen, H. P. Fritton, T. Igo-Kemenes, unpublished). This suggests that the same sequences that have been shown to bind the receptor in vitro and that are able to confer hormone inducibility to heterologous promoters are also implicated in the induction of the gene by hormones in vivo.

Conclusions and Perspectives

Although the identification of DNA regulatory elements for steroid hormones has contributed to a molecular understanding of gene regulation by these hormones, many questions are still open. For instance, the discovery that steroid hormone receptors are phosphorylated raises the possibility that the state of their phosphorylation determines their activity. This could either be by directly influencing the ability of the protein to interact with DNA regulatory elements or by influencing the interaction of the protein with other factors involved in transcriptional regulation.

Another open question is the role of the hormone in the process of induction and in the interaction of the receptor protein with the regulatory sequences. Until recently it was assumed that unoccupied steroid hormone receptors are localized in the cytosol and that after hormone binding they acquire an affinity for DNA and chromatin. However, using monoclonal antibodies and cytochemical techniques, unoccupied hormone receptors have been detected within the nucleus, thus raising the question of whether in vivo the receptors may be already attached to DNA or chromatin before binding the hormone (Welshons et al. 1984; King and Greene 1984). Recently the cDNA for human and rat glucocorticoid receptors have been cloned and the structure of one of them has been published (Miesfeld et al. 1984; Hollenberg et al. 1985; Weinberger et al. 1985a). The human glucocorticoid receptor exhibits a striking homology to the *erbA* oncogene (Weinberger et al. 1985b) and exists in two forms in vivo that differ by only a few amino acids at the carboxy terminal end. The shorter form is not able to bind the steroid but might be responsible for the specific DNA binding observed without steroid (T. Willmann and M. Beato, unpublished). The availability of receptor cDNA and the possibility of producing large amounts of proteins with defined primary structure will greatly facilitate our understanding of the role or receptor modifications and protein interactions in the hormonal regulation of gene activity.

Acknowledgments. The experimental work reported here was supported by the Deutsche Forschungsgemeinschaft and the Fond der Chemischen Industrie.

References

Ahe D von der, Janich S, Scheidereit C, Renkawitz R, Schütz G, Beato M (1985) Glucocorticoid and progesterone receptors bind to the same sites in two hormonally regulated promoters. Nature 313:706–709

Ahe D von der, Renoir JM, Buchou T, Baulieu EE, Beato M (1986) The receptors for glucocorticosteroid and progesterone recognize distinct features with a DNA regulatory element. Proc Natl Acad Sci USA 83:2817–2821

Aiba H (1983) Autoregulation of the *E. coli* crp gene: CRP is a transcriptional repressor of its own gene. Cell 32:141–149

Beato M (1986) Interaction of steroid hormone receptors with DNA. In: Thompson EB, Papaconstantinou J (eds) DNA-protein interactions and gene regulation. Univ of Texas Press (in press)

Birnberg NC, Lissitzky JC, Hinman M, Herbert E (1983) Glucocorticoids regulate proopiomelanocortin gene expression in vivo at the levels of transcription and secretion. Proc Natl Acad Sci USA 80:6982–6986

Brown D (1984) The role of stable complexes that repress and activate eukaryotic genes. Cell 37:359–365

Cato ACB, Geisse S, Wenz M, Westphal HM, Beato M (1984) The nucleotide sequences recognized by the glucocorticoid receptor in the rabbit uteroglobin gene are located far upstream from the initiation of transcription. EMBO J 3:2731–2736

Cato ACB, Miksicek R, Schütz G, Arnemann J, Beato M (1986) The hormone regulatory element of mouse mammary tumour virus mediates progesterone induction. EMBO J (in press)

Chandler VL, Maler BA, Yamamoto KR (1983) DNA sequences bound specifically by glucocorticoid receptor in vitro render a heterologous promoter hormone responsive in vivo. Cell 33:489–499

DeFranco D, Yamamoto K (1986) Two different factors act separately or together to specify functionally distinct activities at a single transcriptional enhancer. Mol Cell Biol 6:993–1001

Eberwine JH, Roberts JL (1984) Glucocorticoid regulation of proopiomelanocortin gene transcription in the rat pituitary. J Biol Chem 259:2166–2170

Herr W (1985) Diethyl pyrocarbonate: a chemical probe for secondary structure in negatively supercoiled DNA. Proc Natl. Acad Sci USA 82:8009–8013

Hollenberg SM, Weinberger C, Ong ES, Cerelli G, Oro A, Lebo R, Thompson EB, Rosenfeld MG, Evans RM (1985) Primary structure and expression of a functional human glucocorticoid receptor cDNA. Nature 318:635–641

Israel A, Cohen SN (1985) Hormonally mediated negative regulation of human pro-opiomelanocortin gene expression after transfection in mouse L-cells. Mol Cell Biol 5:2443–2453

Johnston BH, Rich A (1985) Chemical probes of DNA conformation: detection of Z-DNA at nucleotide resolution. Cell 42:713–724

Karin M, Haslinger A, Holtgreve H, Richards RI, Krauter P, Westphal HM, Beato M (1984) Characterization of DNA sequences through which cadmium and glucocorticoid hormones induce human metallothionein II$_A$-gene. Nature 308:513–519

King WJ, Greene GL (1984) Monoclonal antibodies localize oestrogen receptor in the nuclei of target cells. Nature 307:745–747

Kirchhausen T, Wang JC, Harrison SC (1985) DNA gyrase and its complexes with DNA: Direct observation by electron microscopy. Cell 41:933–943

Liu FL, Wang JC (1978) Micrococcus luteus DNA gyrase: Active components and a model for its supercoiling of DNA. Proc Natl Acad Sci USA 75:2098–2102

Miesfeld R, Okret S, Wikström AC, Wrange Ö, Gustafsson JA, Yamamoto KR (1984) Characterization of a steroid hormone receptor gene and mRNA in wild-type and mutant cells. Nature 312:779–781

Miksicek R, Heber A, Schmid W, Danesch U, Posseckert G, Beato M, Schutz G (1986) Glucocorticoid responsiveness of the transcriptional enhancer of Moloney murine sarcoma virus. Cell 46:283–290

Ponta H, Kennedy N, Skroch P, Hynes NE, Groner B (1985) Hormonal response region in the mouse mammary tumor virus long terminal repeat can be dissociated from the proviral promoter and has enhancer properties. Proc Natl Acad Sci USA 82:1020–1024

Renkawitz R, Schütz G, von der Ahe D, Beato M (1984) Identification of hormone regulatory elements in the promoter region of the chicken lysozyme gene. Cell 37:503–510

Scheidereit C, Beato M (1984) Contacts between receptor and DNA double helix within a glucocorticoid regulatory element of mouse mammary tumor virus. Proc Natl Acad Sci USA 81:3029–3033

Scheidereit C, Geisse S, Westphal HM, Beato M (1983) The glucocorticoid receptor binds to defined nucleotide sequences near the promoter of mouse mammary tumour virus. Nature 304:749–752

Scheidereit C, Westphal HM, Carlson C, Bosshard H, Beato M (1986) Molecular model of the interaction between the glucocorticoid receptor and the regulatory elements of inducible genes. DNA (in press)

Slater EP, Rabenau O, Karin M, Baxter JD, Beato M (1986) Glucocorticoid receptor binding and activation of a heterologous promoter in response to dexamethasone by the first intron of the human growth hormone gene. Mol Cell Biol 5:2984–2992

Takahashi K, Bigneron M, Matthes H, Wildeman A, Zenke M, Chambon P (1986) Requirement of stereospecific alignments for initiation from the simian virus 40 early promoter. Nature 319:121–126

Weinberger C, Hollenberg SM, Ong ES, Harmon JM, Brower ST, Cidlowski J, Thompson EB, Rosenfeld MG, Evans RM (1985a) Identification of human glucocorticoid receptor complementary DNA clones by epitope selection. Science 728:740–742

Weinberger C, Hollenberg SM, Rosenfeld MG, Evans RM (1985b) Domain structure of human glucocorticoid receptor and its relationship to the v-erb A oncogene product. Nature 318:670–672

Weintraub H (1985) Assembly and propagation of repressed and derepressed chromosomal states. Cell 42:705–711

Weintraub H, Larsen A, Groudine M (1981) α-Globin-gene switching during the development of chicken embryos: expression and chromosome structure. Cell 24:333–344

Welshons WV, Lieberman ME, Gorski J (1984) Nuclear localization of unoccupied oestrogen receptors. Nature 307:747–749

Zaret KS, Yamamoto KR (1984) Reversible and persistent changes in chromatin structure accompany activation of a glucocorticoid-dependent enhancer element. Cell 38:29–38

Enhancers as Control Elements for Tissue-Specific Transcription

Uwe Schlokat and Peter Gruss

A multicellular organism consists of an extensive variety of different tissues and cell types. Since all of these cells were originally derived from a single cell (zygote), with few exceptions each cell contains identical genetic information. Thus, the differences between tissues must result from differential expression of genes controlling cellular determination and differentiation. One major control mechanism is exerted at the transcriptional level. Control sequences residing upstream of, and closest to polymerase II transcribed genes have been termed promoters. Undoubtedly promoters, as demonstrated for the immunoglobulin genes, can act as tissue-specific control elements. An additional class of transcriptional control element is represented by the so-called enhancers of transcription (see Gluzman 1985 for review). These peculiar elements, in the presence of a responsive promoter, are able to enhance transcription up to 1000-fold while some also show cell-type specificity.

A number of enhancers for viral and cellular genes (e.g., immunoglobulin, fibroin, elastase and others) have been described (see Gruss 1984 and Gluzman 1985 for references). This review focuses on transcriptional enhancers and the components essential for their tissue-specific activation.

Activity of the SV40-Enhancer Requires *cis* and *trans* Elements

The first enhancer identified was an upstream control element of the DNA tumor virus SV40 (for references see Gruss 1984 and Khoury and Gruss 1983). When present in the viral genome, it drastically stimulates transcriptional activity of the early genes. Most interestingly, this effect is also exerted on completely unrelated promoters such as, for example, that of the β-globin gene. In this case, a potentiation effect of at least two orders of magnitude, independent of orientation and distance — even up to 3 kb away from the promoter sequences — was observed. These striking results clearly distinguished enhancers from the position-dependent promoter elements. As far as its enormous distance effect is concerned, the SV40 enhancer turned out to have less dramatic effects on some other promoters. In the case of the SV40 early and the chicken conalbumin promoters, careful analyses revealed a biphasic decrease of activity with increasing distance between enhancers and promoters (Wasylyk et al. 1984). For maximal potentiation effect, two separate domains within the 72 bp unit — the core element (GTGTGGAAAG) and Sphl motif

Oncogenes and Growth Control
Edited by P. Kahn and T. Graf
© Springer-Verlag Berlin Heidelberg 1986

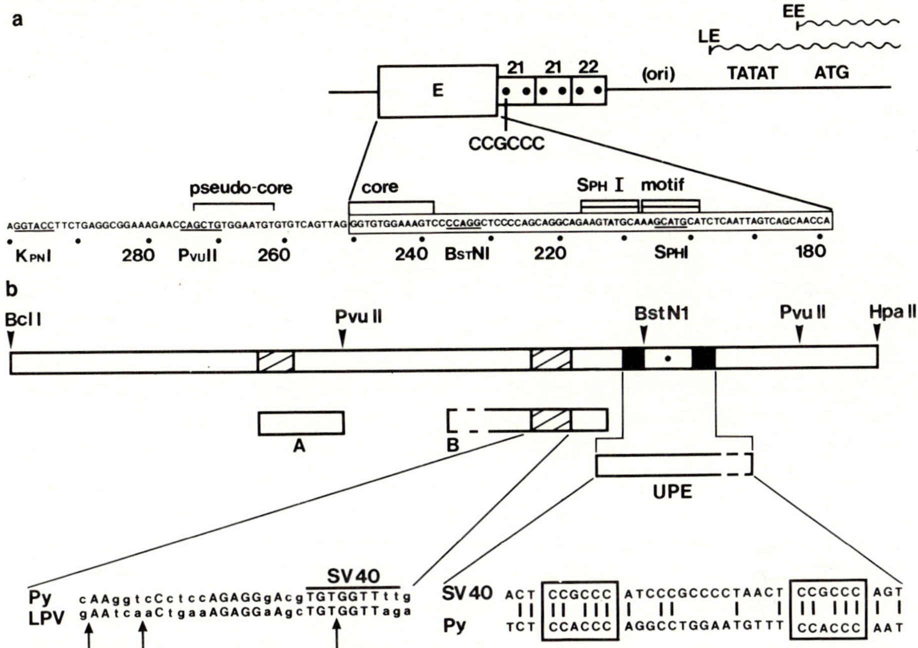

Fig. 1. a Schematic representation of the SV40 early control region. *E* denotes one unit of the 72 bp repeats of the SV40 enhancer, *21* represents 21, 22 bp repeats respectively, *ori* points to the origin of replication, *TATAT* represents the Goldberg-Hogness box, *EE* and *LE* denote the early and late starts of SV40 early mRNA, respectively. Sequences depicted as *core, pseudo-core,* and the *SphI motif* are *cis*-essential for the enhancer activity. **b** Schematic representation of the polyoma (*Py*) control region, *A* and *B* denote enhancer *A* and *B*, respectively. The *dot* between the two solid squares indicates the mutation in the F9-1 mutant. *UPE* represents the upstream promoter element. The regions homologous to the SV40 21 bp repeats are indicated as *solid boxes*. The homology of the *UPE* to SV40 21 bp repeats (see **a**) is pointed out by sequence comparison. The site in enhancer *B* to which a factor is bound is indicated as a *hatched rectangle*. The sequence of the binding site is shown at the bottom of the figure. A comparison to a related sequence in the LPV enhancer is also presented

— as well as the additional "pseudo-core" motif (GTGGAAtG) upstream of the 72 bp repeat are required (Fig. 1a; Zenke et al. 1986). Interestingly, the isolation of revertants indicated that either duplication of the core motif or, likewise, of the SphI-motif is sufficient to restore functional activity (Herr and Clarke 1986). These results demonstrated that at least two domains are required for enhancement function and, moreover, that they can functionally replace one another. Thus, a functionally active enhancer requires defined nucleotide sequences on the same DNA molecule as the promoter. It was therefore tempting to postulate that the stimulating activity might be mediated via an interaction of specific cellular factors with the essential enhancer domains.

A first attempt to identify possible *trans*-acting cellular factors required for enhancer activity employed an in vivo competition assay (Schöler and

Gruss 1984). It quickly became evident that a limited number of these factors are present in all cells examined. In order to elucidate the nature of the factors involved, it was essential to develop suitable in vitro systems. In crude cellular protein extracts (Sassone-Corsi et al. 1984; Sergeant et al. 1984), the SV40 enhancer, when located upstream of suitable promoter, increased transcriptional activity 5- to 15-fold. Using these in vitro transcriptional assays, the effects unique to enhancers could be mimicked qualitatively and subsequent competition experiments confirmed the in vivo results (Wildeman et al. 1984; Sassone-Corsi et al. 1985; Schöler and Gruss 1985). Thus, the factors responsible for this effect must be present in the extracts and can be identified. Recent DNAase I footprinting analyses have indeed revealed protected domains in the SV40 enhancer (Wildeman et al. 1986). Furthermore, experiments with mutants which prevent enhancer function in vivo (Zenke et al. 1986) confirmed the footprint analyses in that they prevented binding of the respective factors in vitro (Wildeman et al. 1986). Recent footprinting experiments using nuclear extracts from cells as diverse in origin as HeLa (Wildeman et al. 1986), BJA-B (a human Burkitt lymphoma cell line; P. Chambon, personal communication and our own unpublished observations) and rat liver cells (S. McKnight, personal communication) have revealed similar, but not always identical footprints. This indicates that some of the factors interacting with the SV40 enhancer in different cell types might be related. Some enhancer-interacting factors have already been partially purified from crude extracts by chromatographic fractionation. In particular, the core-binding factor was separated from other DNA-binding proteins (S. McKnight, personal communication). We can thus conclude that at least two different cellular factors are essential for the activity of the SV40 enhancer in vivo and in vitro, which do not have to be identical in different cell lines.

Cell-Specific Enhancers

In contrast to the SV40 enhancer, which is promiscuously active in most cell types, the enhancers of the papova viruses polyoma, JCV and LPV, as well as of the human T cell leukemia viruses HTLV-I, II, III (see Haseltine et al., this Vol.) exhibit interesting cell type specificities. Polyoma virus replicates and produces progeny in primary or established mouse cell lines while no virus progeny can be detected when embryonal stem cells such as F9 or PCC4 are infected. However, viral mutants were isolated that replicate in both embryonal carcinoma cell lines and in their differentiated derivatives (see Amati 1985 for review). These mutants exhibit point mutations (such as in F9-1) or other alterations within their enhancer/promoter region. In contrast to other papova viruses, polyoma contains two enhancers (A and B; see Amati 1985 for references). In mouse fibroblasts both enhancers function equally well with the polyoma promoter while in embryonal carcinoma cells they are inactive. The point mutation in F9-1 DNA enables enhancer B to potentiate polyoma early gene expression in F9 cells, whereas enhancer A is dispensable. Thus, by introducing a single point mutation, the host range of the virus is altered due to its influence on the enhancer B activity. Recently, a cellular factor was discovered that binds specifically to the polyoma B enhancer sequences. This factor was found to be present in several differentiated and undifferentiated mouse cell lines, and therefore most likely represents a general enhancer-binding factor (Piette et al. 1985; Böhnlein and Gruss 1986; see Fig. 1b).

Another interesting model system for the study of cell-type-specific expression is the human JC virus. Its enhancer (a 98 bp repeat) is tenfold more effective in fetal glial cells than in HeLa cells (see Serfling et al. 1985). Similarly, an enhancer from a lymphotropic papova virus (LPV) produces a transcriptional stimulation of several orders of magnitude in human cells of hematopoietic, but not of epithelial or fibroblastoid origin (Mosthaf et al. 1985). The tropism of the LPV enhancer can be mimicked in in vitro transcription assays. Using these assays, it could also be demonstrated that three enhancers (SV 40, LPV, IgCμ) compete for, and therefore share, common factors (Schöler and Gruss 1985). Furthermore, in the course of these experiments, it was found that the LPV enhancer harbors a sequence motif also present in the polyoma B enhancer (Fig. 1 b).

By far the best-studied tissue-specific cellular enhancers are the immunoglobulin κ- and heavy chain gene (IgCμ) enhancers (see Voss et al. 1986 for review and references). The IgCμ enhancer was discovered to be located in the intron between the joining and constant regions of the immunoglobulin heavy chain gene. This enhancer has been shown to be restricted in its activity to B-cells and to some, but not all, T-cell lines (Mosthaf et al. 1985). In vivo (Mercola et al. 1985) and in vitro competition experiments (Sassone-Corsi et al. 1985; Schöler and Gruss 1985) demonstrated that the enhancement effect was due to the presence of cellular factors. At least some of these factors are DNA-binding proteins, as first demonstrated in vivo using the novel technique of "genomic sequencing" (see Gluzman 1985 for references). The protected regions are outlined in Fig. 2. In contrast to the in vivo data, these DNA-binding factors were shown to be present not only in B-cells, but also in T-cells and HeLa cells (Schlokat et al. 1986; D. Baltimore, personal communication). At least four protected areas are present on the IgCμ enhancer as schematically outlined in Fig. 2, suggesting the presence of four factors. The IgPE-1-binding factor is competed by polyoma and MSV enhancers, and includes the "core" sequence (see Fig. 1a). The IgXP-1-binding factor appears unique for IgCμ sequences, whereas the IgPE-3-binding factor likely interacts with an octanucleotide (Singh et al. 1986) that is found in all immunoglobulin gene promoters examined and other enhancer elements. Finally, interaction of cellular factors with region IgPE-2 can be competed by the LPV enhancer.

A recent mutational analysis of the IgCμ enhancer (Wasylyk and Wasylyk 1986) revealed that in fibroblasts, an element contained within the region from nucleotide 383 to 683 (Fig. 2) has a broader activity spectrum than the

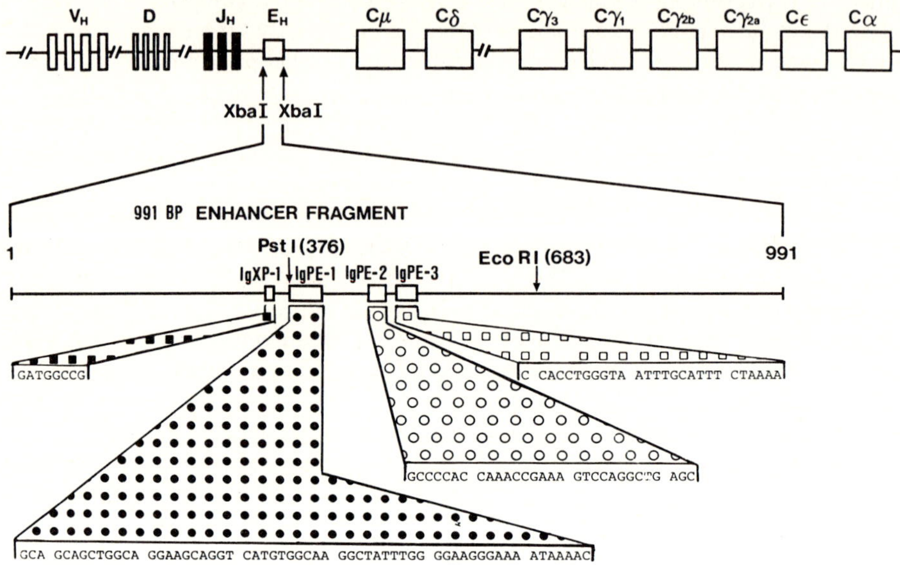

Fig. 2. Schematic representation of the immunoglobulin heavy chain gene and its different components in germline configuration. (Top line: *V* Variable; *D* Diversity; *J* Joining; *C* Constant regions as well as the XbaI enhancer fragment *E* are indicated as *boxes*). The middle diagram shows an enlargement of the 991 bp enhancer fragment with the putative protein binding sites indicated as boxes and *PstI* and *EcoRI* restriction sites at positions 376 and 683 bp marked for orientation. The lower line shows the nucleotide sequences of regions IgXP-1 and IgPE 1 to 3 where binding of cellular factors has been demonstrated

larger 1-kb fragment within which it is contained. This element is about equally as active in lymphoid as in nonlymphoid cell lines. Thus, sequences flanking this region might be responsible for the tissue specificity, suggesting that the central region comprises a ubiquitously active "cassette" (in analogy to SV40, see above), while the adjacent sequences contain the information resulting in restriction of enhancer activity. In this context, region IgXP-1 (Fig. 2) may represent one target for the binding of cell-specific factors.

Many other cellular genes have been identified as harboring tissue-specific enhancers (see Gluzman 1985 for review) including genes from the endocrine and exocrine pancreas, such as insulin and chymotrypsin. In these instances, the enhancer is located in the 5′-flanking sequence in a region shown to bind cellular factors as well (Ohlsson and Edlund 1986). One of these factors seems to be restricted to pancreatic cells.

Another approach which elegantly demonstrates the cell-type specificity of enhancer elements utilizes transgenic mice. Introduction into the mouse germline of upstream control sequences of the rat insulin gene (rI2) linked to the SV40 large T-Ag coding region led to the induction of tumors specific for β-cells of the pancreas (Hanahan 1985 and this Vol.). Similar approaches have been used to reveal the cell specificity of enhancers such as those of the α-fetoprotein gene (Krumlauf et al. 1985), of the mouse mammary tumor virus long terminal repeat (MMTV LTR; Stewart et al. 1984), and of the immunoglobulin genes (Adams et al. 1985).

Inducible Enhancers

The MMTV LTR is stimulated by glucocorticoids (see Yamamoto 1985 and article by Beato, this Vol., for review and references). The stimulation is mediated by hormonal ligands, such as dexamethasone, bound to the glucocorticoid receptor. In the presence of dexamethasone, the glucocorticoid receptor stimulated transcriptional activity through an element termed GRE (glucocorticoid responsive element) that is located within a 122-bp segment of MMTV DNA known to contain regulatory sequences. This element has all the characteristics of an enhancer which, however, is active only in the presence of the ligand. Binding studies as well as linker scanning mutagenesis revealed a consensus hexanucleotide that is essential for binding and enhancement. Recently, GREs have also been found in the Moloney Murine Sarcoma Virus (MoMSV) LTR in a region that carries, in addition, a constitutive enhancer element (DeFranco and Yamamoto 1986). Thus, two different enhancers are present in the MoMSV LTR.

Other well-studied inducible systems include the murine and human metallothionein genes. The human metallothionein-II$_A$ gene (Haslinger and Karin 1985) and the murine metallothionein-I gene (MT-I; Serfling et al. 1985) contain enhancers that can stimulate the expression of the HSV Tk gene and the β-globin gene, respectively. The enhancer element in the human MT gene could, however, be clearly dissociated from the metal responsive element (Haslinger and Karin 1985).

cis and *trans* Repression of Enhancer Activity

Both the MMTV GRE (Ostrowski et al. 1984), as well as β-interferon upstream control sequences (Zinn and Maniatis 1986; Goodbourn et al. 1986), seem to harbor elements that decrease enhancer activity in the absence of inducer. In analogy to prokaryotic repressors (Ptashne et al. 1980), this could be explained by binding of repressor-like molecules to enhancer sequences. In the case of MMTV, this could be the glucocorticoid receptor without its ligand. It remains unclear whether this negative effect can be exerted over large distances or whether the putative repressor-binding site must lie between the enhancer and promoter domains. The region located between 2- and 4-kb upstream of the rat insulin rI1 gene is largely distance-independent. This fragment downregulates the activity of the SV40 and MoMSV heterologous enhancers in transient assays, and was therefore termed "silencer" (Laimins et al. 1986). It was also shown that this sequence belongs to a rat repetitive sequence family (LINES). However, the question of why this sequence does not repress the activity of the endogenous rI1 gene remains unanswered.

Surprisingly, the adenovirus (Ad) E1A products, which are known to stimulate the activity of several genes, can also act, either directly or indirectly, as repressors of several enhancers (see Bos and van der Eb 1985 for references). Notably, Ad2 E1A products negatively interfere with SV40 and polyoma as well as with IgCµ enhancer activity (Hen et al. 1985). Also, Ad5 E1A products have been shown to repress transcription from the SV40 promoter. Interestingly, the E1A gene products of the highly oncogenic Ad12 strain suppress expression of endogenous class I major histocompatibility (MHC) antigen genes, also known to contain enhancers (Gillies et al. 1984). However, whether MHC gene repression in vivo is exerted by a related mechanism remains unclear. It is tempting to speculate that the E1A repression is paralleled by related cellular proteins, a mechanism that was suggested for gene regulation in embryonal carcinoma stem cells (Imperiale et al. 1984; Gorman et al. 1985). In support of this notion is the finding that the polyoma F9-1 mutant enhancer cannot be repressed by Ad2 E1A products, in contrast to the wild-type enhancer (Hen et al. 1986). This result suggests that the activity of F9-1 mutants in EC stem cells is due to an elimination of the repression effect.

Conclusions and Perspectives

Research over the past few years has provided deeper insights into some of the requirements for the functional activity of enhancer elements but has left us with an incomplete understanding of the mechanism by which they function. At least for some enhancers, a detailed analysis of the *cis*-essential sequences has been performed. Mutational analysis in conjuction with sequence comparison revealed the presence of common stretches of sequences (cassettes). At least two or more cassettes, either identical or nonidentical, seem to be required to bring about the enhancement effect. Thus, a small number of cassettes can generate a high degree of specificity, depending upon the respective combination of cassettes. Most interestingly, recent in vitro experiments demonstrated that at least some, if not all, of these cassettes can interact specifically with cellular (probably protein) factors. However, most of the factors identified so far by DNA-binding assays seem to be present in all cell types. The difficulty to identify cell-specific factors could be explained by the assumption that they are less abundant than the general factors and would therefore not be detectable in DNA-binding assays using crude extracts. Alternatively, the cell-specific factors may not be DNA-binding proteins but may exert their effects via protein-protein interaction(s). In vitro assays such as delayed migration of DNA-protein complexes in native gels might shed some light on the molecular basis of cell type specificity. Future research must be directed toward identification of the *trans*-factors (and their genes) involved in cell-specific expression. If, on the other hand, complex structures such as nuclear scaffolds (Mirkowitch et al. 1984) are involved in enhancer function, we still have a long way to go before the mechanism of enhancer function will be completely understood.

Acknowledgments. We thank E. Böhnlein, L. Mosthaf, and H. Schöler for critical discussions and for providing data prior to publication. Moreover, we are indebted to Dr. A. Colberg-Poley and Stephan Voss for critical reading and, finally, to G. Wee and B. John for typing the manuscript. Research performed in our lab was funded by the Bundesministerium für Forschung und Technologie, the Deutsche Forschungsgemeinschaft and the Fonds der Chemischen Industrie.

References

Adams JM, Harris AW, Pinkert CA, Corcoran LM, Alexander WS, Cory S, Palmiter RD, Brinster RL (1985) The c-myc oncogene driven by immunoglobulin enhancers induces lymphoid malignancy in transgenic mice. Nature 318:533–538

Amati P (1985) Polyoma regulatory region: a potential probe for mouse cell differentiation. Cell 43:561–562

Böhnlein E, Gruss P (1986) Interaction of distinct nuclear proteins with sequences controlling the expression of polyoma early genes. Mol Cell Biol 6:1401–1411

Bos JL, van der Eb AJ (1985) Adenovirus region E1A: transcription modulator and transforming agent. TIBS 10:310–313

DeFranco D, Yamamoto KR (1986) Two different factors act separately or together to specify functionally distinct activities at a single transcriptional enhancer. Mol Cell Biol 6:993–1001

Gillies SD, Folson V, Tonegawa S (1984) Cell type specific enhancer element associated with a mouse MHC gene, Eβ. Nature 310:594–597

Gluzman Y (ed) (1985) Eucaryotic transcription: The role of cis- and trans-acting elements in initiation. Current communications in molecular biology. Cold Spring Harbor Lab, Cold Spring Harbor, NY

Goodbourn S, Burstein H, Maniatis T (1986) The human β-interferon gene enhancer is under negative control. Cell 45:601–610

Gorman CM, Rigby PWJ, Lane DP (1985) Negative regulation of viral enhancers in undifferentiated embryonic stem cells. Cell 42:519–526

Gruss P (1984) Magic enhancers. DNA 3:1–5

Hanahan D (1985) Heritable formation of pancreatic β-cell tumours in transgenic mice expressing recombinant insulin/simian virus 40 oncogenes. Nature 315:115–122

Haslinger A, Karin M (1985) Upstream promoter element of the human metallothionein-IIA gene can act like an enhancer element. Proc Natl Acad Sci USA 82:8572–8576

Hen R, Borelli E, Chambon P (1985) Repression of the immunoglobulin heavy chain gene enhancer by the adenovirus-2 E1A products. Science 230:1391–1394

Hen R, Borelli E, Fromental C, Sassone-Corsi P, Chambon P (1986) A mutated polyoma virus enhancer which is active in undifferentiated embryonal carcinoma cells is not repressed by the adenovirus-2 E1A products. Nature 321:249–251

Herr W, Clarke J (1986) The SV40 enhancer is composed of multiple functional elements that can compensate for one another. Cell 45:461–470

Imperiale MJ, Kao HT, Feldman L, Nevins J, Strickland S (1984) Common control of the heat shock gene and early adenovirus genes: evidence for a cellular E1a-like activity. Mol Cell Biol 4:867–874

Khoury G, Gruss P (1983) Enhancer elements. Cell 33:313–314

Krumlauf R, Hammer RE, Tilghman SM, Brinster RL (1985) Developmental Regulation of α-fetoprotein genes in transgenic mice. Mol Cell Biol 5:1639–1648

Laimins L, Holmgren-König M, Khoury G (1986) Transcriptional "silencer" element in rat repetitive sequences associated with the rat insulin 1 gene locus. Proc Natl Acad Sci USA (in press)

Mercola M, Goverman J, Mirell C, Calame K (1985) Immunoglobulin heavy chain enhancer requires one or more tissue-specific factors. Science 227:266 – 270

Mirkovitch J, Mirault ME, Laemmli UK (1984) Organization of the higher-order chromatin loop: Specific DNA attachment sites on nuclear scaffold. Cell 39:223 – 232

Mosthaf L, Pawlita M, Gruss P (1985) A viral enhancer element specifically active in human haematopoietic cells. Nature 315:597 – 600

Ohlsson H, Edlund T (1986) Sequence-specific interactions of nuclear factors with the insulin gene enhancer. Cell 45:35 – 44

Ostrowski MC, Huang AL, Kessel M, Wolford RG, Hager GL (1984) Modulation of enhancer activity by the hormone responsive regulatory element from mouse mammary tumor virus. EMBO J 3:1891 – 1899

Piette J, Kryszke MH, Yaniv M (1985) Specific interaction of cellular factors with the B enhancer of polyomavirus. EMBO J 4:2675 – 2685

Ptashne M, Jeffrey A, Johnson AD, Maurer R, Meyer BJ, Pabo CO, Roberts TM, Sauer RT (1980) How the lambda repressor and cro work. Cell 19:1 – 11

Sassone-Corsi P, Dougherty JP, Wasylyk B, Chambon P (1984) Stimulation of in vitro transcription from heterologous promoters by the SV40 enhancer. Proc Natl Acad Sci USA 81:308 – 312

Sassone-Corsi P, Wildeman A, Chambon P (1985) A transacting factor is responsible for the simian virus 40 enhancer activity in vitro. Nature 313:458 – 463

Schlokat U, Bohmann D, Schöler H, Gruss P (1986) Nuclear factors binding specific sequences within the immunoglobulin enhancer differentially interact with other enhancer elements. (submitted)

Schöler HR, Gruss P (1984) Specific interaction between enhancer-containing molecules and cellular components. Cell 36:403 – 411

Schöler HR, Gruss P (1985) Cell-type specific transcriptional enhancement in vitro requires the presence of trans-acting factors. EMBO J 4:3005 – 3013

Serfling E, Jasin M, Schaffner W (1985) Enhancers and eucaryotic gene transcription. TIG 1:224 – 230

Sergeant A, Bohmann D, Zentgraf H, Weiher H, Keller W (1984) A transcription enhancer acts in vitro over distances of hundreds of base-pairs on both circular and linear templates but not on chromatin-reconstituted DNA. J Mol Biol 180:577 – 600

Singh H, Sen R, Baltimore D, Sharp PA (1986) A nuclear factor that binds to a conserved sequence motif in transcriptional control elements of immunoglobulin genes. Nature 319:154 – 158

Stewart TA, Pattengale PK, Leder P (1984) Spontaneous mammary adenocarcinomas in transgenic mice that carry and express MTV/myc fusion genes. Cell 38:627 – 637

Voss SD, Schlokat U, Gruss P (1986) The role of enhancers in the regulation of cell-type-specific transcriptional control. TIBS II:287 – 289

Wasylyk C, Wasylyk B (1986) The immunoglobulin heavy-chain B-lymphocyte enhancer efficiently stimulates transcription in non-lymphoid cells. EMBO J 5:553 – 560

Wasylyk B, Wasylyk C, Chambon P (1984) Short and long range activation by the SV40 enhancer. Nucleic Acids Res 12:5589 – 5608

Wildeman AG, Sassone-Corsi P, Grundström T, Zenke M, Chambon P (1984) Stimulation of in vitro transcription from the SV40 early promoter by the enhancer involves a specific trans-acting factor. EMBO J 13:3129 – 3133

Wildeman AG, Zenke M, Schatz C, Wintzerith M, Grundström T, Matthes H, Takahashi K, Chambon P (1986) Specific protein binding to the SV40 enhancer in vitro. Mol Cell Biol 6:2098 – 2105

Yamamoto KR (1985) Steroid receptor regulated transcription of specific genes and gene networks. Annu Rev Genet 19:209 – 252

Zenke M, Grundström T, Matthes H, Wintzerith M, Schatz C, Wildeman A, Chambon P (1986) Multiple sequence motifs are involved in SV40 enhancer function. EMBO J 5:387 – 397

Zinn K, Maniatis T (1986) Detection of factors that interact with the human β-interferon regulatory region in vivo by DNAase I footprinting. Cell 45:611 – 618

The Effect of DNA Methylation on DNA-Protein Interactions and on the Regulation of Gene Expression

WALTER DOERFLER

The transcriptional activity of a gene can be controlled in a number of ways and modulated by different factors. The interaction of specific proteins with highly specific sequence elements ("motifs") in the promoter regions of genes plays a pivotal role in this regulation (see Schlokat and Gruss, this Vol., and Beato, this Vol.). In addition, the methylation of promoter sequences can cause the inactivation of genes in a number of eukaryotic systems (for reviews see Razin and Riggs 1980; Doerfler 1981, 1983; Riggs and Jones 1983). This chapter discusses the role of site-specific promoter methylations in the control of gene expression. (For detailed information on the biochemistry and biology of DNA methylation, see Taylor 1984; Adams and Burdon 1985; Trautner 1984; Razin et al. 1984; and Cantoni and Razin 1985).

A Two-Step Model for Gene Activation by DNA Binding Proteins

One reason for the relatively limited understanding of promoter function may lie in the linearity of most analytical approaches chosen so far. It is likely that promoter function in fact depends on DNA-protein interactions that assume complex conformations. Any model trying to explain the shut-off effects of site-specific methylations on promoter activity will have to encompass highly specific DNA-protein interactions. It is a reasonable but still unproven assumption that promoter methylations affect the interactions with proteins essential in gene expression.

Two distinct steps can be proposed in the interaction of DNA-binding proteins with a given regulatory sequence. In a first step, "conformational proteins" bind to specific "conformational sequences" which convert the DNA to an "alerted" state. In a second step, the "alerted" DNA interacts with proteins involved in transcription, and subsequently transcription can be initiated. Conformational sequences could be part of or separate from the promoter, enhancer, and other sequences essential for transcription (transcriptional sequences). The term conformational sequence or conformational protein implies a functional rather than an anatomical unit. Conformational and transcriptional sequences do not have to be adjacent to each other, but may be juxtaposed only after the DNA assumes a certain configuration. In this model methylation is suggested to inactivate a promoter by modulating either one of the two types of sequences or both. For example, it is conceivable that a

Oncogenes and Growth Control
Edited by P. Kahn and T. Graf
© Springer-Verlag Berlin Heidelberg 1986

methylation-inhibited promoter might be reactivated by supplying unmethylated conformational or transcriptional sequences upstream or downstream of the methylated sequences. It should be emphasized that this model is purely hypothetical. Its main purpose is to provide a framework for the discussion that follows and to stimulate the design of new experiments.

Significance of DNA Methylation in the Regulation of Eukaryotic Gene Expression

Patterns of DNA methylation are faithfully inherited, probably by the action of maintenance DNA methyltransferases which, after DNA replication, restore a previously given pattern of methylation. In higher eukaryotes the predominant site of DNA methylation is in 5′-CG-3′ dinucleotide sequences. It is unknown how specific patterns of DNA methylation arise and to what extent they are subject to change. DNA can be methylated de novo in eukaryotic cells (Sutter et al. 1978). A possible mechanism of enzymatic demethylation has been suggested (Razin et al. 1986). The activity of maintenance DNA methyltransferases has to be selectively inhibited during replication to cause specific changes in DNA methylation. From this notion, and from the fact that patterns of DNA methylation cannot easily be changed, it follows that DNA methylation plays a role in long-term gene inactivation.

The following observations have established a role of sequence-specific methylations in the regulation of eukaryotic gene expression.

1. Inverse correlations between promoter methylations and gene expression have been demonstrated for most genes investigated (Sutter and Doerfler 1980; Kruczek and Doerfler 1982; Stein et al. 1983). The apparent exceptions might be explained by the fact that in these genes the sequences important for promoter regulation are not known.

2. Inactive genes could be reactivated by treating cell cultures with 5-aza-deoxycytidine (Constantinides et al. 1977; Taylor and Jones 1979; Groudine et al. 1981). The deoxycytidine analog presumably acts as an inhibitor of DNA methyltransferases and consequently alters patterns of DNA methylation. When critical promoter sites are demethylated in this way, gene expression ensues. In one of the analyzed genes, demethylation correlated not only with gene activation but also with an alteration of chromatin structure as evidenced by an increased DNase-I sensitivity (Groudine et al. 1981).

3. Selective, site-specific methylations in the promoter of several genes have been found to cause gene inactivation (Busslinger et al. 1983; Kruczek and Doerfler 1983; Langner et al. 1984; Keshet et al. 1985; Langner et al. 1986). These experiments were done by the enzymatic methylation of promoters in vitro with subsequent transfer into cells. Gene expression was either measured in transient expression systems or after stable integration of the transferred DNA and selection of DNA-positive cells after double-transfection with a selectable marker gene. The results obtained demonstrated that in

most genes methylation of specific sequences within the promoter and 5′ flanking sequences led to inactivation, whereas methylation in the coding or 3′ flanking sequences had no effect on gene activity. An exception is the thymidine kinase gene (Keshet et al. 1985). Depending on the arrangement of conformational and transcriptional sequences of the gene studied, the pattern of methylation associated with the inactive state is different from gene to gene.

For the promoter of the adenovirus type 12 (Ad12) E1A gene, it has been shown that methylation of specific 5′-CG-3 sequences abolished promoter function. Introduction of N^6-methyldeoxyadenosine into certain E1A promoter sequences (Knebel and Doerfler 1986) and into the thymidine kinase gene also led to their inactivation (Waechter and Baserga 1982). These findings were surprising in view of the fact that this modification is usually not present in mammalian DNA, and suggest that a methyl group introduced into the unusual A position can also be effective in promoter inactivation. In the model proposed above, methylations of sequences critically involved in gene expression either as conformational or transcriptional sequences or both modulate the binding of conformational or transcriptional proteins and eventually block expression. Since such critical sequences have not yet been precisely mapped for any gene, it is not surprising that seemingly disparate results have occasionally been obtained in different systems. A careful analysis of the methylation patterns of genes investigated in their inactivated states is an essential precondition for any meaningful in vitro methylation studies.

Viral Promoters as Tools

This discussion will be limited to the few viral promoters which have been studied in detail. It was initially established that the early gene promoters of integrated Ad12 DNA and the late E2A gene promoter of integrated Ad2 DNA were hypermethylated when the genes were not expressed and undermethylated when they were expressed (Sutter and Doerfler 1980; Vardimon et al. 1980; Kruczek and Doerfler 1982). In these studies, the 5′-CG-3′ sites of these promoters were analyzed using the isoschizomeric restriction endonuclease pair HpaII and MspI (Waalwijk and Flavell 1978). It was found that in particular for the late E2A gene promoter, this inverse correlation was consistent in all 14 5′-CCGG-3′ sites. In promoters where such a striking correlation was not seen, different 5′-CG-3′-containing sequences might be functionally significant.

These observations encouraged us to attempt to inactivate the promoters by in vitro methylation. For this purpose, constructs of the E1A promoter methylated or unmethylated in 5′-CCGG-3′ or 5′-GCGC-3′ sequences, linked to an indicator gene (CAT) were transfected into mammalian cells or injected into *Xenopus* oocytes. The methylated promoter was found to be inactive, while the unmethylated promoter led to expression of the CAT gene (Kruzcek and Doerfler 1983). Similarly, when all 14 5′-CCGG-3′ sequences of the late E2A gene promoter of Ad2 were methylated, the gene was inactive upon microinjection into *Xenopus* oocytes (Vardimon et al. 1982). In this system it could be further demonstrated that methylation of three sites located in the promoter and in 5' sequences of the E2A gene sufficed for inactivation, whereas methylation of the remaining 11 sites located in the 3' region were ineffective (Langner et al. 1984), an effect also observed in mammalian cells (Langner et al. 1986). During DNA replication even completely methylated promoter sequences might become transiently hemimethylated. Therefore, it will be interesting to determine whether the methylation of the three critical sites either in the reading or in the complementary strand of a E2A promoter construct has an effect on its activity. Preliminary results demonstrate that in vitro methylation of the transcribed strand inactivates the E2A promoter (K.-D. Langner, D. Renz, U. Lichtenberg and W. Doerfler, in preparation).

Promoter-inactivation by site-specific methylation was also demonstrated in insect cells for the p10 gene promoter of Autographa californica nuclear polyhedrosis virus (Knebel et al. 1985). Although 5-methyldeoxycytidine is not a prevalent signal in insect cells, it is presumably recognized as a conformational and/or a transcriptional block that leads to the shutdown of the viral promoter. This observation underscores the general importance of 5-methyldeoxycytidine as a long-term regulatory signal within a wide spectrum of species.

Recent studies have shown that the inactivating effect of promoter methylation can be overcome by certain viral sequences. Thus, transactivating functions encoded by the E1A region of integrated Ad2 or Ad5 DNA can reactivate the late E2A promoter of Ad2 after being silenced by 5'-CCGG-3' methylation (Langner et al. 1986). This promoter reactivation was found to be accompanied by transcription initiation in the E2A cap sequence. In another series of investigations, sequences of the human cytomegalovirus enhancer (Boshart et al. 1985) were found to reactivate the methylation-inactivated E2A promoter of Ad2 DNA in mammalian cells (K.-D. Langner, B. Fleckenstein and W. Doerfler, in preparation). It is conceivable that in this case the unmethylated cytomegalovirus enhancer can substitute for conformational or transcriptional sequences in a construct in which the E2A promoter was inactivated by methylation.

Future Perspectives

An immediate goal for further research on the mechanism of gene inactivation by site-specific methylations is to determine the in vivo distribution of methylated sites in regulatory regions surrounding active or inactivated genes. This analysis of methylation patterns should be facilitated by the elegant technique of genomic sequencing (Church and Gilbert 1984). The patterns thus determined could then be simulated by in vitro methylation experiments. It will also be necessary to assess the minimal number of methylated deoxycytidines required for promoter inactivation and their exact locations.

Site-specific promoter methylations can function as signals for long-term gene inactivations and these signals are probably recognized in many eukaryotic species. Since it is assumed that the extent and site-specificity of DNA methylation is governed by DNA methyltransferase(s), it will be interesting to determine how the activity of these enzymes is controlled in growing versus resting cells, in organs at different stages of development and in tumors at different stages of progression. The rationale for these studies is to correlate the expression of a given gene (for example, a proto-oncogene) to specific patterns of methylation in its promoter. Investigations in this area promise to yield exciting new insights about the mechanisms of gene regulation in differentiation and malignant transformation.

Acknowledgments. This paper is dedicated to Albrecht K. Kleinschmidt, University of Ulm, Germany, on the occasion of his 70th birthday. I am indebted to Petra Böhm for excellent editorial work. Research in the author's laboratory has been supported by the Deutsche Forschungsgemeinschaft through SFB74-C1.

References

Adams RLP, Burdon RH (1985) Molecular biology of DNA methylation. Springer, New York Berlin Heidelberg

Boshart M, Weber F, Jahn G, Dorsch-Häsler K, Fleckenstein B, Schaffner W (1985) A very strong enhancer is located upstream of an immediate early gene of human cytomegalovirus. Cell 41:521 – 526

Busslinger M, Hurst J, Flavell RA (1983) DNA methylation and the regulation of globin gene expression. Cell 34:197 – 206

Cantoni GL, Razin A (eds) (1985) Biochemistry and biology of DNA methylation. Liss, New York

Church GM, Gilbert W (1984) Genomic sequencing. Proc Natl Acad Sci USA 81:1991 – 1995

Constantinides PG, Jones PA, Gevers W (1977) Functional striated muscle cells from non-myoblast precursors following 5-azacytidine treatment. Nature 267:364 – 366

Doerfler W (1981) DNA methylation – a regulatory signal in eukaryotic gene expression. J Gen Virol 57:1 – 20

Doerfler W (1983) DNA methylation and gene activity. Annu Rev Biochem 52:93 – 124

Groudine M, Eisenman R, Weintraub H (1981) Chromatin structure of endogenous retroviral genes and activation by an inhibitor of DNA methylation. Nature 292:311 – 317

Keshet I, Yisraeli J, Cedar H (1985) Effect of regional DNA methylation on gene expression. Proc Natl Acad Sci USA 82:2560 – 2564

Knebel D, Doerfler W (1986) N^6-methyldeoxyadenosine residues at specific sites decrease the activity of the E1a promoter of adenovirus type 12 DNA. J Mol Biol 189:371 – 375

Knebel D, Lübbert H, Doerfler W (1985) The promoter of the late p10 gene in the insect nuclear polyhedrosis virus Autographa californica: activation by viral gene products and sensitivity to DNA methylation. EMBO J 4:1301 – 1306

Kruczek I, Doerfler W (1982) The unmethylated state of the promoter/leader and 5′-regions of integrated adenovirus genes correlates with gene expression. EMBO J 1:409 – 414

Kruczek I, Doerfler W (1983) Expression of the chloramphenicol acetyltransferase gene in mammalian cells under the control of adenovirus type 12 promoters: effect of promoter methylation on gene expression. Proc Natl Acad Sci USA 80:7586 – 7590

Langner K-D, Vardimon L, Renz D, Doerfler W (1984) DNA methylations of three 5′-C-C-G-G-3′ sites in the promoter and 5′ region inactivated the E2a gene of adenovirus type 2. Proc Natl Acad Sci USA 81:2950 – 2954

Langner K-D, Weyer U, Doerfler W (1986) Transeffect of the E1 region of adenoviruses on the expression of a prokaryotic gene in mammalian cells: Resistance to 5'-CCGG-3' methylation. Proc Natl Acad Sci USA 83:1598 – 1602

Razin A, Riggs AD (1980) DNA methylation and gene function. Science 210:604 – 610

Razin A, Cedar H, Riggs AD (eds) (1984) DNA methylation. Springer, New York Berlin Heidelberg

Razin A, Szyf M, Kafri T, Roll M, Giloh H, Scarpa S, Carotti D, Cantoni GL (1986) Proc Natl Acad Sci USA 83:2827 – 2831

Riggs AD, Jones PA (1983) 5-methylcytosine, gene regulation, and cancer. Adv Cancer Res 40:1 – 30

Stein R, Sciaki-Gallili N, Razin A, Cedar H (1983) Pattern of methylation of two genes coding for housekeeping functions. Proc Natl Acad Sci USA 80:2422 – 2426

Sutter D, Doerfler W (1980) Methylation of integrated adenovirus type 12 DNA sequences in transformed cells is inversely correlated with viral gene expression. Proc Natl Acad Sci USA 77:253 – 256

Sutter D, Westphal M, Doerfler W (1978) Patterns of integration of viral DNA sequences in the genomes of adenovirus type 12-transformed hamster cells. Cell 14:569 – 585

Taylor JH (1984) DNA methylation and cellular differentiation. Springer, Wien New York

Taylor SM, Jones PA (1979) Multiple new phenotypes induced in 10T1/2 and 3T3 cells treated with 5-azacytidine. Cell 17:771 – 779

Trautner TA (ed) (1984) Methylation of DNA. Springer, Berlin Heidelberg New York

Vardimon L, Neumann R, Kuhlmann I, Sutter D, Doerfler W (1980) DNA methylation and viral gene expression in adenovirus-transformed and -infected cells. Nucl Acids Res 8:2461 – 2473

Vardimon L, Kressmann A, Cedar H, Maechler M, Doerfler W (1982) Expression of a cloned adenovirus gene is inhibited by in vitro methylation. Proc Natl Acad Sci USA 79:1073 – 1077

Waalwijk C, Flavell RA (1978) MspI an isoschizomer of HpaII which cleaves both unmethylated and methylated HpaII sites. Nucl Acid Res 5:3231 – 3236

Waechter DE, Baserga R (1982) Effect of methylation on expression of microinjected genes. Proc Natl Acad Sci USA 79:1106 – 1110

Trans-Acting Elements Encoded in Immediate Early Genes of DNA Tumor Viruses

LENNART PHILIPSON

The adenovirus E1A region is capable of inducing the immortalization of primary rodent fibroblasts and cooperates with other genes such as adenovirus E1B, polyoma middle T or the *ras* gene to induce a fully transformed phenotype (see Land, this Vol., and Cuzin and Mougneau, this Vol.). It is probably the first transcription unit of adenovirus to be activated during lytic infection. Its gene products activate the transcription of other early regions of the viral genome. Several other early viral proteins produced in cells infected by DNA viruses also activate viral genes expressed during the later phases of the lytic cycle. Among these are the SV40 large T-antigen, the polyoma large T-antigen, the herpes virus immediate early (IE) proteins and the E2 gene-product of bovine papilloma virus (Spalholz et al. 1985). The *tat* proteins of the RNA containing human T-cell leukemia virus HTLV I and II also have a transactivating function (see Haseltine et al., this Vol.). All these genes have efficient promoters that are activated by *cis*-elements such as enhancers (see Schlokat and Gruss, this Vol.). The strong early viral promoters thus have a dual function: they are regulated by *cis* elements to ensure initiation of infection and their products *trans*-activate the promoter of other early viral genes to permit the virus to progress through its lytic cycle. Since most studies have been done with the adenovirus E1A region, I will review this gene as a model for the organization and the mechanism of action of *trans*-acting elements.

Organization of the E1A Region

The early region 1A of adenoviruses (Ad) 2 and 5 each express three mRNAs with a relative size of 13S, 12S and 9S (Fig. 1). The 13S and 12S mRNAs of Ad2 and 5 code for two proteins of 289 and 243 amino acids (266 and 235 for Ad12). These two proteins are responsible for the *trans* functions and are identical except for the additional 46 amino acid of the larger protein. A smaller protein with an unidentified function is expressed from the 9S mRNA at late times during the lytic cycle from the same transcriptional unit of Ad2 and Ad5 but it has not been found in Ad12. The 13S and the 12S mRNAs originate by differential splicing of the main transcript. The 13S mRNA does not function as a precursor for the smaller mRNA by subsequent splicing (Svensson et al. 1983). Recently, three different enhancer elements have been identified that are located between nucleotides − 140 and − 500 relative to the

Oncogenes and Growth Control
Edited by P. Kahn and T. Graf
© Springer-Verlag Berlin Heidelberg 1986

 L. Philipson

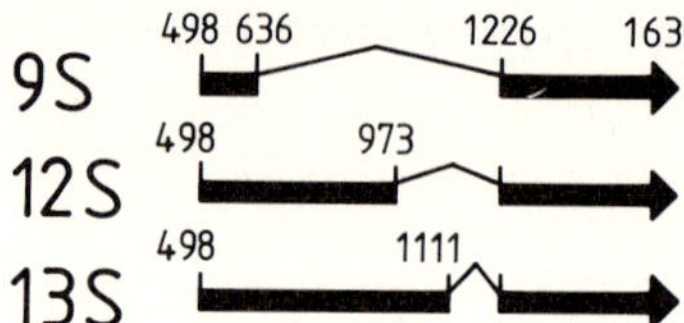

Fig. 1. E1A gene mRNAs

transcription start of E1A (Hearing and Shenk 1983; Hen et al. 1983). A region between nucleotides $+859$ and $+923$ appears also to enhance transcription from a plasmid E1A gene (Osborne et al. 1984), but this was not observed in a deletion analysis of the viral genome (Hearing and Shenk 1985).

The protein products of the 13S and 12S mRNAs, as deduced from the DNA sequence, would be predicted to have 31.9 and 26.5 kDa, but the products actually encountered range in size from 54 – 25 kDa (Harlow et al. 1985), suggesting that they undergo post-translational modifications. The polypeptides expressed in *E. coli* from cDNA clones are homogenous but exhibit apparent molecular weights in excess of the predicted ones (Ferguson et al. 1985), demonstrating an anomalous electrophoretic mobility in SDS gels. In vivo, the E1A proteins have half-lives of around 60 min (Branton and Rowe 1985), indicating that some of the low molecular weight forms result from degradation. Although several other immediate early viral proteins exhibit similar effects as the E1A products, no sequence homology between E1A with the herpes virus IE proteins or with the HTLV *tat*-proteins has been found. Only limited homology has been detected to the papova virus large T-antigens (Stabel et al. 1985).

The phenotypic effects of the E1A gene are complex since it mediates not only activation of several early viral genes, but it can also activate or repress transcription of plasmid-borne and certain chromosomal genes. These effects are discussed in the following two sections.

Transactivation by E1A

The E1A gene products can stimulate transcription from five early transcription units on the adenovirus genome (Jones and Shenk 1979; Berk et al. 1979). This stimulation appears to occur primarily at the level of transcription initiation (Nevins 1981) although claims (Katze et al. 1981) for an effect on mRNA stability have not been excluded. A direct comparison of transactivation by SV40 large T-antigen, adenovirus E1A, and herpes virus IE proteins on eukaryotic promoters in plasmid constructs containing CAT as an indicator gene (Alwine 1985) revealed that T-antigen and the IE protein are promiscuous activators of the SV40 late, the adenovirus E3, the $\alpha2(I)$-collagen promoters and the LTR promoter of Rous sarcoma virus.

In contrast, the E1 proteins activate only the adenovirus E3 promoter and suppress the basal activity of the others. But E1A can also stimulate transcription of nonviral genes such as the rabbit β-globin and the rat preproinsulin gene introduced into mammalian cells by transfection or infection (Green et al. 1983; Gaynor et al. 1984; Svensson and Akusjärvi 1984); the endogenous copies of these genes are, however, unaffected by E1A. On the other hand, at least two chromosomal genes appear to be stimulated: the gene coding for the 70 kDa heat shock protein (Kao and Nevins 1983) and the β-tubulin gene (Stein and Ziff 1984). Thus any theory explaining the mechanism of activation by the E1A products must account for the difference in chromatin structure of transfected plasmids and viral and cellular chromatin. Attempts to identify the sequences required for E1A activation in the responding genes have shown that they are inseparable from elements responsible for basal level expression. An exception is the adenovirus E4 unit in which E1A-mediated activation appears to depend on an upstream element (Gilardi and Perricaudet 1984).

Recently it has been shown that the E1A products as well as the herpes virus IE protein can also enhance *pol* III transcription to the same level as that of *pol* II (Berger and Folk 1985; Gaynor et al. 1985). It is, therefore, tempting to conclude that the E1A proteins do not transactivate by combining directly or indirectly with the specific promoter sequence, but rather induce a change in the nuclear environment favoring both *pol* II and *pol* III transcription from exogenously introduced genes. This environmental change in the nuclei occasionally also triggers chromosomal genes.

Utilizing *E. coli* manufactured proteins specific for the 13S and 12S mRNAs respectively, it was demonstrated that both E1A products are equally effective on a weight basis in activating the adenovirus early transcription units (Ferguson et al. 1985). Similar results have been obtained when comparing the two products with regard to induction of cell growth in quiescent mouse fibroblasts (Stabel et al. 1985), suggesting that also in this assay the 12S and 13S mRNA products are equally effective.

E1A proteins derived from the 13S or 12S mRNA accumulate in the nucleus of the cell. Dissection of the E1A proteins by deletion mutant analyses and expression in *E. coli* demonstrated that the C-terminal end contains a signal for transport across the nuclear membrane (Krippl et al. 1985). A basic sequence containing lys-arg-pro-arg, found in E1A proteins of both Ad5 and Ad12 as well as in polyoma large T antigen, seems to be required for the transport. On the other hand, information for efficient transactivation of the early adenovirus genes appears to be contained in sequences of the first exon corresponding to amino acid 121 to 150 from the N-terminus. A portion of the second exon also contributes to transactivation but a deletion of 87 amino acids at the C-terminus induces nearly wild-type levels of early viral mRNAs (Jones and Shenk 1979). The anomaly with these results is that E1A mutants which lack the C-terminal signal and only slowly accumulate in the nucleus can still transactivate, perhaps suggesting that the primary site of action of the E1A products is not in the nucleus.

Repression by E1A

In addition to their capacity of transcriptional activation, the E1A proteins can repress the SV40 promoter (Borelli et al. 1984; Velcich and Ziff 1985) and possibly their own promoter (Smith et al. 1985). The latter finding was made using a plasmid construct with a change in the reading frame that leads to a truncation of the protein after the synthesis 70 N-terminal amino acids. This plasmid overexpressed the E1A mRNA 30-fold, suggesting that the expression of E1A products is subject to an autoregulatory mechanism. Surprisingly, however, this effect was not observed when viral constructs were used (Hearing and Shenk 1985).

A special case of E1A repression is the recent finding by van der Eb and his colleagues (Schrier et al. 1983) that, in contrast to Ad5 E1A, Ad12 E1A can suppress expression of class I transplantation antigens (and possible other antigens) at the surface of transformed cells. The loss of expression of class I may explain the higher tumorigenicity of Ad12 as compared to Ad5, perhaps by providing a mechanism for these cells to escape surveillance by cytotoxic T cells. On the other hand, others found only partially reduced expression of class I antigens in rat cells by Ad12 infection, and suggested that restriction of class I antigens is of minor significance in determining the tumorigenicity of adenovirurses (Mellow et al. 1984). Recent transfection experiments have, however, established that the Ad12 E1A gene can suppress accumulation of class I gene mRNA, although it is not yet clear whether the control is at the level of transcription initiation or mRNA stability (Vaessen et al. 1986 and Van der Eb, personal communication). The finding that γ-interferon rapidly restimulates the appearance of class I antigens in Ad12-transformed mouse cells irrespective of the class I genotype of the cells (Eager et al. 1985) suggests that Ad12 does not preferentially transform cells that are deficient in expression of class I genes but that it actively suppresses these genes. Transfection and concomitant over-expression of class I antigens in Ad12 transformed cells also eliminates the tumorigenicity of these cells, lending further support to the idea that the ability to suppress class I antigens is of importance in conferring a high tumorigenic potential to Ad12 (Tanaka et al. 1985).

A Complex Model for E1A Function

It is difficult to provide a comprehensive model for the action of the E1A gene in molecular terms due to its pleiotropic effects, involving transactivation of some *pol* II and *pol* III genes, its preferential effects on plasmid-encoded versus chromosomal genes and the repression of selected plasmid and cellular genes. However, it seems to be clear that the E1A gene does not directly interact with DNA sequences of the responding gene, but rather provides a signal in the cell for turning on a defined set of genes and repressing others. Analogous systems have been defined in bacteria where a single gene seems to be responsible for regulating the heat shock response (Landick et al. 1984) which leads to induction of a cascade of transcriptional and translational control elements. Although the 70 kDa heat shock gene is turned on by the E1A product (Kao and Nevins 1983), the heat shock response is probably not relevant for E1A function, since a shift to high incubation temperatures cannot bypass the requirement of E1A expression for the activation of adenovirus early genes. The mammalian cell may, however, have several similar pathways to control transcription that are dependent on external or internal stimuli. A possibly analogous system is the induction of genes expressed in the nucleus of growth-arrested cells stimulated by growth factors (see chapters by Zullo et al.; Bravo and Müller). In this context it is interesting that E1A genes can overcome the requirement for growth factors of quiescent fibroblasts (Stabel et al. 1985).

It would be advantageous for viruses to use a cellular pathway that controls macromolecular synthesis in order to secure expression of their own genes. The fact that various cell lines have an endogenous E1A-like activity (Bellet et al. 1985) which can reduce the E1A requirement for early adenovirus gene expression supports this idea. The finding that tumor antigens encoded by DNA tumor viruses exhibit multifunctional properties mimicking cellular proteins which regulate gene expression raises the possibility that during evolution these viruses originally acquired cellular genes advantageous for viral replication. Due to the restriction imposed by the minimal genome size of DNA tumor viruses, they might have further evolved into efficient *trans*-acting elements in that they kept only the relevant functional domains of the ancestral cellular genes.

References

Alwine JC (1985) Transient gene expression control: Effects of transfected DNA stability and trans-activation by viral early proteins. Mol Cell Biol 5:1034–1042

Bellett AJD, Li P, David ET, Mackey EJ, Braithwaite AW, Cut JR (1985) Control functions of adenovirus transformation region E1A gene products in rat and human cells. Mol Cell Biol 5:1933–1939

Berger S, Folk WR (1985) Differential activation of RNA polymerase III-transcribed genes by the polyoma-virus enhancer and the adenovirus E1A gene product. Nucl Acids Res 13:1413–1428

Berk AJ, Lee F, Harrison T, Williams J, Sharp PA (1979) Pre-early Ad5 gene product regulates synthesis of early viral mRNAs. Cell 17:935–944

Borrelli E, Hen R, Chambon P (1984) Adenovirus-2 E1A products repress enhancer-induced stimulation of transcription. Nature 312:608–612

Branton PE, Rowe DT (1985) Stabilities and interrelations of multiple species of human adenovirus type 5 early region 1 proteins in infected and transformed cells. J Virol 56:633–638

Eager KB, Williams J, Breiding D, Pan S, Knowles B, Appella E, Ricciardi RP (1985) Expression of histocompatibility antigens H-2K, -D, and -L is reduced in adenovirus-12-transformed mouse cells and is restored by interferon γ. Proc Natl Acad Sci USA 82:5525–5529

Ferguson B, Krippl B, Andrisani O, Jones N, Westphal H, Rosenberg M (1985) E1A 13S and 12S mRNA products made in *Escherichia coli* both function as nucleus-localized transcription activators but do not directly bind DNA. Mol Cell Biol 5:2653–2661

Gaynor RB, Hillman D, Berk AJ (1984) Adenovirus early region 1A protein activates transcription of a non-viral gene introduced into mammalian cells by infection or transfection. Proc Natl Acad Sci USA 81:1193–1197

Gaynor RB, Feldman LT, Berk AJ (1985) Transcription of class III genes activated by viral immediate early proteins. Science 230:447 – 450

Gilardi P, Perricaudet M (1984) The E4 transcriptional unit of Ad2: far upstream sequences are required for its transactivation by E1A. Nucl Acids Res 12:7877 – 7888

Green MR, Treisman R, Maniatis T (1983) Transcriptional activation of cloned human β-globin genes by viral immediate-early gene products. Cell 35:137 – 148

Harlow E, Franza BR Jr, Schley C (1985) Monoclonal antibodies specific for adenovirus early region 1A proteins: extensive heterogeneity in early region 1A products. J Virol 55:633 – 646

Hearing P, Shenk T (1983) The adenovirus type 5 E1a transcriptional control region contains a duplicated enhancer element. Cell 33:695 – 703

Hearing P, Shenk T (1985) Sequence-independent autoregulation of the adenovirus type 5 E1a transcription unit. Mol Cell Biol 5:3214 – 3221

Hen R, Borrelli E, Sassone-Corsi P, Chambon P (1983) An enhancer is located 340 bp upstream from the adenovirus-2 E1a capsite. Nucl Acids Res 11:8747 – 8760

Jones N, Shenk T (1979) An adenovirus type 5 early gene function regulates expression of other early viral genes. Proc Natl Acad Sci USA 76:3665 – 3669

Kao H-T, Nevins JR (1983) Transcriptional activation and subsequent control of the human heat shock gene during adenovirus infection. Mol Cell Biol 3:2058 – 2065

Katze MG, Persson H, Philipson L (1981) Control of adenovirus early gene expression: posttranscriptional control mediated by both viral and cellular gene products. Mol Cell Biol 1:807 – 813

Krippl B, Ferguson B, Jones N, Rosenberg M, Westphal H (1985) Mapping of functional domains in adenovirus E1A proteins. Proc Natl Acad Sci USA 82:7480 – 7484

Landick R, Vaughn V, Lau ET, Van Bogelen RA, Erickson JW, Neidhardt FC (1984) Nucleotide sequence of the heat shock regulatory gene of *E. coli* suggests its protein product may be a transcription factor. Cell 38:175 – 182

Mellow GH, Fohring B, Dougherty J, Gallimore PH, Raska K (1984) Tumorigenicity of adenovirus-transformed rat cells and expression of class I major histocompatibility antigen. Virology 134:460 – 465

Nevins JR (1981) Mechanism of activation of early viral transcription by the adenovirus E1A gene product. Cell 26:213 – 220

Osborne TF, Arvidson DN, Tyau ES, Dunsworth-Brown M, Berk AJ (1984) Transcription control region within the protein-coding portion of adenovirus E1A genes. Mol Cell Biol 4:1293 – 1305

Schrier PI, Bernards R, Vaessen RTMJ, Houweling A, Van der Eb AJ (1983) Expression of class I major histocompatibility antigens switched off by highly oncogenic adenovirus 12 in transformed rat cells. Nature 305:771 – 775

Smith DH, Kegler DM, Ziff EB (1985) Vector expression of adenovirus type 5 E1a proteins: evidence for E1A autoregulation. Mol Cell Biol 5:2684 – 2696

Spalholz BA, Yang Y-C, Howley PM (1985) Transactivation of a bovine papilloma virus transcriptional regulatory element by the E2 gene product. Cell 42:183 – 191

Stabel S, Argos P, Philipson L (1985) The release of growth arrest by microinjection of adenovirus E1A DNA. EMBO J 4:2329 – 2336

Stein R, Ziff EB (1984) HeLa cell β-tubulin gene transcription is stimulated by adenovirus 5 in parallel with viral early genes by an E1a-dependent mechanism. Mol Cell Biol 4:2792 – 2801

Svensson C, Akusjärvi G (1984) Adenovirus 2 early region 1A stimulates expression of both viral and cellular genes. EMBO J 3:789 – 794

Svensson C, Pettersson U, Akusjärvi G (1983) Splicing of adenovirus 2 early region 1A mRNAs is non-sequential. J Mol Biol 165:475 – 499

Tanaka K, Isselbacher KJ, Khoury G, Jay G (1985) Reversal of oncogenesis by the expression of a major histocompatibility complex class I gene. Science 228:26 – 30

Vaessen RTMJ, Houweling A, Israel A, Kourilsky P, Van der Eb AJ (1986) Adenovirus E1A-mediated replication of class I MHC expression. EMBO J 5:335 – 341

Velcich A, Ziff E (1985) Adenovirus E1a proteins repress transcription from the SV40 early promoter. Cell 40:705 – 716

Transactivator Genes of HTLV-I, II, and III

WILLIAM A. HASELTINE, JOSEPH SODROSKI, CRAIG ROSEN, WEI CHUN GOH, ANDREW DAYTON, and DANIEL CELANDER

The investigation of the molecular biology of the three prototypes of Human T Lymphotropic Viruses (HTLV) reveals that each virus encodes at least one protein that greatly accelerates the rate of viral gene expression. A brief survey of these transactivating genes, their mechanism of action, and their role in the life cycle of the virus is presented.

HTLV-I and HTLV-II

HTLV-I (Poiesz et al. 1980) and II (Kalyanaraman et al. 1982a) are the etiological agents of T4 + human leukemias. HTLV-I induces a virulent form of adult T cell leukemia/lymphoma after a long period of latency (Poiesz 1980; Kalyanaraman 1982b; Gallo 1984). HTLV-II induces a milder form of the disease, a T-cell variant of hairy cell leukemia that is often accompanied by cutaneous manifestations (Kalyanaraman et al. 1982a, b; Chen et al. 1983). Cocultivation of HTLV-I- and II-infected cells with primary lymphocytes derived either from the peripheral circulation or cord blood results in transformation of T4 + cells (Chen et al. 1983; Miyoshi et al. 1982; Popovic et al. 1983).

In addition to the *gag, pol,* and *env* genes characteristic of all retroviruses, the genome of HTLV-I and II contains a region of 1500 to 1600 nucleotides located between the *env* gene and the 5′ LTR, also called the pX region (Seiki et al. 1983); (Fig. 1). At least two possible reading frames in this region appear to be used. These include the longest open reading frame that encodes a 42 kDa protein for HTLV-I and a 38 kDa protein for HTLV-II, as well as a protein synthesized from a smaller open reading frame that partially overlaps that of the longest open reading frame (Haseltine et al. 1984; Sagata et al. 1984; Lee et al. 1984; Kiyokawa et al. 1986). The genome of the bovine leukemia virus is arranged similarly and encodes at least two proteins in a region corresponding to the pX region of HTLV-I and II (Sagata et al. 1984; A. Burny, personal communication). The product of the longest open reading frame of HTLV-I and II corresponds to the transactivator (*tat*) protein. The rate of expression of HTLV-I genes is accelerated 50 − 200-fold in cells that constitutively express the *tat* protein (Sodroski et al. 1984; Sodroski et al. 1985c; Felber et al. 1985). Recent experiments suggest that an increase in the rate of transcription accounts for the majority of this effect (Felber et al. 1985).

Oncogenes and Growth Control
Edited by P. Kahn and T. Graf
© Springer-Verlag Berlin Heidelberg 1986

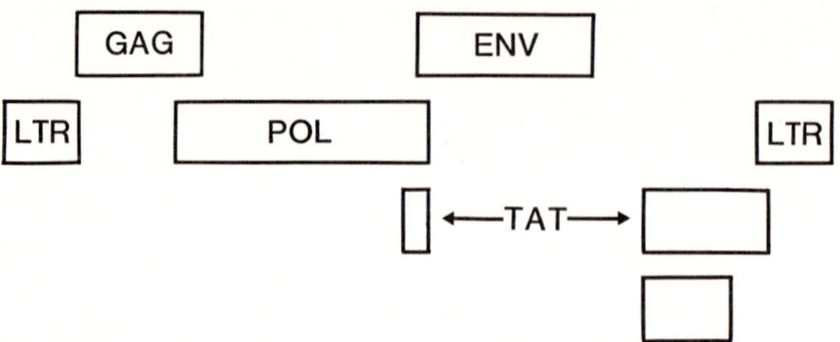

Fig. 1. The genome of HTLV-I. The long terminal repeats (LTR), *gag, pol* and *env* genes are shown. The transactivator gene (tat) is bipartite and made from a doubly spliced messenger RNA. The open box beneath the *tat* second coding exon represents another gene of unknown function that encodes p21 and p27 products (Kiyokawa et al. 1986)

The region of the HTLV-I genome responsive to the transactivator signal has been determined to lie within the long terminal repeat regions (LTRs) that flank the proviral DNA and includes both enhancer and promoter elements important for transcription initiation (Rosen et al. 1985; see also Schlokat and Gruss, this Vol.). Neither the enhancer nor the minimal-sequence promoter of HTLV-I responds to the *tat* I product (Rosen et al. 1985a). However, a sequence that extends about 150 nucleotides 5′ to the start site of RNA synthesis is required for the transactivating response and is called the *trans*-acting responsive element (Rosen et al. 1985a).

The corresponding sequences of the HTLV-II LTR are arranged somewhat differently and also vary in their biological properties in that the enhancer is inactive in most cell types in the absence of the transactivator (Rosen et al. 1986a). Moreover, stimulation of HTLV-II LTR-directed gene expression by either the *tat* I or *tat* II proteins occurs only in very few cell types such as immortalized T4+ cell lines (Sodroski et al. 1985c). Interestingly, gene expression directed by the HTLV-II LTR is not induced even in some cell lines expressing the *tat* II gene in which this transactivator stimulates the expression of genes under the control of the HTLV-I LTR (unpublished observation). It is therefore likely that viral and cellular factors other than the 38 kDa *tat* II protein are required for proper function of the HTLV-II LTR. The LTR of the bovine leukemia virus resembles that of the HTLV-II LTR in this respect (Derse et al. 1984; Rosen et al. 1985b, 1986a).

The observations that HTLV-I and II are capable of in vitro transformation of primary lymphocytes and that both viruses possess transactivator genes not found in other retroviruses led to the hypothesis that cellular genes involved in proliferation of normal T cell are transcriptionally activated by the HTLV-I and II transactivator genes (Sodroski et al. 1984).

To test this hypothesis, the *tat* II gene was introduced into T4+ cell lines and the expression of the genes encoding IL2 and IL2 receptor, known to be induced by antigen activation of T cells, was measured. These experiments demonstrated that mRNA for both genes is induced by the *tat* II protein in

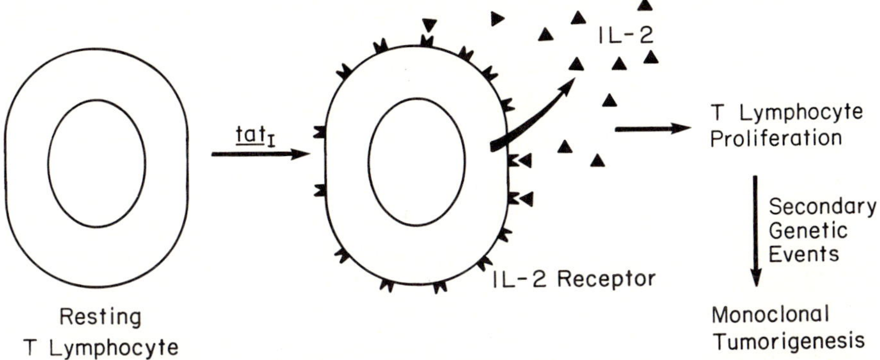

Fig. 2. Autocrine model of T-lymphocyte proliferation. In this model, after HTLV infection, the *tat_I* gene product induces the simultaneous expression of interleukin-2 (IL-2) and its receptor. In the proliferating pool of infected T lymphocytes, secondary genetic events occur that lead to outgrowth of a monoclonal tumor after a long latent period

T4+ cell lines while no expression was observed in B cell lines (Greene et al. 1986). Moreover, functional IL2 and IL2 receptors could be demonstrated in the T4+ cell lines that produce *tat* II protein while cell lines containing the same sequences but in an inactive configuration remained negative. These experiments suggest that the *tat* gene initiates the transformation of T4+ cells by the induction of IL2 and IL2 receptors. Simultaneous induction of these two proteins could lead to proliferation of T4 cells via an autocrine mechanism (Fig. 2). The cell lineage specific induction of IL2 and IL2 receptor expression may provide an explanation for the T cell tropism of the virus.

If a similar pattern of IL2 and IL2 receptor induction occurs after infection of T cells in vivo, why does leukemia occur only with a low incidence and long latent period? We speculate that the same mechanisms that normally limit the expansion of activated T cell populations also limit the expansion of HTLV-I- and HTLV-II-infected T cells. Such autoregulatory mechanism might control the proliferation of the infected T cells until such time as either a secondary change occurs in one of the infected cells or the host suppression mechanisms fail.

What advantage might such an activity have for the virus? Epidemiological studies suggest that transmission of HTLV-I and the related bovine leukemia virus occurs via cell to cell contact (sexual transmission, maternal transmission, milk transmission and transfer of blood either by transfusion or needle) rather than by infection with cell-free virus (Gallo 1984). Difficulties in in vitro infection by cell-free HTLV-I virus support this view. If the infected T4+ cell is the infectious unit, expansion of the infected population by stimulation of a T4+ specific autocrine proliferation pathway would provide an obvious selective advantage to the virus. In this view, lymphomas induced by HTLV-I and II would be consequences of the mode of virus propagation.

HTLV-III

HTLV-III (also known as LAV or HIV) is the etiological agent of the acquired immune deficiency syndrome (AIDS) and other disorders including marked degeneration of the central and peripheral nervous system (Barre-Sinoussi et al. 1983; Seligman et al. 1984; for review and further references, see Broder and Gallo 1984). There is a long and variable period between time of infection and onset of disease symptoms. Recent evidence suggests that the HTLV-III virus may establish latent as well as lytic infections of T4+ cells (Zagury et al. 1986; Folkes et al. 1985).

In addition to the *gag, pol,* and *env* proteins, the genome of HTLV-III encodes at least three other proteins, the 23 kDa *sor* gene product (Lee et al. 1986; Sodroski et al. 1985a), the 14 kDa *tat* III gene product (Arya et al. 1985; Sodroski et al. 1985b) and the 27 kDa *orf* gene product (Allan et al. 1985). Of these, only the *tat* III gene is required for growth and cytopathic activity in T4+ cells (Sodroski et al. 1985a; Dayton et al. 1986). Defects in the *tat* III gene can be complemented in cell lines constructed so as to express constitutively the *tat* III gene product (Dayton et al. 1986).

Viruses deleted for the *tat* III gene show a greater than 500-fold reduction in the expression of the *gag* and *env* gene products as compared to viruses containing the gene. Surprisingly, this defect occurs not at the level of transcription but rather post-transcriptionally, since cells transfected with viruses defective in the *tat* III gene produce the same amount of viral RNA as cells transfected with the intact virus (Rosen et al. 1986b). The LTR of HTLV-III, like that of HTLV-I, contains enhancer and promoter sequences that drive mRNA synthesis in the absence of the transactivator (Rosen et al. 1985d). Neither of these two elements responds to the transactivator. The LTR of HTLV-III also includes a regulatory sequence that acts as a negative "enhancer" (Rosen et al. 1985d).

The sequences responsive to the *tat* gene product (TAR) are located at the 5′ end of the RNA species between nucleotides +1 and +80 (unpublished observation). Presumably all HTLV-III viral mRNAs are subject to *tat* III regulation, as they all contain the same 5′ nontranslated leader sequences derived from the R and U5 regions of the HTLV-III LTR. The level of protein synthesis directed by mRNAs consisting of TAR sequences fused to an indicator gene such as dihydrofolate reductase is greatly increased in the presence of the *tat* III gene product (Rosen et al. 1986b). In such constructs, the TAR sequences are only active if they comprise the 5′ end of the RNA; if located 5′ or 3′ to the +1 position, protein synthesis directed by the mRNA is insensitive to the *tat* III gene product (Rosen et al. 1986b).

What role does the *tat* III gene play in the virus life cycle? Once replication begins, the *tat* III gene product should greatly accelerate the rate of synthesis of virus protein. This process should be autocatalytic, limited only by the amount of viral RNA. The resulting acceleration in virus growth is likely to provide a selective advantage to the virus. The *tat* III gene may also play a role in the maintenance of the latent state observed after infection of nondividing T4+ cells (Zagury et al. 1986; Folkes et al. 1982). Thus, virus structural proteins are not synthesized in such infected cells in the absence of the *tat* III gene product (Dayton et al. 1986; Rosen et al. 1986b), even in the presence of the appropriate viral mRNAs. It is possible that resting, HTLV-III-infected T4+ cells express a cellular factor that suppresses *tat* III activity with the consequence that viral RNA but no virus progeny is produced. Relief from such an antagonist (for example by T cell activation) could induce rapid virus replication. Such a "rapid escape" synthesis might be a mechanism that has evolved as a consequence of the lethal effects of HTLV-III replication in T4+ cells. The ability of the HTLV-III virus to proliferate rapidly as a result of the induction of transactivator function probably plays an important role in the transmission and progression of HTLV-III-induced disease.

References

Allan JS, Coligan JE, Lee TH, McLane MF, Kanki PJ, Groopman JR, Essex M (1985) A new HTLV-III/LAV encoded antigen detected by antibodies from AIDS patients. Science 230:810–813

Arya SK, Guo C, Josephs SF, Wong-Staal F (1985) Trans-activator gene of human T-lymphotropic virus type III (HTLV-III). Science 229:69–73

Barre-Sinoussi F, Chermann JC, Rey F, Nugeyre MT, Chamaret S, Gruest J, Dauguet C, Axler-Blin C, Vezinet-Brum F, Rouzious C, Rozenbaum W, Montagnier L (1983) Isolation of a T-lymphotropic retrovirus from a patient at risk for acquired immune deficiency syndrome (AIDS). Science 220:868–871

Broder SM, Gallo RC (1984) A pathogenic retrovirus (HTLV-III) linked to AIDS. N Eng J Med 311:1292–1297

Chen ISY, Quan SG, Golde DW (1983) Human T-cell leukemia virus type II transforms normal human lymphocytes. Proc Natl Acad Sci USA 80:7006–7009

Dayton AI, Sodroski JG, Rosen CA, Goh WC, Haseltine WA (1986) The *trans*-activator gene of the HTLV-III virus is required for replication. Cell 44:941–947

Derse D, Caradonna SJ, Casey J (1984) Bovine leukemia virus long terminal repeat: A cell type-specific promoter. Science 227:317–320

Felber BK, Paskalis H, Ewing CK, Wong-Staal F, Pavlakis GN (1985) The px protein of HTLV-I is a transcriptional activator of its long terminal repeats. Science 229:675–679

Folkes T, Benn S, Rabson A, Theodore T, Hoggan MD, Martin M, Lightfoote M, Sell K (1985) Characterization of a continuous T-cell line susceptible to the cytopathic effects of the acquired immunodeficiency syndrome (AIDS)-associated retrovirus. Proc Natl Acad Sci USA 82:4539–4543

Gallo RC (1984) Human T-cell leukemia lymphoma virus and T cell malignancies in adults. Cancer Surveys 3:113–158

Greene W, Leonard W, Wano Y, Svetlik P, Pfeffer N, Sodroski J, Rosen CA, Goh WC, Haseltine WA (1986) Transactivator gene of HTLV-II induces IL-2 receptor and IL-2 cellular gene expression. Science 232:877–880

Haseltine WA, Sodroski JG, Patarca R, Briggs D, Perkins D, Wong-Staal F (1984) 3′ terminal sequency of human T-cell leukemia virus type II: evidence for a new coding region. Science 225:419–421

Kalyanaraman VS, Sarngadharan MG, Robert-Guroff M, Miyoshi I, Blayney D, Golde D, Gallo RC (1982a) A new subtype of human T-cell leukemia virus (HTLV-II) associated with a T-cell variant of hairy cell leukemia. Science 218:571–573

Kalyanaraman VS, Sarngadharan MG, Nakao Y, Ito Y, Aoki T, Gallo RC (1982b) Natural antibodies to the structural core protein (p24) of the human T-cell leukemia (lymphoma) retrovirus found in sera of leukemia patients in Japan. Proc Natl Acad Sci USA 79:1653–1657

Kiyokawa T, Seiki M, Iwashita S, Shimizu F, Yoshida M (1985) p27^{x-III} and p21^{x-III}, proteins encoded by the pX sequence of human T cell leukemia virus type I. Proc Natl Acad Sci USA 82:8359–8363

Lee TH, Coligan JE, Allah JS, McLane MF, Groopman JE, Essex M (1986) A new HTLV-III/LAV protein encoded by a gene found in cytopathic retroviruses. Science 231:1546–1549

Miyoshi I, Taguchi H, Fujishita M, Yoshimoto S, Kubonishi I, Ohtsuki Y, Shiraishi Y, Akagi T (1982) Transformation of monkey lymphocytes with adult T-cell leukemia virus. Lancet 1(8273):1016

Poiesz BJ, Ruscetti FW, Gazdar AF, Bunn PA, Minna JD, Gallo RC (1980) Detection and isolation of type C retrovirus particles from fresh and cultured lymphocytes of a patient with cutaneous T-cell lymphoma. Proc Natl Acad Sci USA 77:7415–7419

Popovic M, Lange-Wantzin G, Sarin PS, Mann D, Gallo RC (1983) Transformation of human umbilical cord blood T cells by human T-cell leukemia/lymphoma virus. Proc Natl Acad Sci USA 80:5402–5406

Rosen CA, Sodroski JG, Haseltine WA (1985a) Location of Cis-acting regulatory sequences in the human T-cell leukemia virus type I long terminal repeat. Proc Natl Acad Sci USA 82:6502–6506

Rosen CA, Sodroski JG, Kettman R, Burny A, Haseltine WA (1985b) *Trans*-activation of the bovine leukemia virus long terminal repeat in BLV-infected cells. Science 227:321–323

Rosen CR, Sodroski JG, Campbell K, Haseltine WA (1985c) Construction of recombinant retroviruses that express the human T cell leukemia virus type II and human T cell leukemia virus type III transactivator. J Virol 57:379–384

Rosen C, Sodroski J, Haseltine WA (1985d) Location of cis-acting regulatory sequences in the human T cell lymphotropic virus type III (HTLV-III/LAV) long terminal repeat. Cell 41:813–823

Rosen CA, Sodroski JG, Kettman R, Haseltine WA (1986a) Activation of enhancer sequences in type II human T-cell leukemia virus and bovine leukemia virus long terminal repeats by virus-associated *trans*-acting regulatory factors. J Virol 57:379–384

Rosen CA, Sodroski JG, Goh WC, Dayton AI, Lippke J, Haseltine WA (1986b) Post-transcriptional regulation accounts for the trans-activation of the human T-lymphotropic virus type III. Nature 319:555–559

Sagata N, Yasunaga T, Ohishi K, Tsuzuku-Kawamura J, Onuma M, Ikawa Y (1984) Comparison of the entire genomes of bovine leukemia virus and human T cell leukemia virus and characterization of their unidentified open reading frames. EMBO J 3:3231–3237

Seiki M, Hattori S, Hirayama Y, Yoshida M (1983) Human adult T-cell leukemia virus. Complete nucleotide sequence of the provirus genome integrated in leukemia cell DNA. Proc Natl Acad Sci USA 80:3618–3622

Seligman M, Chess L, Fahey JL, Fauci AS, Lachmann PJ, L'Age-Stehr J, Ngu J, Pinching AJ, Rosen FS, Spira TJ, Wybran J (1984) AIDS – an immunologic re-evaluation. N Engl J Med 311:1286–1292

Sodroski JG, Rosen C, Haseltine WA (1984) *Trans*-acting transcriptional activation of the long terminal repeat of human T lymphotropic viruses in infected cells. Science 225:381–385

Sodroski JG, Goh WC, Rosen CA, Tartar A, Portetelle D, Burny A, Haseltine WA (1985a) Replicative and cytopathic potential of HTLV-III/LAV with *sor* gene deletions. Science 231:1549–1553

Sodroski JG, Rosen CA, Wong-Staal F, Popovic M, Arya S, Gallo RC, Haseltine WA (1985b) *Trans*-acting transcriptional activation of the long terminal repeat of human T cell leukemia virus type III (HTLV-III). Science 227:171–173

Sodroski JG, Rosen CA, Goh WC, Haseltine WA (1985c) A transcriptional activator protein encoded by the x-*lor* region of the human T-cell leukemia virus. Science 228:1430–1434

Zagury D, Bernard J, Cheynier R, Feldman M, Sarin P, Gallo RC (1986) Long-term cultures of HTLV-III-infected cells: a model of cytopathology of T-cell depletion in AIDS. Science (in press)

Involvement of Proto-Oncogenes in Growth Control: The Induction of c-*fos* and c-*myc* by Growth Factors

RODRIGO BRAVO and ROLF MÜLLER

One of the most conspicuous features of retroviral oncogenes is their impact on the growth control machinery of the target cell. For this reason, it has become a generally accepted idea that their normal cellular homologs, the proto-onco-genes, may have crucial physiological functions in metabolic pathways involved in the regulation of cellular proliferation. It was not until 1983 that this hypoth-esis was substantiated by the observation that two proto-oncogenes, c-*sis* and c-*erbB,* are part of the cellular growth factor receptor system (Doolittle et al. 1983; Waterfield et al. 1985; Downward et al. 1984) and articles by Heldin and Westermark; Schlessinger and Beng et al., this Vol.). At the same time, expres-sion of a proto-oncogene encoding a nuclear protein, c-*myc*, was reported to increase rapidly following the treatment of different cell types with growth-in-ducing components (Kelly et al. 1983). This discovery was followed by the find-ing that c-*myc* mRNA accumulation is even preceded by the transcriptional acti-vation of another proto-oncogene coding for a nuclear product, the c-*fos* gene (Greenberg and Ziff 1984; Cochran et al. 1984; Kruijer et al. 1984; Müller et al. 1984). This chapter discusses the results of these studies and several subsequent investigations analyzing the expression of c-*fos* and c-*myc* during cell prolifera-tion. (For reviews of each of these proto-oncogenes, the reader is referred to Jenuwein and Müller, this Vol., and Mölling, this Vol.).

Induction of c-*fos* and c-*myc* mRNAs by Serum and Growth Factors in Quiescent Cells

The involvement of c-*onc* genes in cell proliferation has been investigated by studying their expression following growth stimulation of mouse fibroblasts. In these experiments, quiescent NIH3T3 (i.e., serum-deprived) cells were stim-ulated with fetal calf serum (FCS), which leads to a synchronous entry of the cell population into the S-phase of the cell cycle within 10 – 12 h (Müller et al. 1984). One hour after serum stimulation, the steady-state level of c-*myc* mRNA shows an approximately 20-fold increase followed by a slow decrease, reaching basal levels of expression at approximately 18 h. The induction of c-*fos* expression is even more dramatic, increasing at least 50-fold within 30 min and decreasing rapidly to basal levels within the following 90 min (Kelly et al. 1983; Campisi et al. 1984; Cochran et al. 1984; Greenberg and Ziff 1984; Kruijer et al. 1984; Müller et al. 1984). While the increased levels of c-*fos*

Oncogenes and Growth Control
Edited by P. Kahn and T. Graf
© Springer-Verlag Berlin Heidelberg 1986

mRNA were shown to be largely due to transcriptional activation (Greenberg and Ziff 1984), the concentration of c-*myc* mRNA is controlled by post-transcriptional mechanisms (i.e., by modulation of the stability of c-*myc* mRNA; Blanchard et al. 1985). Although the induction of c-*fos* precedes the accumulation of c-*myc* mRNA it is unlikely that the c-*fos* gene product is involved in the induction of c-*myc*, since both genes can be induced in the presence of protein synthesis inhibitors (Cochran et al. 1984; Greenberg and Ziff 1984; Müller et al. 1984).

Purified growth factors, such as epidermal growth factor (EGF), fibroblasts growth factor (FGF) and platelet-derived growth factor (PDGF), have proven to be as efficient in inducing c-*myc* as 10% FCS (Müller et al. 1984). In contrast, c-*fos* expression is efficiently induced only by FGF and PDGF. Both proto-oncogenes are induced with similar kinetics by PDGF, FGF, and FCS. c-*myc* RNA, however, reaches basal levels more rapidly after stimulation by EGF than after PDGF or FGF stimulation (Müller et al. 1984). No significant increase in c-*myc* RNA has been found in Balb/c 3T3 cells after EGF stimulation (Kelly et al. 1983; Cochran et al. 1984), which is probably due to differences between the cells used.

Induction of c-*fos* Protein by Serum and Growth Factors

As shown in immunoprecipitation experiments, the induction of c-*fos* mRNA is rapidly followed by the synthesis of c-*fos* protein. Maximum synthesis is observed at about 1 h after stimulation, and after another hour c-*fos* protein synthesis is greatly reduced, correlating with the observed levels of c-*fos* mRNA. The growth factor-induced c-*fos* protein is post-transcriptionally modified, as judged from its increased molecular mass. While unmodified c-*fos* protein migrates as a 54 kDa protein on reducing polyacrylamide gels, several forms in the range of 55 – 65 kDa are observed in stimulated 3T3 cells (Müller et al. 1984; Kruijer et al. 1984). Similar post-translational modifications are absent from transforming v-*fos* gene products and may therefore represent a regulatory mechanism to control the oncogenic potential of c-*fos* protein (Müller et al. 1984). The nature of these modifications, however, remains enigmatic.

Steady-state levels of c-*fos* protein following growth factor stimulation were assessed by immunofluorescence studies. These analyses showed that the typical granular staining pattern of c-*fos* protein is seen throughout the nucleus (with the exclusion of the nucleoli) at 30 min following stimulation. Maximal levels of c-*fos* protein are reached after 1 – 2 h of stimulation by either serum or purified growth factors. At 4 h, c-*fos* protein levels significantly decrease. A slight increase in c-*fos* protein can be observed already after 15 min of serum stimulation. This finding is in agreement with the observation that an increase in the rate of transcription of the c-*fos* gene is observed as early as 10 min after serum stimulation (Greenberg and Ziff 1984). Induction of c-*fos* is thus the earliest known effect on the control of gene expression exerted by growth factors.

Induction of c-*fos* and c-*myc* Is Independent of Cell Growth

Although EGF is a potent mitogen for a number of cell types in culture, it inhibits proliferation of the human A431 epidermal carcinoma cells (Gill and Lazer 1981; Barnes 1982 and by Carpenter et al., this Vol.). Subclones which display either no change or an increase in growth rate after response to EGF have been isolated (Bravo 1984). These cells have been useful in analyzing whether c-*myc* and c-*fos* expression is strictly correlated with the induction of cell proliferation.

The results of these studies have shown that A431 cells and the variant subclones respond to EGF treatment in a similar way with an increased expression of c-*fos* and c-*myc* (Bravo et al. 1985a). The kinetics of induction of both proto-oncogenes are very similar to those described for fibroblasts. Surprisingly, although EGF induces low levels of c-*fos* in fibroblasts, this growth factor is a potent inducer of c-*fos* in A431 cells. In spite of some minor variations in the levels of c-*fos* and c-*myc* induction between the different subclones, no correlation could be found between induction of these proto-oncogenes and the ability of EGF to influence proliferative activity in these cells. These findings show that the activation of c-*fos* and c-*myc* does not necessarily lead to an increase in growth rate. The activation of c-*fos* and c-*myc* rather seems to be a direct response to the growth factor: receptor interaction. In addition, the fact that the A431 cells used for EGF treatment were nonsynchronized, actively growing cells, strongly suggests that cells do not need to be arrested in G0 to permit c-*fos* and c-*myc* induction by EGF and possibly other growth factors.

c-*fos* and c-*myc* Expression During the Cell Cycle

Recent analyses have shown that c-*fos* mRNA levels are extremely low or even undetectable throughout the cell cycle in NIH3T3 cells (Bravo et al. 1986). This suggests that a high expression of c-*fos* is not required for the continuous cycling of NIH3T3 cells. However, a role of c-*fos* in the normal proliferation of cells cannot be completely ruled out, since c-*fos* expression in growing cells has been shown to be slightly elevated compared to quiescent cells (Müller et al. 1984).

In contrast to c-*fos*, c-*myc* mRNA and protein is expressed at readily detectable levels at all stages of the cell cycle, in chicken fibroblasts as well as in human lymphocytes, leukemia cells, and HeLa cells, suggesting that c-*myc* product plays a role during normal cell proliferation (Hann et al. 1985; Rabbitts et al. 1985; Thompson et al. 1985). The level of expression during the cell cycle is, however, lower than in growth factor-stimulated cells. Another evidence for a role of the c-*myc* product in normal cell proliferation is that its expression is high in growing asynchronous cells and very low in quiescent cells. Serum deprivation of asynchronous cultures for a period as short as 2 h leads to a dramatic decrease in the level of c-*myc* mRNA, indicating that the constant presence of external growth factors is required to sustain the expression of c-*myc* at high levels (Bravo et al. 1986).

Role of c-*fos* and c-*myc* in Competence Induction

It has been shown that growth factors can act as either "competence factors" or "progression factors" (Stiles et al. 1979). Quiescent mouse fibroblasts briefly exposed to FGF or PDGF become competent to synthesize DNA. The addition of platelet-poor plasma (PPP), which contains progression factors but lacks PDGF, allows these cells (but not untreated cells) to progress through G1 and enter the S-phase (Stiles et al. 1979). The observation that c-*fos* and c-*myc* induction is among the earliest events following growth factor stimulation of quiescent fibroblasts suggests that both proto-oncogene products may play an important role in conferring competence on stimulated cells. There are several lines of evidence that support this hypothesis. First, both c-*fos* and c-*myc* are induced significantly by competence factors, not by the progression factors in PPP (Bravo et al. 1985b). Second, it has been shown that the induction of competence is possible at different stages during the cell cycle (Stiles et al. 1979). Accordingly, c-*fos* and c-*myc* can be induced by PDGF or serum during any phase of the cell cycle except for mitosis (Bravo et al. 1986). Third, wounding of a confluent monolayer of cells induces competence in cells lining the wound, as shown by their mitogenic response to progression factors (Stiles et al. 1979). Likewise, cells lining the wound transiently express c-*fos* protein (Verrier et al. 1986).

Other evidence suggests that c-*fos* and c-*myc* must act in concert to induce the competent state in fibroblasts. First, epidermal growth factor, which induces c-*myc* efficiently but c-*fos* only weakly, is not a competence factor (Müller et al. 1984). Second, cells transformed by *fos* oncogenes are not able to grow in PPP (Bravo and Müller, unpublished observations). Third, cells transfected with c-*myc* show a competence-like state in fibroblasts, but the growth response to progression factors is weak compared to PDGF (Armelin et al. 1984). These results suggest that c-*fos* and c-*myc* may function in concert to promote cell division. The observation that several other unidentified genes are induced by growth factors or serum in quiescent cells (Cochran et al. 1983; Lau and Nathans 1985) suggests that possibly the transition from G0 to G1 is the result of a complex interaction of several genes including c-*fos* and c-*myc*.

The competence state induced by a brief PDGF treatment is very stable, showing a significant decrease in their response to progression factors only 8 h after the withdrawal of the competence factor. However, PDGF-stimulated cells kept in serum-free medium for a similar time show levels of c-*fos* and c-*myc* expression that are equivalent to those observed in quiescent cells (Bravo et al. 1985b). When PPP is added to these cells, c-*fos* and c-*myc* are only weakly induced although but a strong mitogenic effect is observed. The persistence of the competent state cannot be due to the presence of high levels of c-

fos and c-*myc* protein, since the half-life of both products is 30 – 60 min (Müller et al. 1984; Thompson et al. 1985) compared to the half-life of 16 h for competence (Singh et al. 1983; Bravo et al. 1985b). Three major conclusions can be drawn from these observations. First, cells can remain competent in spite of expressing c-*fos* and c-*myc* at levels equivalent to quiescent cells. Second, *competent* cells do not require high levels of c-*fos* and c-*myc* expression to progress through G1. Third, the persistence of competence could reflect the presence of a stable biochemical change triggered by the early induced genes.

References

Armelin HA, Armelin MCS, Kelly K, Stewart T, Leder P, Cochran BH, Stiles CD (1984) Functional role for c-*myc* in mitogenic response to platelet-derived growth factor. Nature 310:655 – 660

Barnes DW (1982) Epidermal growth factor inhibits growth of A431 human epidermoid carcinoma in serum-free cell cultures. J Cell Biol 93:1 – 4

Blanchard J-M, Piechaczyk M, Dani C, Chambard J-C, Franchi A, Pouyssegur J, Jeanteur P (1985) C-*myc* gene is transcribed at high rate in G0 arrested fibroblasts and is post-transcriptionally regulated in response to growth factors. Nature 317:443 – 445

Bravo R (1984) Epidermal growth factor inhibits the synthesis of the nuclear protein cyclin in A431 human carcinoma cells. Proc Natl Acad Sci USA 81:4848 – 4850

Bravo R, Burckhardt J, Curran T, Müller R (1985a) Stimulation and inhibition of growth by EGF in different A431 cell clones is accompanied by the rapid induction of c-*fos* and c-*myc* proto-oncogenes. EMBO J 4:1193 – 1197

Bravo R, Burckhardt J, Müller R (1985b) Persistence of the competent state in mouse fibroblasts is independent of c-*fos* and c-*myc* expression. Exp Cell Res 160:540 – 543

Bravo R, Burckhardt J, Curran T, Müller R (1986) Expression of c-*fos* in NIH3T3 cells is very low but inducible throughout the cell cycle. EMBO J 5:695 – 700

Campisi J, Gray HE, Pardee AB, Dean M, Sonenshein GE (1984) Cell cycle control of c-*myc* and c-*ras* expression is lost following chemical transformation. Cell 36:242 – 247

Cochran BH, Reffel AC, Stiles CD (1983) Molecular cloning of gene sequences regulated by platelet-derived growth factor. Cell 33:939 – 947

Cochran BH, Zullo J, Verma IM, Stiles CD (1984) Expression of the c-*fos* oncogene and a newly discovered r-*fos* is stimulated by platelet-derived growth factor. Science 226:1080 – 1082

Doolittle RF, Hunkapiller MW, Hood LE, Deware SG, Robins KC, Aaronson SA, Antoniades HN (1983) Simian sarcoma virus *onc* gene, v-*sis*, is derived from the gene (or genes) encoding a platelet-derived growth factor. Science 211:275 – 277

Downward J, Yarden Y, Mayes E, Scarce G, Totty N, Stockwell P, Ullrich A, Schlessinger J, Waterfield MD (1984) Close similarity of epidermal growth factor receptor and v-*erbB* oncogene protein sequences. Nature 307:521 – 527

Gill GN, Lazer CS (1981) Increased phosphotyrosine content and inhibition of proliferation in EGF-treated A431 cells. Nature 293:305 – 307

Greenberg ME, Ziff EB (1984) Stimulation of 3T3 cells induces transcription of the c-*fos* proto-oncogene. Nature 311:433 – 438

Hann SR, Thompson CR, Eisenman RE (1985) c-*myc* oncogene protein synthesis is independent of the cell cycle in human and avian cells. Nature 314:366 – 369

Kelly K, Cochran BH, Stiles CD, Leder P (1983) Cell-specific regulation of the c-*myc* gene by lymphocyte mitogens and platelet-derived growth factor. Cell 35:603 – 610

Kruijer W, Cooper JS, Hunter T, Verma IM (1984) Platelet-derived growth factor induces rapid but transient expression of the c-*fos* gene and protein. Nature 312:711 – 716

Lau LF, Nathans D (1985) Identification of a set of genes during the G0/G1 transition of cultured mouse cells. EMBO J 4:3145 – 3151

Müller R, Bravo R, Burckhardt J, Curran T (1984) Induction of c-*fos* gene and protein by growth factors precedes activation of c-*myc*. Nature 312:716 – 720

Rabbitts PH, Watson JV, Lamond A, Forster A, Stinson MA, Evan G, Fischer W, Atherthon E, Sheppard R, Rabbitts TH (1985) Metabolism of c-*myc* gene products: c-*myc* mRNA and protein expression in the cell cycle. EMBO J 4:2009 – 2015

Singh JP, Chaikin MA, Pledger WJ, Scher CD, Stiles CD (1983) Persistence of the mitogenic response to platelet-derived growth factor (competence) does not reflect a long-term interaction between the growth factor and the target cell. J Cell Biol 96:1497 – 1502

Stiles CD, Capone GT, Scher CD, Antoniades HN, Van Wijk JJ, Pledger WJ (1979) Dual control of cell growth by somatomedins and platelet-derived growth factor. Proc Natl Acad Sci USA 76:1279 – 1283

Thompson CB, Challoner PB, Neiman PE, Groudine M (1985) Levels of c-*myc* oncogene mRNA are invariant throughout the cell cycle. Nature 314:363 – 366

Verrier B, Müller D, Bravo R, Müller R (1986) Wounding a fibroblast monolayer results in the rapid induction of the c-*fos* proto-oncogene. EMBO J 5:913 – 917

Waterfield MD, Scrace GT, Whittle N, Stroobant P, Johnson A, Wasteson A, Westermark B, Heldin CH, Huang JS, Deuel T (1983) Platelet-derived growth factor is structurally related to the putative transforming protein p28sis of simian sarcoma virus. Nature 304:35 – 39

Oncogenes and Interferons: Genetic Targets for Animal Cell Growth Factors

JOHN ZULLO, DAVID HALL, BARRETT ROLLINS, and CHARLES D. STILES

Many, if not all, oncogenes are functional components of a mitogenic cascade which is normally controlled by growth factors. Some oncogenes such as *sis* seem to function at the onset of this cascade by directing synthesis of an auto-mitogenic growth factor (see Heldin and Westermark, this Vol.). Others such as *erbB* and *fms* appear to function at subsequent steps of the cascade by directing synthesis of a growth factor receptor or a structurally altered receptor derivative (see Beug et al.; Schlessinger; and Sherr and Stanley, all this Vol.). Still other oncogenes such as *myc, fos* and perhaps *myb* are mutated or rearranged homologs of proto-oncogenes whose expression is normally induced by growth factors (see Bravo and Müller; Mölling; and Jenuwein and Müller, all this Vol. and references therein).

The interface between growth factors and oncogenes has been widely publicized even within the lay community, and no serious tumor biologist can now be unaware of it. The purpose of this article is to draw attention to an interface of another sort − between growth factors and interferons. This newly discovered relationship may ultimately become more exciting and have more practical consequences than the one between growth factors and oncogenes.

Chemistry and Nomenclature of the Interferons

The interferons are a group of small glycopeptides which were first identified and isolated as anti-viral agents. The history, biology, and molecular biology of these agents have been frequently reviewed (Revel et al. 1979; Stewart 1979; Lengyel 1981). Current nomenclature recognizes three interferon subclasses: alpha, beta, and gamma. The alpha and beta interferons are both stimulated in response to viral infection and are predominantly synthesized by lymphocytes and fibroblasts respectively. Gamma interferon is unique in that it is synthesized by lymphocytes in response to antigens or plant lectins (Torrence and DeClerq 1981).

Within the three major classes of interferon, subclasses may be found. In human, for example, there are more than a dozen alpha interferons (Hiscott et al. 1984; Capon et al. 1985; Roscouet et al. 1985). Taniguchi et al. (1980) obtained a cDNA clone for the human beta interferon whose NH_2-terminal amino acid sequence was determined by Knight et al. (1980). Several other genes which have nucleic acid sequence homology and functional homology to the

Oncogenes and Growth Control
Edited by P. Kahn and T. Graf
© Springer-Verlag Berlin Heidelberg 1986

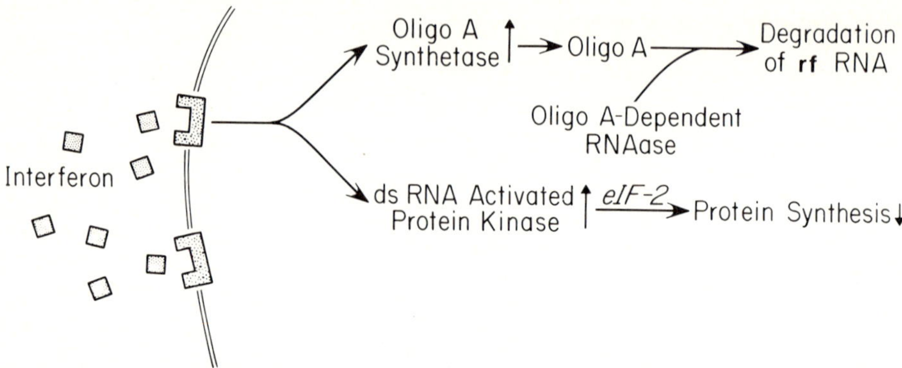

Fig. 1. Two independent metabolic pathways which are regulated by interferon. For explanation, see text. rf, replicative form intermediate of viral RNA

gene isolated by Taniguchi et al. have been described (Friedman-Einat et al. 1982; Sehgal et al. 1983; Sager et al. 1984).

Interferon-Associated Enzymes

Figure 1 summarizes two independent metabolic pathways which are regulated by interferon (Revel et al. 1979; Stewart 1979; Lengyel 1981). Either or both of these pathways may function to mediate the anti-viral state which is induced by interferon. In the pathway shown in the lower part of Fig. 1, formation of the interferon:receptor complex activates a protein kinase which utilizes double-stranded (ds) RNA as a cofactor. Among the targets of this ds RNA-dependent protein kinase is eukaryotic initiation factor-2 (eIF-2). Phosphorylation of eIF-2 inhibits its ability to function in the translational apparatus and consequently protein synthesis is suppressed.

The metabolic pathway shown in the upper part of Fig. 1 involves a transcription-dependent event. Formation of the interferon:receptor complex stimulates expression of the gene encoding the oligoadenylate synthetase enzyme. This enzyme ligates ATP together to form short poly(A)$^+$ tracts with an unconventional $(2'-5')$ phosphodiester linkage. These short tracts of oligoadenylate ("oligo A") serve as a cofactor for an oligo A-dependent ribonuclease. The activated oligo A-dependent ribonuclease cleaves single-stranded RNA at UpAp, UpGp, and UpUp sequences (Wreschner et al. 1981; Floyd-Smith et al. 1981). Revel and his associates have obtained a cDNA clone of the human $(2'-5')$-oligoadenylate synthetase gene (Merlin et al. 1983).

Induction of Interferons and of $(2' - 5')$-oligo A Synthetase by Cell Growth Factors

Recent work in our laboratory has drawn attention to an interesting relationship between platelet-derived growth factor (PDGF), double-stranded ribonucleic acids and interferons (Zullo et al. 1985). The synthetic double-stranded RNA poly(rI):poly(rC) stimulates prompt and dramatic expression of a "competence" gene family which we had previously shown to be regulated by PDGF (Cochran et al. 1983). The c-*myc* and c-*fos* proto-oncogenes are contained within the competence gene family (see Bravo and Müller, this Vol. and references therein). Conversely, PDGF stimulates expression of a beta-fibroblast interferon-like gene.

The proto-oncogenes *myc* and *fos* are induced promptly by PDGF (within 1 h), whereas interferon is induced slowly and continues to accumulate over a 12- to 24-h time period. We also observed that PDGF stimulates expression of the $(2' - 5')$-oligoadenylate synthetase gene. Like interferon, PDGF-mediated induction of $(2' - 5')$-oligoadenylate synthetase occurs over a longer time period than induction of c-*myc*, c-*fos* and other members of the competence gene family.

The observations on PDGF drew our attention to an older and broader body of literature on relationships between interferon, interferon-associated enzymes, and growth factors. A variety of growth factors, including interleukin-1, interleukin-2, colony-stimulating factor-1, epidermal growth factor, fibroblast growth factor, and platelet-derived growth factor, stimulate expression of interferons (Van Damme et al. 1985; Torres et al. 1982; Moore et al. 1984; Johnson and Torres 1985). Like PDGF, all of these mitogens stimulate expression of c-*myc* and c-*fos* very quickly and dramatically in their respective target cells. The response of interferons to these mitogens occurs more gradually than oncogene induction, as was noted in our studies on PDGF.

A related body of literature documents the induction of $(2' - 5')$-oligoadenylate synthetase activity by epidermal growth factor and by serum in fibroblast cell cultures (Lin et al. 1983; Wells and Mallucci 1985). Revel et al. (1979) have shown that agents which promote differentiation of the U937 cell line and of Friend erythroleukemia cells induce interferon and oligoadenylate synthetase (Friedman-Einat et al. 1982; Yarden et al. 1984).

Summary: Feedback Control of Cell Growth by Oncogenes and Interferons

We have suggested that oncogenes and interferon function in a feedback loop to regulate the response to cell growth factors (Zullo et al. 1985). Genes such as c-*myc* and c-*fos* (which are induced quickly by a variety of growth factors) function to initiate a round of cell division. The c-*myc* gene product has been shown to function as an intracellular mediator of growth-factor-induced proliferation (Armelin et al. 1984; Mougneau et al. 1984; Keath et al. 1984; Kacz-

marek et al. 1985; Rapp et al. 1985 and Bravo and Müller, this Vol., and Mölling, this Vol.). The very same agents which induce c-*myc* and c-*fos* induce interferon and $(2' - 5')$-oligoadenylate synthetase activity in their respective target cells, albeit with a slower time course. Interferon and interferon-associated
enzymes may function to terminate the growth response. A broad body of literature is at least consistent with such a scheme.

Acknowledgments. Work from the author's laboratory was supported by grants from the Nation-
al Institute of Health, the American Cancer Society and the Ajinomoto Company of Japan.
David Hall is a Damon Runyon postdoctoral fellow, and Barrett Rollins is an American Cancer
Society postdoctoral fellow.

References

Armelin HA, Armelin MCS, Kelly K, Stewart T, Leder P, Cochran BH, Stiles CD (1984) A functional role for c-*myc* in the mitogenic response to platelet-derived growth factor. Nature
310:655 – 660
Capon DJ, Shepard M, Goeddel DV (1985) Two distinct families of human and bovine interferon
alpha interferon-genes are coordinately expressed and encode functional peptides. Mol Cell
Biol 5:768 – 779
Cochran BH, Reffel AC, Stiles CD (1983) Molecular cloning of gene sequences regulated by platelet-derived growth factor. Cell 33:939 – 947
Floyd-Smith C, Slattery E, Lengyel P (1981) Interferon action: RNA cleavage pattern of $(2' - 5')$-
oligoadenylate-dependent endonuclease. Science 212:1030 – 1032
Friedman-Einat M, Revel M, Kimchi A (1982) Initial characterization of a spontaneous interferon
secreted during growth and differentiation of friend erythroleukemia cells. Mol Cell Biol
2:1472 – 1480
Hiscott J, Cantell K, Weissmann C (1984) Differential expression of human interferon genes.
Nucl Acids Res 12:3727 – 3746
Johnson HM, Torres BA (1985) Peptide growth factors PDGF, EGF and FGF regulate interferon
gamma production. J Immunol 134:2824 – 2826
Kaczmarek L, Hyland JK, Watt R, Rosenberg M, Baserga R (1985) Microinjected c-*myc* as a
competence factor. Science 228:1313 – 1315
Keath EJ, Caimi PG, Cole MD (1984) Fibroblast lines expression activated c-*myc* oncogenes are
tumorigenic in nude mice and syngenic animals. cell 39:339 – 348
Knight E Jr, Hunkapiller MW, Korant BD, Hardy RWF, Hood LE (1980) Human fibroblast interferon: Amino acid analysis and amino terminal amino acid sequence. Science 207:525 – 526
Lengyel P (1981) Enzymology of interferon – a short survey. Methods Enzymol 17:135 – 148
Lin SL, Tso POP, Hollenberg MD (1983) Epidermal growth factor-urogastrone action: Induction
of 2' – 5'-oligoadenylate synthetase activity and enhancement of the mitogenic effect by anti-
interferon antibody. Life Sci 32:1479 – 1488
Merlin G, Chebath J, Benech P, Metz R, Revel M (1983) Molecular cloning and sequence of parti-
al cDNA for interferon-induced $(2' - 5')$-oligo (A) synthetase mRNA for human cells. Proc
Natl Acad Sci USA 80:4904 – 4908
Moore RN, Larsen HS, Horahov DW, Rouse BT (1984) Endogenous regulation of macrophage
proliferative expansion by colony-stimulating factor-induced interferon. Science 223:178 – 181
Mougneau E, Lemieux L, Rassoulzadegan M, Cuzin F (1984) Biological activities of v-*myc* and c-
myc oncogenes in rat fibroblast cells in culture. Proc Natl Acad Sci USA 81:5758 – 5762
Rapp UR, Cleveland JL, Brightman K, Scott A, Ihle JN (1985) Abrogation of IL-3 and IL-2 dependence by recombinant murine retroviruses expressing v-*myc* oncogenes. Nature 317:
434 – 438

Revel M, Kimchi A, Schmidt A, Shulman L, Chernajovsky Y (1979) Studies of interferon action: synthesis, degradation and biological activity of (2' − 5')oligo-isoadenylate. In: Koch C, Richter D (eds) Regulation of macromolecular synthesis by low molecular weight mediators. Academic Press, London New York, pp 3441 − 3459

Roscouet DL, Vodjdani G, Lamaigre-Dubreuil Y, Tovey MG, Latta M, Doly J (1985) Structure of a murine alpha interferon pseudogene with a repetative R-type sequence in the 3' flanking region. Mol Cell Biol 5:1343 − 1348

Sagar AD, Schgal PB, May LT, Inouye M, Slate DL, Shulman L, Ruddle FH (1984) Interferon-beta-related DNA is dispersed in the human genome. Science 223:1312 − 1315

Sehgal PB, May LT, Sagar AD, LaForge KS, Inouye M (1983) Isolation of novel human genomic DNA clones related to human interferon beta cDNA. Proc Natl Acad Sci USA 80:3632 − 3636

Stewart WE (1979) The interferon system. Springer, Berlin Heidelberg New York

Taniguchi T, Ohno S, Fuji-Kutiyama Y, Muramatsu M (1980) The nucleotide sequence of human fibroblast interferon cDNA. Gene 10:11 − 15

Torrence PF, DeClercq E (1981) IF inducers: General survey and classification. Methods Enzymol 78:291 − 299

Torres BA, Farrar WL, Johnson HM (1982) Interleukin 2 regulates immune interferon (IFN$_\gamma$) production by normal and suppressor cell cultures. J Immunol 128:2217 − 2219

Van Damme J, De Ley M, Opdenakker G, Billrau A, De Somer P, Van Beeumen JV (1985) Homogeneous interferon-inducing 22K factor is related to endogenous pyrogen and interleukin 2. Nature 314:266 − 268

Wells V, Mallucci L (1985) Expression of the 2-5A system during the cell cycle. Exp Cell Res 159:27 − 36

Wreschner DH, McCauley SW, Skehel JJ, Kerr IM (1981) Interferon action-sequence specificity of the ppp(A2'p)nA-dependent ribonuclease. Nature 289:414 − 417

Yarden A, Shure-Gottlieb H, Chebath J, Revel M, Kimchi A (1984) Autogenous production of interferon-β switches on HLA genes during differentiation of histiocytic lymphoma U937 cells. EMBO J 3:969 − 973

Zullo J, Cochran BH, Huang A, Stiles CD (1985) Platelet-derived growth factor and double-stranded ribonucleic acids induce the same genes in 3T3 cells. Cell 43:793 − 800

Regulation of c-*myc* Expression in Normal and Transformed Mammalian Cells

PAUL D. FAHRLANDER and KENNETH B. MARCU

The c-*myc* proto-oncogene present in mammalian species is expressed in a wide variety of cell types. It consists of three exons (see Fig. 1) of which the first, approximately 560 nucleotides in length, is noncoding although evolutionarily conserved. The c-*myc* gene product is translated from exons 2 and 3. The transcripts have a size of 2.4 and 2.2 kb that are synthesized under the direction of two promoters, P1 and P2, respectively. Normal c-*myc* RNAs (Dani et al. 1985) and their 62–64 kDa polypeptide products (Hann and Eisenman 1984) are highly unstable in vivo. The c-*myc* protein has been localized within the nucleus and has been proposed to contribute to a cell's competence to enter and progress through the cell cycle (see Moelling, this Vol., and Bravo and Müller, this Vol., and references therein).

Modulation of c-*myc* Activity in Proliferating and Differentiating Cells

Transcriptional and post-transcriptional mechanisms are responsible for alterations in c-*myc* expression. The addition of serum or defined growth factors to growth-arrested fibroblasts results in transient increases in c-*myc* RNA levels as a consequence both of increased transcription and enhanced messenger stability (Greenberg and Ziff 1984; Blanchard et al. 1985, Bravo and Müller,

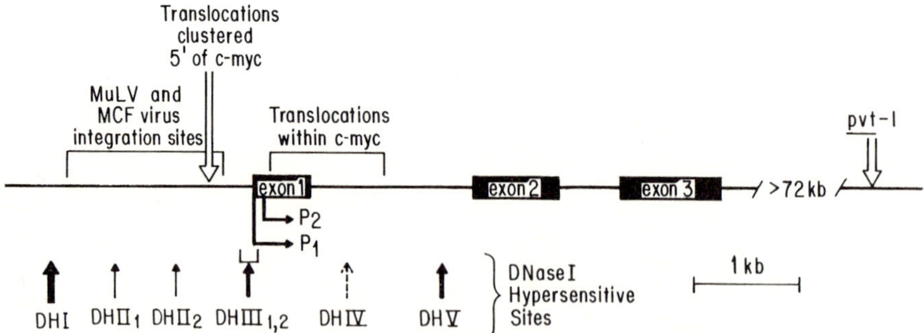

Fig. 1. Diagram of murine c-*myc* showing the structure of the gene, the normal promoters (P1 and P2), and regions of disruption of the locus by chromosome translocation and retroviral insertion. *Arrow sizes* correspond to the relative hypersensitivity to DNase I of the DH sites. *MuLV* and *MCF* are murine retroviruses

Oncogenes and Growth Control
Edited by P. Kahn and T. Graf
© Springer-Verlag Berlin Heidelberg 1986

this Vol.), although the relative contributions of these two factors to c-*myc* induction remain controversial. Recent findings have shown that inducers of protein kinase C such as TPA (Coughlin et al. 1985) as well as double-stranded RNA (Zullo et al. 1985, this Vol.) result in increased steady-state levels of c-*myc* transcripts. Accumulation of c-*myc* RNA is also induced in resting B and T lymphoid cells by mitogenic stimuli (Reed et al. 1985). The mechanisms responsible for the down-regulation of c-*myc* are post-transcriptional in interferon-treated Daudi Burkitt lymphoma cells (Dani et al. 1985) and in teratocarcinoma cells induced to differentiate with retinoic acid (Dony et al. 1985). However, differentiation is not necessarily associated with c-*myc* inactivation. Thus, a short pulse of c-*myc* RNA accumulation has been observed during the G_1 phase of the cell cycle in mouse erythroleukemia cells induced to differentiate (Lachman et al. 1985).

Mechanisms of c-*myc* Deregulation in Transformed Cells

Structural alterations involving c-*myc* and concomitant dysregulation of this proto-oncogene's normal pattern of expression are common features of a variety of tumors and transformed cells. The structural changes which have been identified thus far are chromosome translocation, retroviral insertion, and DNA amplification. The precise means by which most of these aberrations lead to c-*myc* activation remain the subject of intense investigation and considerable controversy. In addition, the contribution(s) of the c-*myc* gene product (which seems to be qualitatively normal in most tumors exhibiting abnormal c-*myc* expression) to the proliferation of normal cells or to the development of the transformed phenotype is by no means clear.

Among murine plasmacytomas and Burkitt lymphomas of humans (both B-cell neoplasias), c-*myc* is generally rearranged *via* reciprocal chromosome translocation with the immunoglobulin heavy or light chain loci. It has been suggested that expression of such translocated genes is grossly abnormal in these transformed counterparts of terminally differentiated B lymphocytes whose c-*myc* alleles may normally be completely silent (Nishikura et al. 1983). It also seems likely that such chromosomal abnormalities (i.e., c-*myc* juxtaposed with the immunoglobulin locus) are growth-selected on the basis of some B cell-specific influence on the transcription of the translocated c-*myc* gene. Recently, it has been shown that the c-*myc* gene under the control of immunoglobulin enhancers is highly tumorigenic for B cells in transgenic mice (Adams et al. 1985 and Hanahan, this Vol.).

The chromosome 15 breakpoints which have been described for c-*myc* translocations in plasma cell tumors fall into three basic categories: (a) those within exon 1 or intron 1 of the normal c-*myc* transcription unit; (b) breakpoints several hundred base pairs upstream of the gene; and (c) breakpoints at a large distance from the gene, such that no rearrangement of c-*myc* is detectable by Southern blot analysis.

In the first category, truncated RNAs are synthesized from normally silent promoters in intron 1. The absence of exon 1 and the presence of intron 1 sequences in such RNAs significantly increases their steady-state accumulation (Piechaczyk et al. 1985). This is despite the fact that c-*myc* genes which employ their normal promoters are transcribed somewhat more efficiently than broken c-*myc* alleles that have fused to immunoglobulin heavy chain constant region (C_H) switch regions. No transcription of the untranslocated allele can be detected in this class of plasma cell tumor (Fahrlander et al. 1985a). Together, these observations establish that activation of c-*myc* by chromosome translocation can result in both abnormal transcriptional and post-transcriptional regulation. Furthermore, the separation of normal and translocated c-*myc* alleles on their resident chromosomes by somatic cell hybrid techniques has elegantly shown that the untranslocated gene in a Burkitt lymphoma is not expressed, as judged by analysis of the steady-state RNA pool (Nishikura et al. 1983). Similar conclusions have been reached by exploiting mutations in the translocated c-*myc* gene's exons in certain Burkitt lymphomas, thus allowing their RNAs to be distinguished from any transcripts arising from the normal allele (Taub et al. 1984).

Examples with breakpoints upstream of c-*myc* exhibit a loss of normal c-*myc* regulatory sequences by a partial shift in promoter usage in favor of P1, the more upstream promoter (Taub et al. 1984; Fahrlander et al. 1985b; Yang et al. 1985). In two plasma cell tumors of this category, the immunoglobulin heavy chain (IgH) enhancer element has been found upstream of c-*myc* (Fahrlander et al. 1985b; Yang et al. 1985), while the majority possess C_H genes in this position. This implies that IgH enhancer and C_H sequences are capable of exerting comparable *cis* effects on translocated c-*myc* loci.

Less obvious disruption of c-*myc* transcriptional control occurs in tumors exhibiting a breakpoint at a large distance from c-*myc*. These include the so-called plasmacytoma variant translocations, which involve the κ light chain locus on chromosome 6 and a site on chromosome 15 termed the *pvt*-1 locus (Cory et al. 1985). The latter is a region spanning approximately 4.5 kb of DNA more than 72 kb downstream (Banerjee et al. 1985) of c-*myc*. It is not yet known whether in those plasmacytomas which have undergone a 12;15 translocation at a large distance from c-*myc* the *pvt*-1 locus is also involved.

Activation of c-*myc* also appears to be common in murine T cell leukemogenesis. Chromosome 15 is frequently trisomic in these malignancies (see Klein, this Vol.) and it has been shown that approximately 45% of early T cell lymphomas and up to 65% of late-developing tumors have proviral insertions near c-*myc* (O'Donnell et al. 1985). Disruption of the c-*myc* locus occurs in a manner qualitatively analogous to that described above for plasma cell tumors. Proviral insertions have been mapped: (a) within the normal c-*myc* transcription unit (O'Donnell et al. 1985); (b) in a cluster approximately 0.7 to 1.3 kb upstream of the gene (O'Donnell et al. 1985); and (c) within the *pvt*-1 locus (Graham et al. 1985). It is possible that alterations in c-*myc* transcription will be observed for these T lymphomas paralleling those found in the plasma cell tumors.

Mechanisms of c-*myc* Gene Regulation

Based largely on studies of structurally altered c-*myc* genes, it is becoming apparent that regulation of this gene's transcription involves an extraordinarily complex series of control elements, lying perhaps both within the gene and in its 5′-flanking region. It is likely that these elements regulate c-*myc* at the transcriptional level by sequence-specific interactions with protein factors. In this context, a binding site for nuclear factor 1 has been identified approximately 1.4 kb upstream of c-*myc* and has been associated with a DNaseI-hypersensitive (DH) site in c-*myc* chromatin (see DH site II_1 in Fig. 1; Siebenlist et al. 1984). A number of other DH sites have been mapped in the 5′-flanking region as well as within the gene itself. One of these, site I, is located approximately 1.8 kb upstream of c-*myc*, and was proposed to be the site of action of a *trans*-acting repressor of c-*myc* transcriptional activity (Siebenlist et al. 1984). Subsequent data conflicted with the repressor model and indicated that the chromatin structure associated with the presence of DH site I was necessary but perhaps not sufficient for transcription of c-*myc* genes whose 5′-flanking region was in germline configuration (Fahrlander et al. 1985a). These data also indicate that any negative modulator(s) of c-*myc* transcription probably lies downstream of DH site I. The noncoding first exon has been postulated to bind a repressor but experimental evidence to support this suggestion is not as yet available.

Other DH sites (III_1 and III_2) were found to be associated with the dual promoters, P1 and P2 (Siebenlist et al. 1984; Fahrlander et al. 1985a). Two additional DH sites (IV and V) are located within the gene's first intron. Site IV was found in an A-MuLV-transformed fibroblast cell line (54c12) which contains approximately 40 copies of the c-*myc* locus (Nepveu et al. 1985). The presence of DH site IV may reflect an alteration of chromatin structure uniquely associated with the c-*myc* amplification unit and which may be partially responsible for changes in the gene's expression. Finally, DH site V was found in every murine cell line tested but only sporadically in human cells (Siebenlist et al. 1984). It is located in the downstream portion of intron 1 (Fahrlander et al. 1985a) and may identify a transcriptional control element responsible for the activity of the intron promoters in truncated c-*myc* genes. The chromatin structure of this region is not altered upon translocation, although the start sites it contains appear to function only when the gene's normal promoters are removed and replaced by immunoglobulin sequences. This lends further credence to the hypothesis that immunoglobulin sequences exert a stimulatory *cis* effect on translocated c-*myc* genes.

Interestingly, the frequency of gene breakage and activation of intron promoters is greater in the plasma cytomas than among Burkitt lymphomas. This difference may be explained on the basis of differences in chromatin structure associated with particular stages of B cell development. It is possible that chromatin structure influences the nature of translocations needed for constitutively deregulated c-*myc* expression. A similar developmental argument could explain the relative proclivity of c-*myc* to recombine with switch regions in plasmacytomas compared with Burkitt lymphomas and the more frequent accumulation of mutations in the translocated c-*myc* genes of the latter malignancy.

Negative regulation of the normal c-*myc* allele is not only a common feature of lymphoid malignancies (see above) but is also apparent in other cell types transformed by *myc*-containing retroviruses. Thus, myeloid and fibroblast cell lines transformed by v-*myc*-expressing murine retroviruses do not accumulate c-*myc* RNAs (Rapp et al. 1985). In contrast to the negative regulation of the normal c-*myc* allele in lymphoid tumors, this effect on endogenous c-*myc* activity may require extremely high levels of exogenous *myc* expression. This idea is supported by the observation that transcription of the endogenous *myc* was detectable in tumorigenic fibroblast cells which expressed only moderate amounts of transfected *myc* (Keath et al. 1984). Finally, normal c-*myc* RNAs were absent in immature and mature B lymphoid malignancies which developed in transgenic mice harboring c-*myc* genes regulated by immunoglobulin heavy or light chain enhancers (Adams et al. 1985). These observations have given support to the proposal that c-*myc* expression may be autoregulated. If autoregulation is a normal feature of c-*myc* control, it will be important to assess: (a) whether this is a direct or indirect effect; (b) the threshold level of the c-*myc* gene product required and whether this level varies from one cell type to another; and (c) whether transcriptional or post-transcriptional mechanisms are involved.

Future Prospects

The observation that the c-*myc* gene is under complex regulation is hardly surprising considering the likely importance of this gene in normal growth processes, its down-regulation in differentiated cells, and its contribution to malignant transformation. Many issues concerning c-*myc* control remain to be resolved. What are the targets within c-*myc* which respond to receptor-transduced stimuli? Is there a c-*myc* repressor? Does c-*myc* activate a repressor factor? What would be its targets within the c-*myc* locus? Might such positive and negative factors be functionally related to the numerous DNase I-hypersensitive sites within and upstream of c-*myc*? It would seem prudent at this stage of the *myc* story to refrain from excessive model building, and instead allow further molecular studies the time to answer these and other questions concerning this pivotal proto-oncogene.

References

Adams JM, Harris AW, Pinkert CA, Corcoran LM, Alexander WS, Cory S, Palmiter RD, Brinster RL (1985) The c-*myc* oncogene driven by immunoglobulin enhancers induces lymphoid malignancy in transgenic mice. Nature 318:533 – 538

Banerjee M, Wiener F, Spira J, Babonits M, Nilsson MG, Sumegi J, Klein G (1985) Mapping of the c-*myc*, *pvt*-1 and immunoglobulin kappa genes in relation to the mouse plasmacytoma-associated variant (6;15) translocation breakpoint. EMBO J 4:3183 – 3188

Blanchard JM, Piechaczyk M, Dani C, Chambard JC, Franchi A, Pouyssegur J, Jeanteur P (1985) c-*myc* gene is transcribed at high rate in G_0-arrested fibroblasts and is post-transcriptionally regulated in response to growth factors. Nature 317:443 – 445

Cory S, Graham M, Webb E, Corcoran LM, Adams JM (1985) Variant (6;15) translocations in murine plasmacytomas involve a chromosome 15 locus at least 72 kb from the c-*myc* oncogene. EMBO J 4:675 – 681

Coughlin SR, Lee WMF, Williams PW, Giels GM, Williams LT (1985) c-*myc* gene expression is stimulated by agents that activate protein kinase C and does not account for the mitogenic effect of PDGF. Cell 43:243 – 251

Dani C, Mechti N, Piechaczyk M, Lebleu B, Jeanteur P, Blanchard JM (1985) Increased rate of degradation of c-*myc* RNA in interferon-treated Daudi cells. Proc Natl Acad Sci USA 82:4896 – 4899

Dony C, Kessel M, Gruss P (1985) Post-transcriptional control of c-*myc* and p53 expression during differentiation of the embryonal carcinoma cell line F9. Nature 317:636 – 639

Fahrlander PD, Piechaczyk M, Marcu KB (1985a) Chromatin structure of the murine c-*myc* locus: implications for the regulation of normal and chromosomally translocated genes. EMBO J 4:3195 – 3202

Fahrlander PD, Sumegi J, Yang JQ, Wiener F, Marcu KB, Klein G (1985b) Activation of the c-*myc* oncogene by the immunoglobulin heavy-chain enhancer after multiple switch region-mediated chromosome rearrangements in a murine plasmacytoma. Proc Natl Acad Sci USA 82:3746 – 3750

Graham M, Cory S, Adams JM (1985) Murine T-lymphomas with retroviral inserts in the chromosomal 15 locus for plasmacytoma variant translocations. Nature 314:740 – 743

Greenberg M, Ziff EB (1984) Stimulation of mouse 3T3 cells induces transcription of the c-*fos* oncogene. Nature 311:438 – 442

Hann SR, Eisenman RN (1984) Proteins encoded by the human c-*myc* oncogene: differential expression in neoplastic cells. Mol Cell Biol 4:2486 – 2497

Keath EG, Caimi PG, Cole MD (1984) Fibroblast lines expressing activated c-*myc* oncogenes are tumorigenic in nude mice and syngeneic animals. Cell 39:339 – 348

Lachman HM, Hatton KS, Skoultchi AI, Schildkraut CL (1985) c-*myc* mRNA levels in the cell cycle change in mouse erythroleukemia cells following inducer treatment. Proc Natl Acad Sci USA 82:5323 – 5327

Nepveu A, Fahrlander PD, Yang JQ, Marcu KB (1985) Amplification and altered expression of the c-*myc* oncogene in A-MuLV-transformed fibroblasts. Nature 317:440 – 443

Nishikura K, Ar-Rushdi A, Erikson J, Watt R, Rovera G, Croce CM (1983) Differential expression of the normal and of the tranlocated human c-*myc* oncogenes in B cells. Proc Natl Acad Sci USA 80:4822 – 4826

O'Donnell PV, Fleissner E, Lonial H, Koehne CF, Reicin A (1985) Early clonality and high-frequency proviral integration into the c-*myc* locus in AKR leukemias. J Virol 55:500 – 503

Piechaczyk M, Yang JQ, Blanchard JM, Jeanteur P, Marcu KB (1985) Post-transcriptional mechanisms are responsible for accumulation of truncated c-*myc* RNAs in murine plasma cell tumors. Cell 42:589 – 597

Rapp UR, Cleveland JL, Brightman K, Scott A, Ihle JN (1985) Abrogation of IL-3 and IL-2 dependence by recombinant murine retroviruses expressing v-*myc* oncogenes. Nature 317: 434 – 438

Reed JC, Nowell PC, Hoover R (1985) Regulation of c-*myc* mRNA levels in normal human lymphocytes by modulators of cell proliferation. Proc Natl Acad Sci USA 82:4221 – 4224

Siebenlist U, Hennighausen L, Battey J, Leder P (1984) Chromatin structure and protein binding in the putative regulatory region of the c-*myc* gene in Burkitt lymphomas. Cell 37:381 – 391

Taub R, Moulding C, Battey J, Murphy W, Vasicek T, Lenoir GM, Leder P (1984) Activation and somatic mutation of the translocated c-*myc* gene in Burkitt lymphoma cells. Cell 36:339 – 348

Yang JQ, Bauer S, Mushinski JF, Marcu KB (1985) Chromosome translocations clustered 5′ of the murine c-*myc* gene qualitatively affect promoter usage: implications for the site of normal c-*myc* regulation. EMBO J 4:1441 – 1447

Zullo JN, Cochran BH, Huang AC, Stiles CD (1985) Platelet-derived growth factor and double-stranded ribonucleic acids stimulate expression of the same genes in 3T3 cells. Cell 43:793 – 800

Properties of the *myc* and *myb* Gene Products

KARIN MOELLING

The *myc* and *myb* oncogenes were originally discovered as the transforming genes in certain avian retrovirus strains and have recently also been implicated in the genesis of human hematopoietic neoplasms. Their gene products were the first transforming proteins found to be localized in the nucleus and are therefore possibly involved in the regulation of gene expression. In this article some of the properties of viral and cellular *myc* and *myb* proteins will be reviewed and their possible roles during normal and neoplastic processes discussed.

The Cellular and Viral *myc* Proteins

The c-*myc* gene consists of three exons, only two of which, exon 2 and exon 3, code for a protein and are transduced by retroviruses (see Fahrlander and Marcu, this Vol.). Exons 2 and 3 code for proteins with apparent molecular weights of about 64 and 55 kDa (for review and references, see Moelling 1985). These values are larger than the 49 kDa predicted from the nucleotide sequence of the gene. Even though both proteins are phosphorylated, this discrepancy cannot be ascribed only to post-transcriptional modifications since c-*myc* protein synthesized in bacteria is unphosphorylated and also has a molecular weight of 64000 (Watt et al. 1985). This aberrant electrophoretic mobility might be due to clustering of proline residues in the N-terminal half of the protein that could result in an unusual electrophoretic mobility.

The majority of information on *myc* proteins was obtained from the study of cells transformed by the v-*myc* containing retroviruses. Properties of the v-*myc* proteins closely resemble those of c-*myc* proteins and will therefore not always be separated in the following discussion.

As evidenced by cell fractionation and indirect immunofluorescence microscopy, the v-*myc* protein is located in the nucleus of infected avian fibroblasts and macrophages; (Donner et al. 1982) in human tumor cell lines c-*myc* protein is also nuclear (Hann and Eisenman 1984; Beimling et al. 1985). The *myc* protein is synthesized in the cytoplasm in about 7 min, is rapidly transported to the nucleus in less than 1 h and exhibits a half-life of only about 15 min (Hann and Eisenman 1984; Beimling et al. 1985). The estimated amount of v-*myc* protein in MC29 virus-transformed fibroblasts is 0.1% of the total cellular protein in MC29 transformed fibroblasts and 0.001% in a

Oncogenes and Growth Control
Edited by P. Kahn and T. Graf
© Springer-Verlag Berlin Heidelberg 1986

small cell lung cancer cell line (N417d) harboring an amplified c-*myc* gene (see Schwab, this Vol.). Phosphorylation of the *myc* protein remains unchanged throughout the cell cycle and is alike in normal and transformed cells (Hann et al. 1985). It is unclear at present what directs *myc* protein to the nucleus. Correct or complete phosphorylation is not a prerequisite for nuclear localization in fibroblasts. An amino acid sequence serving as a signal for transport to the nucleus, such as shown for SV40 T-antigen (Kalderon et al. 1984), has not been identified for the *myc* protein. Interestingly, the p230$^{gag\text{-}pol\text{-}myc}$ protein of OK10 virus is predominantly located in the cytoplasm, presumably due to its *pol* moiety (Moelling 1985). Immunofluorescence data indicate that v-*myc* protein is excluded from nucleoli and that during mitosis it is primarily cytoplasmic rather than nuclear (Winqvist et al. 1984). A slowly growing MC29 variant also exhibits an unusually high level of cytoplasmic v-*myc* protein (unpublished observations). Recently, absence or presence of *myc* protein in nucleoli has been reported to depend on experimental procedures (Persson et al. 1986).

Further characterization of the association of *myc* protein with the nucleus has indicated that it belongs to a discrete subset of nuclear proteins that can be extracted at low salt concentration and that are rendered irreversibly insoluble by exposure of isolated nuclei at 37 °C. According to this analysis, c-*myc* protein is not linked to either the chromatin or to the nuclear matrix (Evan and Hancock 1985). In contrast, some of the v-*myc* protein has been found to be associated with chromatin (Bunte et al. 1982).

The nuclear localization and short half-life of the c-*myc* protein suggest a regulatory mode of action, possibly by interaction with DNA. The purified v- and c-*myc* proteins bind to double-stranded DNA in vitro (Donner et al. 1982; Persson and Leder 1984; Beimling et al. 1985). Bacterially expressed c-*myc* protein does not exhibit any sequence-specific DNA binding. However, this could be due to denaturation or to the fact that the protein is not phosphorylated (Watt et al. 1985). Recent evidence indicates a preferential binding of v- and c-*myc* proteins to cellular sequences upstream of c-*myc*, but not to the viral LTR or to exon 1 (Moelling et al. 1986). The *myc* proteins from three nontransforming MC29 deletion mutants exhibit a reduced ability to bind to DNA (Donner et al. 1983), indicating a correlation between DNA-binding of the *myc* protein with transformation.

Regulation of c-*myc* Expression

C-*myc* mRNA and protein are expressed at similar levels through all stages of the cell cycle including G_0 (Hann et al. 1985; Rabbitts 1985). C-*myc* mRNA is, however, so rapidly turned over during G_0 that no *myc* protein can be identified during this phase. This indicates a regulation of c-*myc* mRNA at the post-transcriptional level (Blanchard et al. 1985). C-*myc* mRNA is extremely unstable during all stages of the cell cycle in normal and transformed cells (its half-life time is about 10 min) (Dani et al. 1985). The rate of c-*myc* protein turnover in the various stages of the cycle is also high (20 to 30 min), indicating that most of the protein and mRNA synthesized results from de novo synthesis. During transition of cells from G_0 to G_1 induced by external stimuli such as growth factors in fibroblasts or mitogens in lymphocytes, c-*myc* mRNA and protein are transiently highly overexpressed (Kelly et al. 1984). Their levels decrease again when the culture resumes quiescence (see article by Bravo and Müller, this Vol.). Likewise, levels of c-*myc* decrease during terminal differentiation in a number of cell systems (see Wagner and Müller, this Vol.).

Inhibition of protein synthesis in most but not all cell types results in a dramatic stabilization of c-*myc* mRNA and protein (Kelly et al. 1983; Dani et al. 1984). This suggests that the controlling element is itself a protein − a nuclease? − that is responsible for the rapid turnover. The finding that the translational efficiency of c-*myc* mRNA is 10-fold higher from transcripts starting at P_2, as compared to P_1 (Darveau et al. 1985) raised the possibility that these two transcripts differ in their stability. As shown recently, this is not the case. However, the absence of exon 1 leads to a higher stability of the c-*myc* mRNA (Butnick et al. 1985; Rabbitts 1985). The above observations might be explained by the recent finding that interferon treatment leads to an increased rate of c-*myc* mRNA degradation (Dani et al. 1985). This raises the possibility that c-*myc* transcripts are regulated post-transcriptionally through their capacity to form double-stranded RNA regions. In support for this notion is the finding that in interferon-treated cells c-*myc* mRNA activates the $(2'-5')$ A synthetase/RNase L endonuclease system apparently by means of its double-strandedness (Dani et al. 1984). These data are compatible with the notion that overexpression of c-*myc* mRNA after cycloheximide treatment is due to the elimination of a short-lived $(2'-5')$ A synthetase/RNase L system which may normally be responsible for c-*myc* mRNA degradation.

Involvement of c-*myc* in Tumor Formation

The c-*myc* oncogene is one of the most convincing examples of an oncogene implicated in tumor etiology (see Klein, this Vol.). Four types of abnormality involving c-*myc* indicate its role in tumor formation: (1) provirus insertion adjacent to c-*myc* in chicken B-cell lymphomas; (2) c-*myc* gene amplification in some human tumors (see Schwab, this Vol.); (3) chromosomal translocations involving the complete c-*myc* gene in various Burkitt's lymphomas (BL) or mouse plasmacytomas; (4) chromosomal translocations and rearrangements of the c-*myc* gene, leading to a truncation of exon 1 (see Fahrlander and Marcu, this Vol.). A result common to these various rearrangements is the constitutive expression of relatively high levels of c-*myc* mRNA and protein throughout all stages of the cell cycle, a situation that might induce a constant proliferation stimulus. Tumor cells differ from normal cells in that they generally do not return to a resting phase. There is indeed a good correlation between the constitutive expression of c-*myc* and transformation (Rabbitts

1985). However, c-*myc* expression may not be sufficient for preventing cells from entering the resting phase. Thus, two partially transformed cell lines which can still be returned to a quiescent state, continue to express c-*myc* (Campisi et al. 1984). Overexpression of c-*myc* is probably also not sufficient for tumor formation. Recent studies with c-*myc* DNA constructs joined to promoters from either Ig genes or mammary tumor virus and introduced into the germ line of mice showed that tumors arise only with a relatively long period of latency in a monoclonal fashion (Adams et al. 1985) (see also Hanahan, this Vol.).

In certain tumors, such as the variant form of small cell cancer of the lung, the amounts of c-*myc* mRNA as well as protein correlate with malignancy of the tumor (Nau et al. 1985). In neuroblastomas, N-*myc* mRNA levels correlate with progression of the disease (see Schwab, this Vol.). It will be interesting to determine whether the level of c-*myc* expression is also a useful tumor or progression marker in other types of human neoplasias.

Possible Function of the c-*myc* Gene Product

The human c-*myc* gene product is assumed to play a regulatory role at the level of gene expression due to its nuclear localization, short half-life and its ability to bind to DNA in vitro. An earlier model suggested a repressor function for the c-*myc* protein possibly acting directly or indirectly on exon 1. This model has been excluded, since the efficiency of translation of c-*myc* mRNA is independent of the presence or absence of exon 1 (Butnick et al. 1985; Cole 1985). The effect of the v-*myc* gene transfected into cells is controversial with respect to its influence on the endogenous c-*myc* expression. While v-*myc* transfection into fibroblasts leaves c-*myc* expression unaltered (Cole 1985), in haematopoietic cells it leads to its shut off (Rapp et al. 1985). In the latter case the introduced v-*myc* gene is, however, expressed at a 100-fold excess over the endogenous one and it is therefore not possible to draw general conclusions regarding an autoregulatory feedback inhibition. If an autoregulatory mechanism of c-*myc* exists, it most likely involves cellular sequences located upstream of exon 1. Seven potential regulatory sites have been determined by DNase I hypersensitivity studies in upstream cellular sequences and within c-*myc*, including the two promoters P_1 and P_2 in exon 1 (Siebenlist et al. 1984). Induction of an inhibition of c-*myc* transcription, e.g., by DMSO-induced differentiation of HL60 cells, results in the loss of several DNase I-hypersensitive sites (Dyson et al. 1985). Whether the c-*myc* protein itself is involved in regulating these sites is still unknown. The c-*myc* gene product has been shown − either directly or indirectly − to introduce expression of indicator genes from a Drosophila heat shock promoter (hsp70) (Kingston et al. 1984), but it is unknown whether mammalian hsp70 is regulated in a similar fashion.

So far DNA-*myc* protein interaction studies have been hampered by lack of material or problems of protein denaturation. It is unknown whether the *myc* protein requires other cellular proteins to exert its specific function. While the SV40 T-antigen recognizes specific DNA sequences without any such factors, the E1A protein of adenovirus does not bind to DNA at all, except when supplemented by a host factor (see Philipson, this Vol., and Schlokat and Gruss, this Vol.).

The *myb* Protein

The c-*myb* oncogene has been implicated in the formation of lymphoid tumors in mice and has also been found to be amplified in a human myelogenous leukemia cell line (Pelicci et al. 1984). It contains 7 exons, of which about 6 1/2 are transduced as the v-*myb* gene of avian myeloblastosis virus (AMV) and about 5 1/2 as the v-*myb* gene of E26 virus. The c-*myb* protein in avian and human tumor cells has an apparent molecular weight of 75 kDa whereas the viral counterparts from AMV has 48 kDa and that from E26 (a fusion protein with *gag* and *ets* sequences) of 135 kDa (Klempnauer et al. 1984; Evan et al. 1984; Moelling et al. 1985). Both v-*myb* proteins have carboxyterminal deletions compared to c-*myb*. In AMV-infected fibroblasts (which are not transformed, in contrast to fibroblasts expressing v-*myc*) and in AMV-transformed myeloblasts the v-*myb* protein is localized in the nucleus. In the latter cells it is also localized to a lesser extent in the perinuclear region of the cytoplasm (Klempnauer et al. 1984). Like the *myc* proteins, v-*myb* protein is not associated with the nuclear matrix but belongs to a group of proteins easily extractable from the nucleus (Evan and Hancock 1985).

V- and c-*myb* proteins bind to DNA in vitro (Moelling et al. 1985; Klempnauer and Sippel 1986), but with unknown specificity. Studies using temperature-sensitive (*ts*) mutants of E26 in v-*myb* indicate that the DNA binding properties of mutant $p135^{gag\text{-}myb\text{-}ets}$ proteins are more heat-labile in in vitro assays than that from wild-type virus (Moelling et al. 1985). Upon induction of differentiation by TPA of AMV-transformed myeloblasts into macrophages, v-*myb* relocates to the cytoplasm (Klempnauer et al. 1984). This result conflicts with the localization of the v-*myb* protein from E26 virus, which remains nuclear in *ts* mutant-infected cells that are induced to differentiate when shifted to the non-permissive temperature (Moelling et al. 1985).

In a recent study, Thompson et al. (1986) analyzed the expression of c-*myb* during fibroblast growth. In contrast to c-*myc*, the c-*myb* proto-oncogene shows a transient increase in expression during the cell cycle that is highest during S phase. Quiescent cells also show a delayed response to serum growth factors in their c-*myb* expression as compared to c-*myc*. These results support the notion that the products of c-*myc* and c-*myb* fulfill different functions in growth control.

In conclusion, in spite of the knowledge of several biochemical properties of the *myc* and *myb* gene products, their functions are still unclear. Nuclear localization and in vitro DNA binding properties suggest a role in DNA-protein interactions such as required in gene regulation. More work is necessary to prove or disprove this concept.

References

Adams JM, Harris AW, Pinkert CA, Corcoran CM, Alexander WS, Cory S, Palmiter RD, Brinster RL (1985) The c-*myc* oncogene driven by immunoglobulin enhancers induces lymphoid malignancy in transgenic mice. Nature 318:533–538

Beimling P, Benter T, Sander T, Moelling K (1985) Isolation and characterization of the human cellular *myc* gene product. Biochemistry 24:6349–6355

Blanchard J-M, Piechaczyk M, Dani C, Chambard J-C, Franchi A, Pouyssegur J, Jeanteur P (1985) c-*myc* gene is transcribed at high rate G_0- arrested fibroblasts and is post-transcriptionally regulated in response to growth factors. Nature 317:443–445

Bunte T, Greiser-Wilke I, Donner P, Moelling K (1982) Association of *gag-myc* proteins from avian myelocytomatosis virus wild-type and mutants with chromatin. EMBO J 1:919–927

Butnick NZ, Miyamoto C, Chizzonite R, Cullen BR, Ju G, Skalka AM (1985) Regulation of the human c-*myc* gene: 5' noncoding sequences do not affect translation. Mol Cell Biol 5:3009–3016

Campisi J, Gray HE, Pardee AB, Dean M, Sonenshein GE (1984) Cell-cycle control of c-*myc* but not c-*ras* expression is lost following chemical transformation. Cell 36:241–247

Cole MD (1985) Regulation and activation of c-*myc*. Nature 318:510–511

Dani C, Blanchard JM, Piechaczyk M, El Sabouty S, Marty L, Jeanteur P (1984) Extreme instability of *myc* mRNA in normal and transformed human cells. Proc Natl Acad Sci USA 81:7046–7050

Dani C, Mechti N, Piechaczyk M, Lebleu B, Jeanteur P, Blanchard JM (1985) Increased rate of degradation of c-*myc* RNA in interferon-treated Daudi cells. Proc Natl Acad Sci USA 82:4896–4899

Darveau A, Pelletier J, Sonenberg N (1985) Differential efficiencies of in vitro translation of mouse c-*myc* transcripts differing in the 5' untranslated region. Proc Natl Acad Sci USA 82:2315–2319

Donner P, Greiser-Wilke I, Moelling K (1982) Nuclear localization and DNA binding of the transforming gene product of avian myelocytomosis virus. Nature 296:262–266

Donner P, Bunte T, Greiser-Wilke I, Moelling K (1983) Decreased DNA-binding ability of purified transformation-specific proteins from deletion mutants of the acute avian leukemia virus MC29. Proc Natl Acad Sci USA 80:2861–2865

Dyson PJ, Littlewood TD, Forster A, Rabbitts TH (1985) Chromatin structure of transcriptionally active and inactive human c-*myc* alleles. EMBO J 4:2885–2891

Evan GI, Hancock DC (1985) Studies on the interaction of the human c-*myc* protein with cell nuclei: $p62^{c\text{-}myc}$ as a member of a discrete subset of nuclear proteins. Cell 43:253–261

Evan GI, Lewis GK, Bishop JM (1984) Isolation of monoclonal antibodies specific for products of avian oncogene *myb*. Mol Cell Biol 4:2843–2850

Hann SR, Eisenman RN (1984) Proteins encoded by the human c-*myc* oncogene differential expression in neoplastic cells. Mol Cell Biol 4:2486–2497

Hann SR, Thompson CB, Eisenman RN (1985) c-*myc* oncogene protein synthesis is independent of the cell cycle in human and avian cells. Nature 314:366–369

Kalderon D, Roberts BL, Richardson WD, Smith A (1984) A short amino acid sequence able to specify nuclear location. Cell 39:499–509

Kelly K, Cochran BH, Stiles CD, Leder P (1984) Cell-specific regulation of the c-*myc* gene by lymphocyte mitogens and platelet-derived growth factor. Cell 35:603–610

Kingston RE, Baldwin AS, Sharp PA (1984) Regulation of heat shock protein 70 gene expression by c-*myc*. Nature 312:280–282

Klempnauer KH, Sippel A (1986) Subnuclear localization of proteins encoded by the oncogene v-*myb* and its cellular homolog c-*myb*. Mol Cell Biol 6:62 – 69

Klempnauer KH, Symonds G, Evan GI, Bishop JM (1984) Subcellular localization of proteins encoded by oncogenes of avian myeloblastosis virus and avian leukemia virus E26 and by the chicken c-*myb* gene. Cell 37:537 – 547

Moelling K (1985) Fusion proteins in retroviral transformation. In: Weinhouse S, Klein G (eds) Advances in cancer research, vol 43. Academic Press, London New York, pp 205 – 239

Moelling K, Pfaff E, Beug H, Beimling P, Bunte T, Schaller HE, Graf T (1985) DNA-binding activity is associated with purified *myb* proteins from AMV and E26 viruses and is temperature-sensitive for E26 *ts* mutants. Cell 40:983 – 900

Moelling K, Lorenz U, Beimling P, Heimann B, Thomas G, Bading H, Sander T (1986) Biochemical properties of oncogene-coded proteins. In: Gallwitz D, Tanner W (eds) Cell cycle and oncogenes, 37th Mosbach Colloquium. Springer, Berlin Heidelberg New York, (in press)

Nau MM, Brooks BJ, Battey J, Sansville E, Gazdar AF, Kirsch IR, McBride OW, Bertness V, Hollis GF, Minna JD (1985) L-*myc*, a *myc*-related gene amplified and expressed in human small cell lung cancer. Nature 318:69 – 73

Pelicci P-G, Lanfrancone L, Braithwaite MD, Wolman SR, Dalla-Favera R (1984) Amplification of the c-*myb* oncogene in a case of human acute myelogenous leukemia. Science 224: 1117 – 1121

Persson H, Leder P (1984) Nuclear localization and DNA binding properties of protein expressed by human c-*myc* oncogene. Science 225:718 – 721

Persson H, Gray H, Godeau F, Braunhut S, Bellvé AR (1986) Multiple growth-associated nuclear proteins immunoprecipitated by antisera raised against human c-*myc* peptide antigens. Mol Cell Biol 6:942 – 949

Rabbitts TH (1985) The c-*myc* proto-oncogene: involvement in chromosomal abnormalities. Trends Gen Dec 327 – 331

Rapp UR, Cleveland JL, Brightman K, Scott A, Ihle JN (1985) Abrogation of IL-3 and IL-2 dependence by recombinant murine retroviruses expressing v-*myc* oncogenes. Nature 317: 434 – 438

Siebenlist U, Henninghausen L, Battey J, Leder P (1984) Chromatin structure and protein binding in the putative regulatory region of the c-*myc* gene in Burkitt lymphoma. Cell 37:381 – 391

Thompson CB, Challoner PB, Neiman PE, Groudine M (1986) Expression of the c-*myb* proto-oncogene during cellular proliferation. Nature 319:374 – 380

Watt RA, Shatzman AR, Rosenberg M (1985) Expression and characterization of the human c-*myc* DNA-binding protein. Mol Cell Biol 5:448 – 456

Winqvist R, Saksela K, Alitalo K (1984) The *myc* proteins are not associated with chromatin in mitotic cells. EMBO J 3:2947 – 2950

The *fos* Oncogene and Transformation

Thomas Jenuwein and Rolf Müller

To date, the *fos* oncogene has been detected in two different mouse osteosarcoma viruses: the Finkel-Biskis-Jinkins mouse osteosarcoma virus (FBJ-MSV), isolated from a spontaneous osteosarcoma in a CF-1 mouse (Finkel et al. 1966; Finkel and Biskis 1968) and the Finkel-Biskis-Reilly mouse osteosarcoma virus (FBR-MSV), isolated from a radiation (^{90}Sr)-induced osteosarcoma in a X/Gf mouse (Finkel et al. 1973). Both FBJ-MSV and FBR-MSV induce chondro-osseous sarcomas of similar histological and pathological appearance (Ward and Young 1976).

The complete nucleotide sequences of the FBJ-MSV and FBR-MSV *fos* oncogenes and of the c-*fos* proto-oncogene have been determined and the primary structures of their encoded proteins deduced (Curran et al. 1983; Van Beveren et al. 1983; van Straaten et al. 1983; Van Beveren et al. 1984; Curran et al. 1984). The predicted size of the c-*fos* gene product is 380 amino acids, which corresponds to a molecular mass of approximately 40 kDa. Presumably due to its high content of proline and charged amino acids, it migrates as a 55-kDa protein (p55$^{c\text{-}fos}$) on SDS-polyacrylamide gels (SDS-PAGE) (Curran and Teich 1982a; Curran et al. 1984). The FBJ-MSV encodes a *fos* protein (pp55$^{v\text{-}fos}$) (Fig. 1) that is very similar to the c-*fos* gene product except for its entirely different C-terminus (49 amino acids), a consequence of a 104 base pair (bp) out-of-frame deletion in the v-*fos* gene (Van Beveren et al. 1983; van Straaten et al. 1983). In addition, the FBJ-MSV gene product contains five point mutations compared with c-*fos* (Van Beveren et al. 1983). In contrast, the *fos* gene of FBR-MSV has retained only 60% of the c-*fos* coding sequence and has undergone several extensive alterations (Fig. 1). These include truncations of the coding sequence at both termini, two in-frame deletions close to the C-terminus, four single amino acid substitutions, and fusion to heterologous sequences at both the 5′ (*gag*) and the 3′ end. The C-terminal 8 amino acids of this fusion product are derived from cellular sequences (termed *fox*) which are unrelated to *fos* (Van Beveren et al. 1984). The transforming protein of FBR-MSV is therefore expressed as a *gag-fos-fox* fusion protein (p75$^{gag\text{-}fos\text{-}fox}$).

Both c-*fos* and v-*fos* proteins are phosphorylated, which in p55$^{v\text{-}fos}$ occurs at least partially on threonine and serine residues. However, only p55$^{c\text{-}fos}$ undergoes additional intensive post-translational modifications which result in a smear of c-*fos* proteins ranging from 55 kDa to more than 62 kDa on SDS-PAGE (Curran et al. 1984). The nature of these post-translational modifica-

Oncogenes and Growth Control
Edited by P. Kahn and T. Graf
© Springer-Verlag Berlin Heidelberg 1986

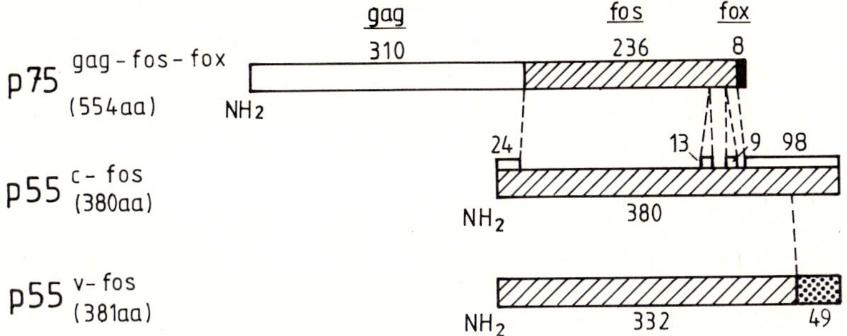

Fig. 1. Comparison of the mouse c-*fos* (p55$^{c\text{-}fos}$) gene product with FBJ-MSV (p55$^{v\text{-}fos}$) and FBR-MSV (p75$^{gag\text{-}fos\text{-}fox}$) transforming porteins (Van Beveren et al. 1983; van Straaten et al. 1983; Van Beveren et al. 1984). *Numbers* refer to amino acids (aa). The sizes of truncations and deletions in the v-*fos* proteins are indicated. FBR-MSV has an N-terminal truncation of 24 amino acids as well as 2 internal deletions of 13 and 9 amino acids, and a C-terminal truncation of 98 amino acids. Amino-acid substitutions in FBR-MSV compared with c-*fos* correspond to positions 64, 138, 268, 279, and 280 of the c-*fos* protein. The FBR-MSV *fos* protein is fused with *gag* (310 amino acids) and *fox* (8 amino acids) sequences. The C-terminal 49 amino acids of FBJ-MSV-encoded p55$^{v\text{-}fos}$ are different from those in p55$^{c\text{-}fos}$ due to an out-of-frame deletion of 104 nucleotides. The five point mutations in FBJ-MSV compared with c-*fos* reside at amino acid positions 15, 67, 110, 175, and 291

tions remains obscure. Cell fractionation, immunoprecipitation, and immunofluorescence data have shown that all *fos* gene products are located in the nucleus and are complexed with a cellular protein, p39 (Curran et al. 1984; Curran and Teich 1982b; Curran et al. 1985). The possible role of this complex is unknown.

Activation of the Transforming Potential of c-*fos*

Analysis of molecularly cloned v-*fos* and c-*fos* genes indicated that the former has the capacity to transform fibroblasts, as expected, but the latter does not (Miller et al. 1985). Using c-*fos*/v-*fos* chimeric molecules it was shown that the transforming potential of the c-*fos* gene can be activated without altering its coding region but rather by addition of the FBJ-MSV long terminal repeat 3′ to the gene and the deletion of a 67 bp inhibitory sequence in the 3′ flanking region 123–189 bp upstream from the poly(A) addition site. This modified c-*fos* gene transforms fibroblasts (although at 50-fold lower efficiencies than v-*fos*) and is no longer regulated by growth factors but is constitutively expressed (Miller et al. 1984; Meijlink et al. 1985; Treisman 1985). The 67-bp element probably interacts with C-terminal coding sequences, since a hybrid molecule with the C-terminus of v-*fos* but retaining the 3′ noncoding region of c-*fos* transforms fibroblasts. Likewise, another construct with a deletion comprising the last 24 amino acids of the c-*fos* carboxy terminus transforms fibroblasts despite the presence of the 3' noncoding region (Miller et al. 1984; Meijlink et al. 1985). The evidence suggests that the interaction between coding and noncoding sequences acts at the post-transcriptional level (Miller et al. 1984; Meijlink et al. 1985). Consistent with these results is the observation that both these sequence elements are deleted in the viral *fos* oncogenes.

Pulse-chase analysis of amnion cells and serum-stimulated fibroblasts showed that the majority of the c-*fos* protein is highly modified under physiological conditions (Curran et al. 1984; Kruijer et al. 1984; Müller et al. 1984). In contrast, both virally encoded transforming *fos* proteins are barely modified and have half-lives (≈ 2 h) which are approximately fourfold longer than that of the c-*fos* protein (30 min) (Curran and Teich 1982a; Curran et al. 1984; Curran and Verma 1984; Kruijer et al. 1984; Müller et al. 1984). The post-translational modifications may therefore influence not only the turnover rate of the different *fos* gene products, but also their transforming ability. This interpretation is based on three observations: (1) normal amnion cells express c-*fos* constitutively at levels as high as those in *fos*-transformed fibroblasts (Müller et al. 1984); (2) modification of the c-*fos* protein in fibroblasts transformed by the deregulated expression of a chimeric c-*fos* gene is not efficient (Curran et al. 1984); and (3) the rapid, transient induction of the c-*fos* protein in various cell types following stimulation by different growth factors leads to expression levels similar to those observed in *fos*-transformed cells, but the stimulated cells maintain a normal phenotype (Kruijer et al. 1984; Müller et al. 1984). Transformation by c-*fos* therefore seems to be due to the constitutive expression of (partially modified) c-*fos* protein in an inappropriate cell type.

The Role of Viral *fos* Oncogenes in Neoplastic Transformation

The oncogenic potential of the v-*fos*-containing retroviruses FBJ-MSV and FBR-MSV was studied by infection and transfection of the rat fibroblast cell line 208F and of early passage mouse connective tissue cell cultures (bone cells, cartilage cells, muscle cells and fibroblasts (Jenuwein et al. 1985). Focus formation in established 208F rat fibroblasts and in nonestablished mouse connective tissue cells was considerably more efficient with FBR-MSV than with FBJ-MSV, as reflected by the number of foci obtained, as well as by the time required for expression of the transformed phenotype. Furthermore, approximately 40% (20/52) of the foci induced in nonestablished cells by the FBR-MSV transforming protein showed an extended life-span, whereas establishment of cell clones transformed by FBJ-MSV occurred much less frequently ($<5\%$). Transformation with Harvey sarcoma virus, which was included in these studies for comparative purposes, was completely unable to rescue cells from senescence (0/30).

The ability of *fos* to promote the rescue from senescence of low passage mouse embryo fibroblasts (MEF) was investigated further in a different assay MEFs were transfected with recombinant DNA molecules carrying the gene encoding resistance to neomycin (neo) and to its analog G418 as a dominant selectable marker covalently linked to either FBR-MSV (pFBR-*neo*) or FBJ-MSV (pFBJ-*neo*). Since cells which have not taken up DNA die in the presence of G418, the cell density on the dish decreases dramatically during the course of the selection. Under these conditions, nonestablished mouse fibroblasts are unable to grow. This assay therefore represents a double selection which allows the immediate identification of transfected clones containing oncogenes that promote cell growth at low cell density, a characteristic of established cell lines.

Both pFBR-*neo* and pFBJ-*neo* gave rise to morphologically transformed, G418-resistant colonies in this assay. However, only G418-resistant colonies transformed by pFBR-*neo* were rescued from senescence, while those transformed with pFBJ-*neo*-transformed colonies rapidly senesced. The deregulated expression in MEFs of an engineered *fos* protein with the C-terminus of c-*fos* also gave rise to cells which proliferated at low cell density, but none of these cultures was able to overcome crisis (T. J., unpublished observations). Thus it appears that all *fos* proteins can contribute to certain events in the rescue of cells from senescence, but further steps in the establishment of cell lines occur with much greater probability in cells transformed by the FBR-MSV gene product.

Nonestablished mouse connective tissue cells transformed by FBR-MSV closely resemble cell lines derived from *fos*-induced osteosarcomas or *fos*-transformed established cell lines, as assayed by the expression of various transformation parameters. However, only a fraction of FBR-MSV transformed early passage mouse connective tissue cell clones grow beyond the normal life span, and in cultures that do so only a subpopulation of cells escapes senescence (Jenuwein et al. 1985). The establishment of *fos*-transformed MEFs may therefore require more than one step. Our observations suggest that p75$^{gag-fos-fox}$ facilitates this process but does not directly induce the rescue of nonestablished mouse cells from senescence. Although morphological transformation seems to be induced by the *fos* oncogene product, it may be the combination of p75$^{gag-fos-fox}$ expression and a random second event that leads to establishment.

The extensive structural alterations of the *fos* gene in FBR-MSV compared to that of FBJ-MSV (Fig. 1) appear to account for its enhanced transforming and "immortalizing" potential, since the expression levels of *fos* RNA and protein are very similar with both viruses (Curran and Teich 1982a; Curran et al. 1984; Curran and Verma 1984; Jenuwein et al. 1985). Recent analyses have shown that the *gag* sequences in p75$^{gag-fos-fox}$ do not affect the transforming potential of the protein (Miller et al. 1985 and T. J., unpublished observations), that the amino-terminal *fos* sequences are dispensable for transformation, and that the primary sequence of the carboxy terminus (amino acid positions 280 to 340) is important in determining the transforming capacity of *fos* proteins (T. J. unpublished observations).

Concluding Remarks

Neoplastic transformation is considered as a multistage process (see Chap. VI). It has been proposed that two oncogenes are required in concert for the malignant conversion of nonestablished rodent cells and that such genes belong to two different complementation groups, one of which induces transformation (for example *ras*) and the other of which triggers "immortalization" (for example *myc*) (Land et al. 1983; Ruley 1983; Land, this Vol.). More recent analyses, however, indicate that under certain conditions the same gene product may carry out both of these functions. This has been shown for the genes encoding *ras* (Spandidos and Wilkie 1984), polyoma virus large-T (PyLT) (Jenuwein et al. 1985), p53 (Eliyahu et al. 1985), adenovirus E1a (J. Lewis, personal communication) and, as described here, *fos* oncogenes (Jenuwein et al. 1985). *Fos* and *ras* oncogenes above therefore behave in a similar way in that they seem to be primarily transforming genes but, depending on various other factors (e.g., the particular construct, expression levels, culture conditions and cell type) they can also display properties of "immortalizing" genes. Conversely, *myc*, PyLT, and E1a are primarly "immortalizing" genes, but are also able to modulate certain transformation parameters. It has been shown, for instance, that PyLT largely abrogates growth-factor requirements in fibroblasts (Rassoulzadegan et al. 1983 and Mougneau and Cuzin, this Vol.) and enhances *fos*-induced morphological transformation (Jenuwein et al. 1985). This suggests that, although the separation of oncogenes into the two categories may be a useful operational definition, transformation and establishment may be functionally related.

References

Van Beveren C, van Straaten F, Curran T, Müller R, Verma IM (1983) Analysis of FBJ-MuSV provirus and c-*fos* (mouse) geen reveals that viral and cellular *fos* gene products have different carboxy termini. Cell 32:1241–1255

Van Beveren C, Enami S, Curran T, Verma IM (1984) FBR murine osteosarcoma virus. II. Nucleotide sequence of the provirus reveals that the genome contains sequences acquired from two cellular genes. Virology 135:229–243

Curran T, Teich NM (1982a) Candidate product of the FBJ murine osteosarcoma virus oncogene: Characterization of a 55,000-Dalton phosphoprotein. J Virol 42:114–122

Curran T, Teich NM (1982b) Identification of a 39,000-Dalton protein in cells transformed by the FBJ murine osteosarcoma virus. Virology 116:221–235

Curran T, Verma IM (1984) FBR murine osteosarcoma virus. I. Molecular analysis and characterization of a 75,000-Da *gag-fos* fusion product. Virology 135:218–228

Curran T, MacConnell WP, van Straaten F, Verma IM (1983) Structure of the FBJ murine osteosarcoma virus genome: Molecular cloning of its associated helper virus and the cellular homolog of the c-*fos* gene from mouse and human cells. Mol Cell Biol 3:914–921

Curran T, Miller DA, Zokas L, Verma IM (1984) Viral and cellular *fos* proteins: a comparative analysis. Cell 36:259–268

Curran T, Van Beveren C, Ling N, Verma IM (1985) Viral and cellular *fos* proteins are complexed with a 39,000-Dalton cellular protein. Mol Cell Biol 5:167–172

Eliyahu D, Michalovitz D, Oren M (1985) Overproduction of p53 antigen makes established cells highly tumorigenic. Nature 316:158–160

Finkel MP, Biskis BO (1968) Experimental induction of osteosarcomas. Prog Exp Tumor Res 10:73–111

Finkel MP, Biskis BO, Jinkins PB (1966) Virus induction of osteosarcomas in mice. Science 151:698–701

Finkel MP, Reilly CA, Biskis BO, Greco IL (1973) Bone tumor viruses. Colston Pap 24:353–366

Jenuwein T, Müller D, Curran T, Müller R (1985) Extended life span and tumorigenicity of nonestablished mouse connective tissue cells transformed by the *fos* oncogene of FBR-MuSV. Cell 41:629–637

Kruijer W, Cooper JA, Hunter T, Verma IM (1984) Platelet-derived growth factor induces rapid but transient expression of the c-*fos* gene and protein. Nature 312:711–716

Land H, Parada LF, Weinberg RA (1983) Tumorigenic conversion of primary embryo fibroblasts requires at least two cooperating oncogenes. Nature 304:596–602

Meijlink F, Curran T, Miller DA, Verma IM (1985) Removal of a 67-base-pair sequence in the noncoding region of protooncogene *fos* converts it to a transforming gene. Proc Natl Acad Sci USA 82:4987–4991

Miller DA, Curran T, Verma IM (1984) c-*fos* protein can induce cellular transformation: a novel mechanism of activation of a cellular oncogene. Cell 36:51–60

Miller DA, Vermas IM, Curran T (1985) Deletion of the *gag* region from FBR murine osteosarcoma virus does not affect its enhanced transforming activity. J Virol 55:521–526

Müller R, Bravo R, Burckhardt J, Curran T (1984) Induction of c-*fos* gene and protein by growth factors precedes activation of c-*myc*. Nature 312:716–720

Rassoulzadegan M, Naghashfar Z, Cowie A, Carr A, Grisoni M, Kamen R, Cuzin F (1983) Expression of the large T protein of polyoma virus promotes the establishment in culture of "normal" rodent fibroblast cell lines. Proc Natl Acad Sci USA 80:4354–4358

Ruley HE (1983) Adenovirus early region 1A enables viral and cellular transforming genes to transform primary cells in culture. Nature 304:602–606

Spandidos DA, Wilkie NM (1984) Malignant transformation of early passage rodent cells by a single mutated human oncogene. Nature 310:469–475

van Straaten F, Müller R, Curran T, Van Beveren C, Verma IM (1983) Complete nucleotide sequence of a human c-*onc* gene: Deduced amino acid sequence of the human c-*fos* protein. Proc Natl Acad Sci USA 80:3183–3187

Treisman R (1985) Transient accumulation of c-*fos* RNA following serum stimulation requires a conserved 5′ element and c-*fos* 3′ sequences. Cell 42:889–902

Ward JM, Young DM (1976) Histogenesis and morphology of periosteal sarcomas induced by FBJ virus in NIH Swiss mice. Cancer Res 36:3985–3992

p53: Molecular Properties and Biological Activities

MOSHE OREN

The p53 cellular tumor antigen is a phosphoprotein which is overproduced in a wide variety of neoplastic cells (reviewed by Klein 1982; Crawford 1983; Rotter and Wolf 1984). High p53 mRNA levels are also seen in certain non-malignant cells, including embryonic tissue and exponentially growing, non-transformed cultured cell lines, but expression is low in most adult organs (Dippold et al. 1981; Rogel et al. 1985). This chapter reviews the evidence which suggests that p53 plays a role both in normal cell proliferation and in neoplastic transformation. It also discusses the question of how p53 expression levels are regulated.

Role of p53 in the Proliferation of Nontransformed Cells in Culture

Mitogenic stimulation of growth-arrested cells results in the rapid induction of p53 mRNA and protein, as shown for fibroblasts (Reich and Levine 1984) and T-lymphocytes (Milner and Milner 1981). In the latter case, the picture is apparently more complex, involving a qualitative rather than quantitative change in p53 (Milner 1984). The obligatory role of p53 in the transition from a resting to a growing state was convincingly demonstrated by a series of microinjection experiments. Introduction of anti-p53 monoclonal antibodies into the nuclei of growth-arrested fibroblasts blocked their ability to be mitogenically stimulated by serum (Mercer et al. 1982, 1984). This effect was only observable in nontransformed cells and was limited to the first few hours following the addition of serum. These findings suggested that p53 may be involved in the induction of competence (see discussion of competence by Bravo and Müller, this Vol.) and are therefore consistent with the idea that constitutive overexpression of this protein may alter normal growth control.

More direct evidence for this idea was recently provided by utilizing p53-overexpression plasmids containing a retroviral LTR linked to cloned mouse p53 DNA. When injected into the nuclei of serum-starved fibroblasts, these plasmids induced a mitogenic response in the presence of platelet-poor plasma (Kaczmarek et al. 1985). This effect was not obtained with similar plasmids from which the major portion of the p53 protein-coding region had been deleted. Interestingly, however, a deletion mutant missing the C-terminal half of p53 was as efficient in this assay as the intact plasmid, indicating that the competence-inducing activity of this protein resides in its N-terminal half.

Oncogenes and Growth Control
Edited by P. Kahn and T. Graf
© Springer-Verlag Berlin Heidelberg 1986

Thus, just like the protein product of the *myc* gene (see Bravo and Müller, this Vol.) p53 can also probably bypass the need for PDGF in the induction of competence.

Role of p53 in Immortalization and Transformation

The finding that p53 accumulates in various transformed cells, as well as the possibility that it regulates a crucial stage in the control of normal cell proliferation, are consistent with the idea that overproduction of this protein may contribute towards neoplastic conversion. This notion was directly tested in a number of experimental systems and, as summarized below, the results suggest that p53 possesses a wide range of biological activities.

When p53-overexpressing plasmids are stably introduced into primary rodent cells, continuous lines of immortalized cells can be obtained (Jenkins et al. 1984). These cells possess an apparently normal morphology and are still subject to density-dependent growth arrest in culture (Jenkins et al. 1984; D. Eliyahu, unpublished). Immortalization with normal p53 can only be achieved when expression is driven by very strong transcriptional control elements (Jenkins et al. 1985). Together with the fact that all p53-immortalized lines express very high levels of the protein (D. Eliyahu, unpublished), these results indicate that the mere constitutive production of p53 is insufficient to induce immortalization. Rather, there appears to be a requirement for markedly elevated levels, indicating the existence of a quantitative threshold above which biological activity becomes aberrant.

When cells immortalized by p53 are transfected with activated Ha-*ras*, they become neoplastically transformed (Jenkins et al. 1984). This can also be directly achieved by cotransfecting primary embryonic fibroblasts with p53 and Ha-*ras* (Eliyahu et al. 1984; Parada et al. 1984). The resultant cells are morphologically altered and are capable of growing over a growth-arrested monolayer of normal cells, giving rise to visible foci. Furthermore, upon injection into syngeneic animals, the progeny of such transformants elicit rapidly growing, lethal tumors. Such studies established p53 as an oncogene of the *myc* complementation group (see Land, this Vol.). It is noteworthy that p53 seems to be five to tenfold more efficient in cotransformation than it is in immortalization (D. Eliyahu, unpublished). Since gross overproduction of *ras* can by itself lead to immortalization (Spandidos and Wilkie 1984), one could argue that in the cotransfection experiments p53 and *ras* collaborate at the level of immortalization to yield a larger number of foci. However, this does not seem to be the case, since many of these foci are actually not immortalized and still undergo senescence (Eliyahu et al. 1984). Hence, it appears that the ability of p53 to participate in transformation is separable from its immortalizing capacity, possibly reflecting two independent activities of the protein.

This conclusion is strongly supported by recent work (Jenkins et al. 1985) showing that constructs containing deleted p53 genes effectively immortalized primary cells but were unable to render such cells subject to transformation by *ras*. In agreement with these findings, no cotransformation with *ras* could be obtained with a deletion mutant encoding the N-terminal half of p53 (D. Eliyahu et al., unpublished), whereas this mutant was as efficient as the intact gene in the induction of competence (Kaczmarek et al. 1985). These results indicate that p53 possesses distinct functional domains, each of which may be exerting its biological effects via a different molecular mechanism.

The apparent distinction between the immortalizing and transforming activities of p53 suggests that overproduction of this protein may also alter the biological properties of established cell lines. Compelling evidence for this notion was obtained in a number of experimental systems. One example is provided by the Abelson MuLV-transformed mouse cell line L12, which is unique in that it makes no detectable p53 owing to the integration of a provirus into the active p53 gene (Rotter and Wolf 1984). Unlike other Abelson-MuLV transformants, L12 cells are incapable of eliciting lethal tumors in syngeneic mice. To determine whether these two unusual properties of L12 are causally related, cells were stably transfected with a p53-overexpressing plasmid, which resulted, as expected, in the restoration of p53 overproduction. The injection of these transfected cells into syngeneic mice led to the efficient formation of lethal tumors, strongly implicating p53 in this process (Wolf et al. 1984). More recently, similar conclusions were derived from a study of Friend erythroleukemia virus-transformed lines (Mowat et al. 1985). In some of those lines, p53 production was abolished due to p53 gene rearrangements. All of these p53-nonproducers were fully transformed in vitro but had a reduced tumorigenic potential, again attesting to the crucial role of p53.

The two systems described above employed transformed cells. When nontransformed, established Rat-1 fibroblasts were made to stably overproduce p53, they still retained a nontransformed phenotype in culture. However, injection of these cells into nude mice led to the formation of rapidly growing, lethal tumors (Eliyahu et al. 1985). Interestingly, such cells were not tumorigenic in syngeneic rats. Taken together, the in vivo experiments demonstrate that p53 overexpression may lead to a series of alterations which eventually result in the loss of normal growth control and the acquisition of neoplastic properties.

Regulation of p53 Levels in Normal and Transformed Cells

It is clear that multiple mechanisms are involved in regulating p53 levels. In some cases, there is strong evidence for the regulation of cytoplasmic p53 mRNA concentrations. For example, nondifferentiated F9 embryonal carcinoma cells contain substantial amounts of p53 mRNA. When these cells are induced to differentiate into apparently nontransformed derivatives, p53 mRNA levels decrease dramatically (Reich et al. 1983). In this case, the decrease is due to a destabilization of p53 mRNA by a completely unknown mechanism and not to changes in the rate of transcription (Dony et al. 1985). However, it is equally possible that in some other cases p53 expression may be regulated at the transcriptional level. Since there is only a single functional p53 gene in all species studied so far (reviewed in Oren 1985), detailed studies comparing the transcription rates of these genes in different cell types (e.g., normal versus transformed) and under different culture conditions (e.g., resting versus growing cells) should resolve this question.

An alternative approach is to analyze in detail the transcriptional control elements of the p53 genes. Such analysis has already led to the interesting finding that the p53 promoter may contain a region which negatively regulates transcription, suggesting that a repressor-like molecule may interact with the DNA at this site (Bienz-Tadmor et al. 1985). Similar observations have recently been made for a number of oncogenes and other growth-related genes, raising the possibility that many such genes may be subject to negative transcriptional regulation. In this case, gene activation under both neoplastic and normal growth conditions might be achieved by depletion of the putative repressor molecule or by its displacement from the DNA. As far as p53 is concerned, this hypothesis appears even more attractive in light of the finding that the negative effect of the "repressor-binding" region is greatly abolished by co-transfection with the adenovirus E1A region (B. Bienz, unpublished), which has been implicated in the modulation of gene expression in many systems (see Philipson, this Vol.).

Although modulation of p53 mRNA levels can be of great importance, it is equally clear that cellular p53 concentrations are also subject to extensive post-translational regulation (Reich et al. 1983). This is particularly so because p53 is a very short-lived protein, as expected for molecules proposed to possess crucial regulatory functions. Modulation of protein turnover rates can be a very efficient way of altering cellular p53 concentrations rapidly and efficiently in response to specific stimuli. Hence, it is important to understand the molecular factors involved in the rapid degradation of p53. In particular, one would like to identify the protease(s) responsible for this process and to determine whether they also recognize other short-lived proteins and particularly other oncogene products. Another unanswered question is what in the p53 molecule makes it a target for rapid degradation. A possible clue was recently provided by the demonstration that certain in-frame deletions result in the production of a much more stable p53 molecule. This stabilization also endows the mutant molecule with an enhanced immortalizing potential (Jenkins et al. 1985).

In a number of neoplastic cell types, there is a remarkable stabilization of p53. The best-documented case is that of SV40-transformed cells, in which the half-life of p53 increases from less than 30 min to more than 24 h (Oren et al. 1981). This is most likely due to the formation of a tight complex between p53 and the SV40 large-T antigen (reviewed by Klein 1982; Crawford 1983), which somehow renders p53 much less vulnerable to rapid proteolytic degradation. Increased stabilities of p53 have also been observed in cells transformed by nonviral agents such as chemical carcinogens. Interestingly, such cells exhibit a specific complex between p53 and a cellular protein, HSP70 (Pinhasi-Kimhi et al. 1986). It is tempting to speculate that this complex, like the one with the SV40 T antigen, contributes to the stabilization of p53.

Mechanism of Action of p53

Probably the major problem with our current understanding of p53 is the total lack of conclusive evidence concerning its mode of action. One attractive possibility is suggested by the fact that p53 is a DNA-binding protein (Lane and Gannon 1983) which tends to accumulate in the nuclei of cells overproducing it (Rotter and Wolf 1984). Several nuclear oncogene products, particularly those of viral origin, have been shown to serve as transactivating factors which modulate gene expression. It is possible that p53 has a similar activity and that it regulates the expression of a set of target genes associated with cellular growth processes. This hypothesis is supported by the demonstration that cotransfection with p53 expression plasmids may increase the promoter activity of the LTR of an intracisternal A particle gene, a retrovirus-like genetic element (Luria et al. 1986). Obviously, more evidence is required to determine conclusively whether p53 functions by regulating specific gene expression. Still, if this assumption turns out to be correct, it should be possible to gain much more insight into the relationship between the molecular properties of p53 and the diverse array of biological effects that it seems to exert under both normal and neoplastic conditions.

Acknowledgments. Supported by the Minerva Foundation, Munich, Germany.

References

Bienz-Tadmor B, Zakut-Houri R, Libresco S, Givol D, Oren M (1985) The 5′ region of the p53 gene: evolutionary conservation and evidence for a negative regulatory element. EMBO J 4:3209–3213

Crawford LV (1983) The 53,000-dalton cellular protein and its role in transformation. Int Rev Exp Pathol 25:1–50

Dippold WG, Jay G, DeLeo AB, Khoury G, Old LJ (1981) p53 transformation-related protein. Detection by monoclonal antibody in mouse and human cells. Proc Natl Acad Sci USA 78:1695–1699

Dony C, Kessel M, Gruss P (1985) Post-transcriptional control of *myc* and p53 expression during differentiation of the embryonal carcinoma cell line F9. Nature 317:636–639

Eliyahu D, Raz A, Gruss P, Givol D, Oren M (1984) Participation of p53 cellular tumor antigen in transformation of normal embryonic cells. Nature 312:646–649

Eliyahu D, Michalovitz D, Oren M (1985) Overproduction of p53 antigen makes established cells highly tumorigenic. Nature 316:158–160

Jenkins JR, Rudge K, Currie GA (1984) Cellular immortalization by a cDNA clone encoding the transformation-associated phosphoprotein p53. Nature 312:651–654

Jenkins JR, Rudge K, Chumakov P, Currie GA (1985) The cellular oncogene p53 can be activated by mutagenesis. Nature 317:816–818

Kaczmarek L, Oren M, Baserga R (1986) Co-operation between the p53 protein tumor antigen and platelet-poor plasma in the induction of cellular DNA synthesis. Exp Cell Res 162: 268 – 272

Klein G (ed) (1982) Advances in viral oncology, vol 2. The transformation-associated cellular p53 protein. Raven, New York

Lane DP, Gannon J (1983) Cellular proteins involved in SV40 transformation. Cell Biol Int Rep 7:513 – 514

Luria S, Horowitz M (1986) The long terminal repeat of the intracisternal A particle as a target for transactivation by oncogene products. J Virol 57:998 – 1003

Mercer WE, Nelson D, DeLeo AB, Old LJ, Baserga R (1982) Microinjection of monoclonal antibodies to protein p53 inhibits serum-induced DNA synthesis in 3T3 cells. Proc Natl Acad Sci USA 79:6309 – 6312

Mercer WE, Avignolo C, Baserga R (1984) Role of the p53 protein in cell proliferation as studied by microinjection of monoclonal antibodies. Mol Cell Biol 4:276 – 281

Milner S (1984) Different forms of p53 detected by monoclonal antibodies in non-dividing and dividing lymphocytes. Nature 310:143 – 145

Milner J, Milner S (1981) SV40-53K antigen: a possible role for p53 in normal cells. Virology 112:785 – 788

Mowat M, Cheng A, Kimura N, Bernstein A, Benchimol S (1985) Rearrangements of the cellular p53 gene in erythroleukemic cells transformed by Friend Virus. Nature 314:633 – 636

Oren M (1985) The p53 cellular tumor antigen: gene structure, expression and protein properties. Biochem Biophys Acta Rev Cancer 823:67 – 68

Oren M, Maltzman W, Levine AJ (1981) Post-translational regulation of the 54K cellular tumor antigen in normal and transformed cells. Mol Cell Biol 1:101 – 110

Parada LF, Land H, Weinberg RA, Wolf D, Rotter V (1984) Cooperation between gene encoding p53 tumor antigen and *ras* in cellular transformation. Nature 312:649 – 651

Pinhasi-Kimhi O, Michalovitz D, Ben-Zeev A, Oren M (1986) Specific interaction between the p53 cellular tumor antigen and major heat shock proteins. Nature 320:182 – 185

Reich NC, Levine AJ (1984) Growth regulation of a cellular tumor antigen, p53 in transformed cells. Nature 308:199 – 201

Reich NC, Oren M, Levine AJ (1983) Two distinct mechanisms regulate the levels of a cellular tumor antigen, p53. Mol Cell Biol 3:2143 – 2150

Rogel A, Popliker M, Webb CG, Oren M (1985) p53 cellular tumor antigen: analysis of mRNA levels in normal adult tissues, embryos and tumors. Mol Cell Biol 5:2851 – 2855

Rotter V, Wolf D (1984) Biological and molecular analysis of p53 cellular encoded tumor antigen. Adv Cancer Res 43:113 – 141

Spandidos DA, Wilkie NM (1984) Malignant transformation of early passage rodent cells by a single mutated human oncogene. Nature 310:469 – 475

Wolf D, Harris N, Rotter V (1984) Reconstitution of p53 expression in a nonproducer Ab-MuLV-transformed cell line by transfection of a functional p53 gene. Cell 38:119 – 126

V Malignant Transformation as a Multistep Process

The discussions of oncogenesis in the preceding sections have focused on the action of individual oncogenes and their effects on cellular phenotypes; indeed, the fact that retroviruses such as the Rous sarcoma virus (RSV) transform fibroblasts in vitro and rapidly cause fatal tumors was initially interpreted to mean that they induce malignant transformation in a single step. However, as the articles in this section describe, more recent evidence suggests that a single oncogene is generally not sufficient to induce full neoplastic transformation, strongly supporting the view long held by epidemiologists that neoplasia results from the accumulation of multiple oncogenic events.

This section begins by discussing oncogene cooperativity in the transformation of cultured cells by a DNA tumor virus, the polyoma virus. The transforming capacity of polyoma resides in its early region, which encodes the proteins known as large T, middle T, and small T. While continuous cell lines of rodent fibroblasts can be transformed by middle T alone, transformation of primary fibroblasts requires the combined action of large T, which confers upon the cells an indefinite lifespan ("establishment" or "immortalization") and middle T, which induces a transformed phenotype. Transformation by polyoma virus also involves cooperativity between a viral oncogene and a proto-oncogene: middle T protein forms a complex with the product of the c-*src* gene, a protein present at extremely low levels in fibroblasts. This association leads to a dramatic enhancement in the tyrosine kinase activity of the c-*src* protein, a change which is necessary although not sufficient for cell transformation. Subsequent articles extend the concept of viral oncogene cooperativity to activated cellular and retroviral oncogenes. Two classes can be distinguished: those which can functionally replace polyoma large T in the transformation of primary rat fibroblasts, for example, the *myc* oncogene; and those which replace polyoma middle T, such as the *ras* oncogene. This classification is not strictly applicable to hematopoietic cells, probably because of their rather different physiology. In primary avian hematopoietic cultures, certain oncogenes induce cells of a particular lineage to undergo extensive self-renewal, while other oncogenes "enhance" the transformed phenotype by blocking differentiation in the transformed cells or by abolishing their growth factor requirements. Interestingly, the same oncogene may exhibit radically different phenotypic effects in cells from different lineages.

The next few articles consider the role of oncogenes in various stages of tumorigenesis. Induction of certain B-cell leukemias in humans and in mice involves activation of the *myc* gene via translocation to a site which activates its constitutive expression. Various co-factors which are thought to extend the lifespan of the target B-cell and/or to provide it with chronic stimulation may increase the probability that the relevant translocation will occur. However, activation of c-*myc* is not sufficient to induce leukemia, and secondary events, possibly including additional translocations, are required. In the induction of skin tumors by chemical carcinogens, mutations in proto-oncogenes of the *ras* family appear to be a crucial event. It is interesting that the specific carcinogen with which the tumor was initiated dictates not only the type of activating point mutation but also its precise location within the *ras* gene.

The last part of the section deals with phenomena related to tumor progression, the process in which neoplasms gradually acquire an increasingly malignant phenotype. One contributing factor could be changes in the dosage of critical genes; such changes might arise as a consequence of the chromosomal instability that is characteristic of many advanced neoplasms. For example, amplification of proto-oncogenes is a common feature of human tumors and has been shown in some instances to result in overexpression of the corresponding proto-oncogene. The finding that more than 90% of the proto-oncogene amplifications which have been observed in human tumors involve *myc*-related loci is consistent with the fact that *myc* genes are activated by deregulation of their expression and do not require alterations in the coding sequence. A highly provocative but still largely speculative concept that also relates to the effect of gene dosage is that of "anti-oncogenes", the normal versions of recessive alleles which predispose their host to cancer, and "suppressor genes", which extinguish the transformed phenotype in somatic cell hybrids between normal and transformed cells. It is possible that the phenotype of the tumor cell is partly determined by the relative expression levels of oncogenes and of anti-oncogenes/suppressor genes. Clearly, the identification of such genes would have an enormous impact on cancer research.

Oncogene Cooperativity in Stepwise Transformation of Rodent Embryo Fibroblasts by Polyoma Virus

Evelyne Mougneau and François Cuzin

More than 50 years ago, Peyton Rous and his colleagues described a tumor progression pathway initiated by infections of rabbits with Shope papilloma virus (Rous and Beard 1935). Experiments performed almost 30 years later by Vogt and Dulbecco (1963) led them to conclude that transformation by polyomavirus is also a multistep process. It is now clear that most, if not all, DNA tumor viruses use more than one oncogene to fully transform rodent embryo fibroblasts in culture. Here, we discuss mechanisms of cooperativity among polyoma virus genes in the transformation of rat embryo fibroblasts (REF).

The transforming ("early") region of polyomavirus encodes three proteins (T antigens): the nuclear large T (*plt*) protein, the membrane-associated middle T (*pmt*) protein, and the cytosolic small T (*pst*) protein. Another well-studied papovavirus is SV40, which exhibits extensive sequence homology with polyomavirus in the genes coding for structural proteins and in the carboxy-terminal portion of the early region, which is thought to encode DNA replication functions (Soeda et al. 1980). The early region of SV40 encodes only two proteins, the large and the small T proteins. (For a detailed description of the genetic structure of polyoma and SV40 viruses see Ito 1980; Cuzin 1984 and references therein).

A multistep transformation process can be most easily explained by the separate and successive "firing" of different oncogenes. However, cooperativity between oncogenes can also reflect direct interaction between their products, thus requiring that they are expressed simultaneously. These two types of oncogene cooperativity, exemplified by polyoma (and SV40) virus genes (see also Cuzin 1984) and by certain cellular or retrovirus-transduced cellular oncogenes, are discussed below.

Cooperativity of Viral and Cellular Oncogenes by a Stepwise Mechanism

The first evidence that the *plt* and *pmt* genes can exert both separate and complementary oncogenic functions stems from the observation by Treisman et al. (1981) that transfer of *pmt* alone induced the transformation of an established rat fibroblast cell line. Further studies (Rassoulzadegan et al. 1982) indicated that these cells were only partially transformed, as they lacked the independence from serum factors characteristic of polyoma virus-induced transformants and reverted to a normal phenotype in low serum medium. How-

Oncogenes and Growth Control
Edited by P. Kahn and T. Graf
© Springer-Verlag Berlin Heidelberg 1986

ever, cells grown under these conditions could be further transformed by the *plt* gene, which induced them to form foci in monolayers and colonies in agar suspension. A stepwise mechanism of cooperation between these two genes was further demonstrated by the observation that transfer of *pmt* alone did not induce focus formation in primary REF cultures, while the *plt* gene induced an indefinite growth potential in these cells ("establishment" or "immortalization"). Once established by the *plt* gene, REF cell lines could be efficiently transformed by the *pmt* gene. Continuous activity of the large T protein was required to maintain the established phenotype, as demonstrated by temperature shift experiments using a temperature-sensitive mutant in the *plt* gene (Rassoulzadegan et al. 1983).

A similar two-step mechanism was demonstrated for adenovirus transformation (Van den Elsen et al. 1982) and has recently been suggested for papillomaviruses (Schiller et al. 1984; Androphy et al. 1985; Yang et al. 1985). This two-step model was extended to cellular oncogenes such as *myc* and *ras* and to viral oncogenes such as the adenovirus E1A gene. These studies showed that *plt* could be functionally replaced by either *myc* or E1A and that *pmt* could be replaced by *ras*, thus establishing two functionally distinct groups of oncogenes that encode nuclear or membrane associated proteins, respectively (see Land, this Vol. and references therein). *Myc*- and *plt*-immortalized cell lines and spontaneously established 3T3 cells (Todaro and Green 1963) share a series of properties, such as growth at low saturation densities and colony-forming ability at low cell density. However, the oncogene-immortalized cells exhibit a number of other properties, most strikingly serum-independent growth in culture and a tendency to transform spontaneously. The frequency of transformation in these cells could be further increased after treatment with TPA (Connan et al. 1985). Expression of nuclear oncogenes thus appears to induce a "high risk state" for transformation. This "high risk state" might be due to an increased genetic instability, since we have recently found that cell lines established by the *plt* and *myc* oncogenes exhibit aneuploid chromosome numbers and increased rates of sister chromatid exchange (Cerni et al. 1986).

It remains to be seen whether the effects induced by *plt, myc,* and E1A oncogenes in primary fibroblasts can be considered in a more general sense as an early transformation step. In fact, recent studies in our laboratory suggest that the early stages of transformation induced by papillomaviruses differ in several ways from those induced by polyoma virus (Cuzin and Meneguzzi 1986).

Cooperativity of Viral and Cellular Oncogenes by a Direct Interaction of Gene Products

The clearest examples of this type of cooperativity are the activation of cellular oncogenes by direct interaction of their gene products with viral proteins. For example, it appears that the transforming capacity of *pmt* gene is indirect and is mediated by the association of the *pmt* protein with the pp60$^{c\text{-}src}$ protein. This leads to an altered tyrosine kinase activity of pp60 c-*src* (see Cheng et al., this Vol. and references therein). Another example is the association of the SV40 large T protein with the cellular p53 protein, probably also leading to the activation of the latter (see Oren, this Vol.). These examples illustrate the notion that the mechanisms of transformation can differ between two types of papovavirus.

Recent studies in our laboratory indicate that a similar type of direct cooperativity between oncogene products also occurs during polyoma virus-induced transformation. This conclusion is based on the following observations. A combination of the polyoma *plt* and *pmt* genes is sufficient for complete malignant transformation of established REF lines such as FR3T3 or RAT-1. In contrast, co-transfer of these two genes into primary REF cultures did not lead to focus formation; rather, transformation of REFs requires that all three viral oncogenes are expressed. A biological effect of the *pst* gene could also be demonstrated in cell lines established from REFs after co-transfer of *plt* and *pmt*. These lines have a normal phenotype but transform "spontaneously" after 12 – 15 generations. However, after introduction of the *pst* gene, they transform without delay (L. Lemieux, M. Rassoulzadegan and F. Cuzin, in preparation). An apparent paradox was the finding that cell lines established by transfer of *plt* in REF cells can then be efficiently transformed by *pmt* alone. Further analysis suggested that transformation by *pmt* requires a cellular function that is constitutively expressed in all established lines, but depends on the expression of small T protein in REF cells. These observations, together with the fact that co-transfer of *plt* and *ras* oncogenes into primary REFs leads to immediate transformation (see above), suggest that the requirement for *pst* involves the *pmt* protein and that *ras* can substitute for signals which result from *pst* and *pmt* acting in concert.

In conclusion, we have seen that even in simple virus models multiple events are necessary to induce a fully transformed phenotype and that complex mechanisms of cooperativity have evolved between viral and cellular oncogenes.

Acknowledgments. We are indebted to L. Carbone, C. Godefroid, M. J. Gonzalez and F. Tillier for skilled technical help. This work was made possible by grants from the Centre National de la Recherche Scientifique and the Association pour la Recherche Médicale (France).

References

Androphy EJ, Schiller JT, Lowy DR (1985) Identification of the protein encoded by the E6 transforming gene of bovine papillomavirus. Science 230:442 – 445

Cerni C, Mougneau E, Zerlin M, Julius M, Marcu KB, Cuzin F (1986) C-*myc* and functionally related oncogenes induce both high rats of sister chromatid exchange and abnormal karyotypes in rat fibroblasts. In: Melchers F, Potter M (eds) Current topics microbiol. and immunol., vol 132. Springer, Berlin Heidelberg New York, pp 193 – 201

Connan G, Rassoulzadegan M, Cuzin F (1985) Focus formation in rat fibroblasts exposed to a tumor promoter after transfer of polyoma *plt* and *myc* oncogenes. Nature 314:277 – 279

Cuzin F (1984) The polyoma virus oncogenes: coordinated functions of three distinct proteins in the transformation of rodent cells in culture. Biochim Biophys Acta 781:193 – 204

Cuzin F, Meneguzzi M (1986) Stepwise transformation and cooperative interactions involving oncogenes of DNA tumor viruses. Adv Viral Onc (in press)

Ito Y (1980) Organization and expression of the genome of polyoma virus. In: Klein G (ed) Viral oncology. Raven, New York, pp 447 – 480

Rassoulzadegan M, Cowie A, Carr A, Glaichenhaus N, Kamen R, Cuzin F (1982) The roles of individual polyoma virus early proteins in oncogenic transformation. Nature 300:713 – 718

Rassoulzadegan M, Naghashfar Z, Cowie A, Carr A, Grisoni M, Kamen R, Cuzin F (1983) Expression of the large T protein of polyoma virus promotes the establishment in culture of "normal" rodent fibroblast cells. Proc Natl Acad Sci USA 80:4354 – 4358

Rous P, Beard JW (1935) The progression to carcinomas of virus-induced papillomas (Shope). J Exp Med 62:523 – 548

Schiller JT, Vass WC, Lowy DR (1984) Identification of a second transforming region in bovine papilloma virus DNA. Proc Natl Acad Sci USA 81:7880 – 7884

Soeda E, Arrand JR, Smolar N, Walsh JE, Griffin BE (1980) Coding potential and regulatory signals of the polyoma virus genome. Nature 283:445 – 453

Todaro GJ, Green H (1963) Quantitative studies on the growth of mouse embryo cells in culture and their development into established lines. J Cell Biol 17:299 – 313

Treisman R, Novak U, Favaloro J, Kamen R (1981) Transformation of rat cells by an altered polyoma virus genome expressing only the middle-T protein. Nature 292:595 – 600

Van den Elsen P, de Pater S, Houweling A, van der Veer J, van der Eb A (1982) The relationship between region *E1a* and *E1b* of human adenoviruses in cell transformation. Gene 18:175 – 185

Vogt M, Dulbecco R (1963) Properties of cell transformed by polyoma virus. Cold Spring Harbor Symp Quant Biol 28:367 – 374

Yang Y-C, Okayama H, Howley PM (1985) Bovine papillomavirus contains multiple transforming genes. Proc Natl Acad Sci USA 82:1030 – 1034

Role of the Middle T: pp60$^{c\text{-}src}$ Complex in Cellular Transformation by Polyoma Virus

Seng H. Cheng, William Markland, and Alan E. Smith

The polyoma virus of mice has been extensively studied over the last two decades, primarily because of its ability to induce neoplastic transformation of mammalian cells in vivo and in vitro (for reviews, see Griffin and Dilworth 1983; Smith and Ely 1983). The small size of the viral genome and the ease with which it can be manipulated make polyoma an attractive system in which to study the mechanisms whereby normal cell growth is transformed to a state of uncontrolled proliferation that culminates in neoplasia. The early region of polyoma has the ability to induce and maintain the transformed state and encodes three proteins, designated large T, middle T and small T. The middle T protein by itself is sufficient to transform established cells (Treisman et al. 1981), although a portion of the large T gene and, under certain circumstances, small T, are also required to transform primary cells (Rassoulzadegan et al. 1982; see article by Mougneau and Cuzin, this Vol.). Here we discuss the activities associated with middle T and the possible mechanisms of action of this viral oncogene.

Biochemical Properties of Polyoma Virus Middle T Antigen

Much of the work on the biochemistry of middle T has centered on the findings that the viral protein has an associated kinase activity which phosphorylates tyrosine residues in vitro (Eckhart et al. 1979; Smith et al. 1979) and that it is present in membrane fractions of cells. Although early studies suggested that middle T antigen is associated with the plasma membrane, more recent reports show the majority to be perinuclear and in the endoplasmic reticulum (Griffin and Dilworth 1983). The protein does not contain a hydrophobic N-terminal signal sequence typical of other secretory proteins and is not known to traverse the lipid bilayer of the plasma membrane (Schaffhausen et al. 1982).

Binding of middle T to cell membranes is believed to be mediated by a hydrophobic sequence of 20 nonpolar residues located near the C-terminus of the molecule (Carmichael et al. 1982; Markland et al. 1986). Mutants which lack sequences from this region are unable to interact with membranes and are defective in both transforming and associated kinase activities (Carmichael et al. 1982). Small changes in this region leave the membrane association and kinase activity intact but abolish transforming capacity (Markland et al. 1986).

Oncogenes and Growth Control
Edited by P. Kahn and T. Graf
© Springer-Verlag Berlin Heidelberg 1986

These findings suggest that middle T needs to associate with membranes to acquire kinase activity. Furthermore, analysis of other mutants has shown that there is an excellent correlation between the presence of the middle T-associated tyrosine kinase and transforming capacity. Indeed, there are no transformation-competent mutants of polyoma which lack kinase activity (Smith and Ely 1983; Courtneidge and Smith 1984; Cheng et al. 1986; Markland et al. 1986).

Several retroviral transforming proteins, including that of Rous sarcoma virus (the pp60$^{v\text{-}src}$ protein), as well as a number of cellular growth factor receptors have been shown to possess tyrosine kinase activity (see Hunter, this Vol.). However, polyoma virus is the only example to date of a DNA tumor virus with a transforming protein associated with kinase activity. Because both middle T and the transforming proteins of some retroviruses are membrane-associated, it first appeared that their modes of action might be at least superficially analogous. However, whereas the kinase activity is clearly intrinsic to the retroviral proteins, all attempts to demonstrate this for middle T were unsuccessful. Furthermore, there is little detectable difference in the phosphotyrosine content of middle T or of cellular proteins in polyoma-transformed cells (Sefton et al. 1980), although this does not exclude the possibility of subtle but critical changes in phosphotyrosine metabolism. The finding that many of the phosphorylated proteins seen in cells transformed by pp60$^{v\text{-}src}$ appear to play little or no direct role in transformation, despite an overall increase in the phosphotyrosine levels (Cooper and Hunter 1983), is consistent with the idea that both middle T and pp60$^{v\text{-}src}$ transform by triggering phosphorylation of some minor and as yet unidentified component. This component could be a protein or an intermediate in the phosphatidylinositol pathway (Whitman et al. 1985) which stimulates cellular proliferation by mobilizing the second messengers diacylglycerol and triphosphoinositide (see by Berridge, this Vol.).

Middle T Antigen Associates with the Proto-Oncogene pp60$^{c\text{-}src}$

Failure to demonstrate that the tyrosine kinase activity of middle T was intrinsic to the protein raised the possibility that the activity resided instead in a protein which associates with middle T. Experiments designed to test this hypothesis (Courtneidge and Smith 1983, 1984) revealed that some of the pp60$^{c\text{-}src}$ from lysates of polyoma-transformed cells migrated more rapidly on sucrose gradients than is usual for this protein. Furthermore, middle T and pp60$^{c\text{-}src}$ can be co-purified in a form that is resistant to high salt concentrations, which is suggestive of a complex. The subpopulation of middle T molecules which does not co-purify with pp60$^{c\text{-}src}$ lacks an associated kinase activity. Finally, a survey of middle T mutants revealed a good correlation between the presence of the kinase activity and of a middle T: pp60$^{c\text{-}src}$ complex. These data, together with the reports that middle T synthesized in vitro (Schaffhausen et al.

1982) or expressed in bacteria (Schaffhausen et al. 1985) does not possess kinase activity, established that the kinase activity is at least in part a property of pp60$^{c\text{-}src}$ present in a complex with middle T. The interaction between the two proteins may be facilitated by the fact that, like middle T, the subcellular distribution of pp60$^{c\text{-}src}$ is perinuclear and at the plasma membrane (Resh and Erikson 1985).

It remains to be resolved whether all the middle T-associated kinase activity is due to pp60$^{c\text{-}src}$ or whether other tyrosine kinases are also present in the complex. The complex has a molecular weight of approximately 220,000, which is sufficient to contain more than one molecule of each protein or possibly other molecular species. Recently, it has been shown that middle-T also forms a complex with a cellular protein of 61 kDa (Grussenmeyer et al. 1985).

Role of the Middle T: pp60$^{c\text{-}src}$ Complex in Cell Transformation

The idea that pp60$^{c\text{-}src}$ participates in the induction of a transformed phenotype is consistent with several findings. For instance, expression vectors which overproduce pp60$^{c\text{-}src}$ elicit a weakly transformed phenotype in transfected established cells (Johnson et al. 1985). Furthermore, the closely related pp60$^{v\text{-}src}$ protein has potent transforming capacity. This suggested that perhaps pp60$^{c\text{-}src}$ in complex with middle T is somehow altered so that it mimics this effect. However, it should be noted that there is a class of middle T mutants which retains associated pp60$^{c\text{-}src}$ kinase activity but is nevertheless transformation-defective, suggesting that the existence of an active complex per se is necessary but not sufficient for transformation (Courtneidge and Smith 1984).

Further examination of the middle T: pp60$^{c\text{-}src}$ complex revealed that the interaction between transformation-competent middle T antigens and pp60$^{c\text{-}src}$ stimulated the specific activity of the pp60$^{c\text{-}src}$ tyrosine kinase (Bolen et al. 1984; Courtneidge 1985). The increased level of kinase activity is probably not a secondary property associated with transformation since only the complexed form of pp60$^{c\text{-}src}$ has enhanced activity. Furthermore, the increased pp60$^{c\text{-}src}$ kinase activity appears to be a feature only of those middle T variants which retain transforming activity and there is a clear correlation between the transforming activity for a given middle T mutant and the degree to which it enhances the kinase activity (Bolen et al. 1984; Cheng et al. 1986).

In other studies, expression of the transformed phenotype induced by plasmids expressing middle T under the control of the dexamethasone-inducible promoter (Raptis et al. 1985) correlated directly with the level of middle T-associated kinase activity rather than with the total amount of middle T expressed in induced cells. This finding implies that not all the middle-T and pp60$^{c\text{-}src}$ are in a complex but that it is the fraction present in the active complex which is important for transformation.

The biological significance of the complex was also demonstrated using mutants which affect the ability of pp60$^{c\text{-}src}$ to associate stably with middle T. A single point mutation (Trp 180) in middle T, which affects the ability to associate stably and actively with pp60$^{c\text{-}src}$, completely abolished the ability of this variant to transform (Cheng et al. 1986). This represents the smallest mutation described to date for middle T that inactivates both its transforming and associated kinase activities.

Mechanism of Transformation

Analysis of the pp60$^{c\text{-}src}$ complexed with middle T indicates that it has an extra tyrosine phosphorylation site within the N-terminal portion of the molecule and that there appears to be a direct correlation between transforming activity and phosphorylation at this site (Yonemoto et al. 1985). This offers a possible explanation for the observed activation: pp60$^{c\text{-}src}$, itself a phosphoprotein, may be regulated by phosphorylation. Precedents for phosphorylations regulating enzyme activity are well documented (see Hunter, this Vol.). Furthermore, a highly phosphorylated form of the closely related pp60$^{v\text{-}src}$ indeed has greater kinase activity than its less phosphorylated counterpart. The idea that phosphorylation on the N-terminus of pp60$^{c\text{-}src}$ is a cause rather than a consequence of its enhanced activity is consistent with the finding that increased pp60$^{c\text{-}src}$ kinase activity in four neuroblastoma cell lines is associated with an N-terminal phosphorylated form of the *src* gene product (Bolen et al. 1985). In addition, treatment of cells with platelet-derived growth factor (PDGF) results in phosphorylation of pp60$^{c\text{-}src}$ and an increase in its kinase activity (Ralston and Bishop 1985), which in turn could account for the mitogenic effect of the hormone.

These in vitro data suggest that the activation of pp60$^{c\text{-}src}$ in complex with middle T may be mediated by an anomalous phosphorylation at the N-terminus of the molecule and that role of middle T may be to effect and/or stabilize this activated form of pp60$^{c\text{-}src}$. The similarity of the cellular response to PDGF, to activated middle-T: pp60$^{c\text{-}src}$ and to modified pp60$^{c\text{-}src}$ in neuroblastoma cells supports the hypothesis that the action of the enhanced pp60$^{c\text{-}src}$ tyrosine kinase represents a pathway which leads ultimately to increased cell proliferation.

A further modification which may modulate the tyrosine kinase of pp60$^{c\text{-}src}$ in complex with middle T involves phosphorylation of a tyrosine at the C-terminus of the molecule (Courtneidge 1985). This C-terminal tyrosine is phosphorylated in pp60$^{c\text{-}src}$ but not in pp60$^{v\text{-}src}$. When pp60$^{c\text{-}src}$ is isolated under conditions which maintain the phosphorylation of this tyrosine, the kinase has a lower specific activity, but it is activated if dephosphorylation is allowed to occur. This suggests that the enhanced specific activity of pp60$^{c\text{-}src}$ complexed to middle T might be a consequence of a masking action of middle T which prevents phosphorylation of this site. It is conceivable that middle T directly modulates this regulatory site, perhaps by disturbing its accessibility to kinases and phosphatases. The activation of $pp60^{c\text{-}src}$ by middle T binding is so far rare in that it is mediated by protein:protein interaction. If other growth-regulating proteins use a similar strategy, one might predict that they interact with other proto-oncogene products or related cellular proteins such as growth factor receptors.

It is not yet understood how the activated middle $T:pp60^{c\text{-}src}$ complex alters the growth properties of the cells, although phosphorylation of some growth-regulating substance is likely to be involved (see Hunter, this Vol.). Further work should seek to establish the identity of the relevant substrates for the tyrosine kinase and their functions in cell growth.

References

Bolen JB, Thiele CJ, Israel MA, Yonemoto W, Lipsich LA, Brugge JB (1984) Enhancement of cellular *src* gene product associated tyrosyl kinase activity following polyoma virus infection and transformation. Cell 38:767–777

Bolen JB, Rosen N, Israel MA (1985) Increased $pp60^{c\text{-}src}$ tyrosyl kinase activity in human neuroblastomas is associated with amino terminal tyrosine phosphorylation of the *src* gene product. Proc Natl Acad Sci USA 82:7275–7279

Carmichael GG, Schaffhausen BS, Dorsky DI, Oliver DB, Benjamin TL (1982) The carboxy terminus of polyoma middle-T antigen is required for attachment to membranes, associated protein kinase activities and cell transformation. Proc Natl Acad Sci USA 79:3579–3583

Cheng SH, Markland W, Markham AF, Smith AE (1986) Mutants around the NG59 lesion indicate an active association of polyoma virus middle-T antigen with $pp60^{c\text{-}src}$ is required for cell transformation. EMBO J 5:325–334

Cooper JA, Hunter T (1983) Regulation of cell growth and transformation by tyrosine-specific protein kinases: The search for important cellular substrate proteins. In: Vogt PK, Koprowski H (eds) Current topics in microbiology and immunology, vol 107. Springer, Berlin Heidelberg New York, pp 125–161

Courtneidge SA (1985) Activation of the $pp60^{c\text{-}src}$ kinase by middle-T antigen binding or by dephosphorylation. EMBO J 4:1471–1477

Courtneidge SA, Smith AE (1983) Polyoma virus transforming protein associates with the product of the c-*src* cellular gene. Nature 303:435–439

Courtneidge SA, Smith AE (1984) The complex of polyoma virus middle-T antigen and $pp60^{c\text{-}src}$. EMBO J 3:585–591

Eckhart W, Hutchinson MA, Hunter T (1979) An activity phosphorylating tyrosine in polyoma T-antigen immunoprecipitates. Cell 18:925–933

Griffin BE, Dilworth SM (1983) Polyomavirus: an overview of its unique properties. Adv Cancer Res 39:183–268

Grussenmeyer T, Scheidtmann KH, Hutchinson MA, Eckhart W, Walter G (1985) Complexes of polyoma virus medium T antigen and cellular proteins. Proc Natl Acad Sci USA 82:7952–7954

Johnson PJ, Coussens PM, Danko AV, Shalloway D (1985) Overexpressed $pp60^{c\text{-}src}$ can induce focus information without complete transformation of NIH/3T3 cells. Mol Cell Biol 5:1073–1083

Markland W, Cheng SH, Roberts BL, Harvey R, Smith AE (1986) Transformation-associated domains of polyoma virus middle-T antigen. Cancer Cells. Cold Spring Harbor Lab, Cold Spring Harbor (in press)

Ralston R, Bishop JM (1985) The product of the proto-oncogene c-*src* is modified during the cellular response to platelet-derived growth factor. Proc Natl Acad Sci USA 82:7845–7849

Raptis L, Lamfrom H, Benjamin TL (1985) Regulation of cellular phenotype and expression of polyomavirus middle-T antigen in rat fibroblasts. Mol Cell Biol 5:2476 – 2485

Rassoulzadegan M, Cowie A, Carr A, Glaichenhaus N, Kamen R, Cuzin F (1982) The role of the individual polyoma virus early proteins in oncogenic transformation. Nature 300:713 – 718

Resh MD, Erikson RL (1985) Highly specific antibody to Rous sarcoma virus *src* gene product recognizes a novel population of pp60$^{v\text{-}src}$ and pp60$^{c\text{-}src}$ molecules. J Cell Biol 100:409 – 417

Schaffhausen BS, Dorai H, Arakee G, Benjamin TL (1982) Polyoma virus middle-T antigen: Relationship to cell membranes and apparent lack of ATP-binding activity. Mol Cell Biol 2:1187 – 1189

Schaffhausen BS, Benjamin TL, Lodge J, Kaplan D, Roberts TM (1985) Expression of polyoma early gene products in *E. coli*. Nucl Acids Res 13:501 – 519

Sefton BM, Hunter T, Beemon K, Eckhart W (1980) Evidence that the phosphorylation of tyrosine is essential for cellular transformation by Rous sarcoma virus. Cell 20:807 – 816

Smith AE, Ely BK (1983) The biochemical basis of transformation by polyoma virus. Adv Viral Oncol 3:3 – 30

Smith AE, Smith R, Griffin B, Fried M (1979) Protein kinase activity associated with polyoma virus middle-T antigen in vitro. Cell 18:915 – 924

Treisman RH, Novak U, Favaloro J, Kamen R (1981) Transformation of rat cells by an altered polyoma virus genome expressing only the middle-T protein. Nature 292:959 – 600

Whitman M, Kaplan DR, Schaffhausen B, Cantley L, Roberts TM (1985) Association of phosphatidylinositol kinase activity with polyoma middle-T competent for transformation. Nature 315:239 – 242

Yonemoto W, Yarvis-Morar M, Brugge JS, Bolen JB, Israel M (1985) Tyrosine phosphorylation within the amino-terminal domain of pp60$^{c\text{-}src}$ molecules associated with polyoma virus middle-sized tumor antigen. Proc Natl Acad Sci USA 82:4568 – 4572

Oncogenes Cooperate, but How?

HARTMUT LAND

The multistage nature of carcinogenesis is demonstrated by a large number of pathological and epidemiological analyses, as well as by a variety of experimental tumor induction systems (for review, see Farber and Cameron 1980). The discovery of cellular oncogenes established the notion that the origins of human cancer are of a genetic nature (Cooper 1982; Weinberg 1982). The subsequent observation that two different oncogenes are required to convert normal cells to a tumorigenic phenotype led to the hypothesis that carcinogenesis may be due to the consecutive activation of functionally different oncogenes (reviewed in Land et al. 1983). The first pair of cellular oncogenes found to cooperate were Ha-*ras* and *myc*. Introduction of either activated *ras* or *myc* genes alone into rat embryo fibroblasts (REFs) does not lead to complete transformation, while REFs harboring both oncogenes give rise to fibrosarcomas.

As summarized in Table 1, replacement of *ras* or *myc* with other oncogenes has allowed us to establish two functionally different classes of oncogenes (reviewed in Weinberg 1985). The gene products of the first class are located at the plasma membrane, while those of the second class are localized in the cell nucleus.

This article reviews some of the properties of these oncogenes and discusses several models concerning the molecular mechanisms of oncogene cooperativity.

Properties of *ras* Genes

Activated *ras* oncogenes have been isolated from a variety of different human neoplasias (see Marshall; Balmain, this Vol.). When introduced into established rodent fibroblast cultures, they induce full transformation, while they mainly induce anchorage-independent growth in primary cultures. The three members of the *ras* gene family encode proteins of 21 kDa that bind GTP and exhibit a GTPase activity, similar to the G-proteins of the adenylate cyclase system (see Masters and Bourne, this Vol.). Oncogenic activation of the Ha-*ras* gene usually occurs by point mutations in the coding domain and correlates with a reduction in the protein's ability to hydrolyze GTP (see Fasano; Marshall, this Vol.).

Oncogenes and Growth Control
Edited by P. Kahn and T. Graf
© Springer-Verlag Berlin Heidelberg 1986

Upon microinjection into resting fibroblasts, Ha-*ras* p21 protein induces DNA synthesis (Feramisco et al. 1984). Injection of *ras*-specific monoclonal antibodies inhibits the growth factor-induced entrance into S-phase, presumably because the cellular *ras* proteins are inactivated (Mulcahy et al. 1985). Such antibodies are also able to block the transforming activities of oncogenes that encode membrane associated products such as *src, fms* and *fes*, while they do not affect the transforming ability of cytoplasmic oncogenes such as *mos* and *raf* (Smith et al. 1986). These observations, as well as the functional analogy to the G proteins, suggest that *ras* proteins act as transducers or modulators of receptor-mediated signals controlling cell proliferation. In this model the oncogenic mutations would constitutively activate the *ras* protein, presumably by prolonging the half-life of the p21-GTP complex. However, it is not clear whether *ras* proteins interact directly with receptor-like molecules, and if so, in a specific manner. In addition, the downstream targets of *ras* proteins are not yet identified. It has been suggested that phospholipase C may be activated by *ras* proteins, leading to an increased phosphatidylinositol turnover and an accumulation of inositol triphosphate and diacylglycerol. These second messengers in turn induce intracellular calcium release and activate protein kinase C (see Berridge, this Vol.).

Another effect of the *ras* oncogene is the induction of tumor growth factor production (see Derynck, this Vol., Moses and Leof, this Vol.). It has been proposed that these factors stimulate growth by an autocrine mechanism and that this is an important factor for tumorigenesis (Sporn and Todaro 1980; see also Kahn et al., this Vol.).

Properties of *myc* Genes

Myc genes have been found to be rearranged in a number of human tumors (see Fahrlander and Marcu, this Vol.). However, *myc* gene rearrangements predominantly affect the regulation of expression rather than the structure of the gene as in *ras* activation. In the absence of other oncogenes, *myc* is able to induce in vitro establishment ("immortalization") but not tumorigenicity of REFs. *Myc* oncogenes enhance the frequency of in vitro establishment by 5 to 6 orders of magnitude (Mougneau et al. 1984; Ruley et al. 1984; Land et al. 1986), but not in cells from species such as chicken or human that show no spontaneous immortalization. Using *ts* mutants of polyoma virus it has been demonstrated that the expression of the viral large T antigen is permanently required for establishment (see Mougneau and Cuzin, this Vol.). Due to the lack of *ts* mutants, similar experiments have not been done with *myc*, but it was found that the frequency of in vitro establishment correlates with *myc* expression levels (Land et al. 1986). In addition to the direct effects of these genes, other factors might be involved in maintaining an established phenotype. Thus, rodent cells, known to have unstable karyotypes in cell culture, are further destabilized after expression of *myc* or polyoma large T genes in

Table 1. Classes of cooperating oncogenes

Class 1	Class 2
Ha-*ras*	*myc*
Ki-*ras*	N-*myc*
N-*ras*	*myb*
src	p53
Polyoma middle T	Adeno E1A
	Polyoma large T
	SV40 large T

that they show an enhanced frequency of sister chromatide exchange (see Mougneau and Cuzin, this Vol.).

Besides the induction of establishment, *myc* exhibits other biological effects. For example, although cellular *myc* oncogenes do not induce focus formation in rodent cell lines, some clones expressing *myc* show an increased rate of proliferation and are tumorigenic (Keath et al. 1984; H. L., unpublished data). In addition, in cell systems in which terminal differentiation can be studied in vitro, v-*myc* blocks maturation and supports self-renewal without necessarily inducing unlimited growth (see Kahn et al., this Vol.; Falcone et al. 1985). Thus, it is possible that at least in certain stages of normal development the expression of c-*myc* regulates the rate of cell division and/or blocks maturation. This function of *myc* may well be based on the same mechanism as the process of establishment of rodent embryo fibroblasts.

The nuclear localization of *myc* protein and its DNA binding capacity in vitro led to speculations that *myc* is involved in regulation of gene expression (see Moelling, this Vol.). In addition, since the adenovirus E1A gene can act as a transcriptional activator (see Philipson, this Vol.; Schlokat and Gruss, this Vol.) and since *myc* and E1A belong to the same functional group (Table 1), it has been suggested that *myc* also operates as a transcriptional activator. However, the ability of E1A to establish cells in vitro can be experimentally separated from its capacity to activate transcription (Zerler et al. 1986). It remains to be seen whether the establishment function of E1A is related to its activity to suppress viral and cellular enhancers (Borelli et al. 1984; Velcich and Ziff 1985; Hen et al. 1985; Philipson, this Vol.).

How Do *ras* and *myc* Cooperate?

One of the early responses during the mitogenic activation of resting cells by platelet-derived growth factor (PDGF) is the induction of c-*myc* expression (Kelly et al. 1983; Bravo and Müller, this Vol.). Induction of c-*myc* sequences under the control of a regulatable promoter can partly relieve cells from the dependence on PDGF for stimulation of proliferation (Armelin et al. 1984).

In addition, when *myc* protein is injected into resting fibroblasts, it is able to induce DNA synthesis (Kaczmarek et al. 1985). TPA is also able to activate c-*myc* expression (Kelly et al. 1983) and it is now generally assumed that the induction of c-*myc* expression by both PDGF and TPA is mediated through protein kinase C (see Berridge, this Vol.; Parker and Ullrich, this Vol. and references therein). Indeed, focus formation can be induced in *ras*-transfected REF cultures after application of TPA (Dotto et al. 1985), a result which otherwise can only be achieved by cotransfection with *myc*.

Src and *ras* also appear to emit signals leading to stimulation of protein kinase C (see Berridge, this Vol.). In a simplistic model this could mean that *ras* and *myc*-like oncogenes act in series in the same signalling pathway, where *ras* would function near the signal reception and *myc* close to the signal response. However, this model does not fit some of the available data. First, transfection experiments have led us to organize *ras*-like and *myc*-like genes into biologically distinct classes. Second, the mitogenic activity of PDGF can be transmitted through a protein kinase C-independent pathway, besides the pathway already mentioned. In addition, c-*myc* expression can be stimulated by epidermal growth factor (EGF) in a protein kinase C-independent manner (Coughlin et al. 1985). Third, it is not known whether the activation of inositol phosphate turnover by p21 *ras*, leading to the activation of protein kinase C, is the only function of this protein relevant to cell division and transformation, or whether *ras* is also involved in protein kinase C-independent signal pathways. In conclusion, a model of *ras* and *myc* oncogenes acting in series appears to be extremely oversimplified.

Another potentially important element in oncogene cooperativity is the ability of *ras* to induce the production of TGFs. An autocrine loop induced by *ras* in *myc*-transformed chicken macrophages has indeed been demonstrated (see Kahn et al., this Vol.). *Myc* on the other hand has been found to increase the sensitivity of a fibroblast cell line to EGF (Stern et al. 1986) and to abrogate growth factor dependence via a non-autocrine mechanism of mouse hematopoietic cells (Rapp et al. 1985).

The available data leave us with a complex puzzle to explain *ras* and *myc* cooperativity. A picture emerges where both *ras* and *myc* oncogenes show distinct as well as common properties. This feature is also reflected by the behaviour of growth factors such as EGF and PDGF. Growth signals from different receptors seem to generate groups of signals, some of which may overlap. Such as system of interdigitated signals allows a sophisticated fine tuning of proliferative processes and the formation of a safety net that can prevent simple failures from irreversibly disrupting growth control. This network might be the reason why multiple genetic alterations are required to fully transform a normal cell.

Responsiveness to Oncogenes Is Cell-Type-Specific and Can Be Altered During Differentiation

During the development from an embryo to an adult organism, the mechanisms that govern growth control have to allow for periods of active proliferation, differentiation, and homeostasis (see Jacobovits, this Vol.). If one assumes that the changes in the control of proliferation are linked to alterations in signalling networks, it would follow that the cellular responsiveness to oncogenes might change from cell type to cell type and during different stages of differentiation. The observation that *ras* and *src* oncogenes can induce not only cell transformation but also the differentiation of a neuronal cell line (see Wagner and Müller; Rohrschneider, this Vol. and references therein) supports this notion.

In P19 cells, a teratocarcinoma cell line that is able to differentiate into neurons, astrocytes and fibroblast-like cells, the presence of a *ras* oncogene does not affect differentiation. Interestingly, however, transformation seems to be lineage-specific in this system. After inducing differentiation with retinoic acid, only fibroblastic cells show a transformed morphology, while all other cell types appear normal, in spite of measurable *ras* expression. In contrast, when the *ras* oncogene is transferred into fibroblasts derived from differentiated P19 cells, transformation is not observed (Bell et al. 1986). These results suggest that transformation occurs only if the oncogene is already expressed during an undifferentiated stage of the cell and underscores the importance of identifying the target cells for oncogenic transformation in vivo.

Studies on normal keratinocyte cultures demonstrate that during their development cells can also change their response to growth factor-like substances. In undifferentiated basal cells TPA acts like a mitogen while high Ca^{2+} concentrations cause differentiation; following Ca^{2+}-induced differentiation, the cells perceive TPA like a hormonal signal by responding with accelerated differentiation. A *ras* oncogene introduced into undifferentiated cells blocks differentiation in that they react to TPA with a proliferative response even in the presence of high concentrations of Ca^{2+} (Yuspa et al. 1985). These data may explain why both initiation and promotion events are required for the induction of mouse skin papillomas (keratinocyte-derived tumors). Keratinocytes "initiated" by expressing a *ras* oncogene would be blocked in their maturation. As a consequence, differentiation signals are unable to reprogram the cells' response to mitogens. Therefore, following administration of TPA, the initiated cells start to proliferate in an uncontrolled manner.

The above observations illustrate that, depending on the cell type or the stage of cellular development, an oncogene can induce differentiation, transformation, a block in maturation, or show no effect at all. Future work on the function of oncogenes and their role in tumorigenesis will have to take this into account.

Environmental Conditions Can Modify Cellular Responsiveness Towards Oncogenes

During experiments testing the effect of *ras* on REFs, we observed that normal cells strongly inhibit the clonal expansion of REFs containing a *ras* oncogene. In contrast, under conditions where untransformed cells that normally surround the transfected cells are selectively killed, transformed colonies were obtained, many of which developed into tumorigenic cell lines (Land et al. 1986). This suggests that the environment of a transformed cell plays an important role in modulating the cellular response to an oncogenic signal. In support of this hypothesis is the recent observation that the inability of transformed cells to grow on a monolayer of untransformed cells correlates with the formation of gap junctions (Metha et al. 1986). The phenomenon of transformation inhibition by normal cells may also be relevant during virus-induced tumorigenesis. Thus, infection of mice with helper-free replication defective virus containing the Ha-*ras* gene does not induce tumors, while sarcomas develop at the site of injection if helper virus is added simultaneously (C. Tabin and R. A. Weinberg, personal communication). The spreading of the virus might create a threshold number of transformed cells that could support the progression of single cells into a tumor. An alternative explanation for this observation is that *ras*-transformed cells do not have a sufficiently long lifespan to develop into a tumor.

Another way to enhance the tumorigenic potential of an oncogenic retrovirus is the wounding of an infected animal. In young chickens infected with the *src* gene carrying Rous sarcoma virus, tumor growth occurs predominantly at the site of injection or at a distant experimental skin lesion (Dolberg et al. 1985). A possible explanation is that at the site of the wound cell growth is induced by the release of growth factors such as PDGF and TGF-β, thus creating a permissive microenvironment for the action of the oncogenic virus. The intercellular influences seen in tissue culture may reflect much stronger cell to cell interactions existing in organized tissues. The elimination of normal cells in experiments studying cellular transformation might remove critical factors affecting the cellular response to an oncogene.

Outlook

Strong evidence indicating oncogene cooperativity has been obtained in a number of different experimental animal systems. However, the relevance of oncogene cooperativity for tumorigenesis in humans has so far not been proven beyond doubt. Only very few human tumors have been demonstrated to contain two different activated oncogenes (Murray et al. 1983; Taya et al. 1984). The development of new types of oncogene bioassays might be necessary to determine whether the activation of multiple oncogenes plays a significant role in human cancer.

References

Armelin HA, Armelin MCS, Kelly K, Stewart T, Leder P, Cochran BH, Stiles CD (1984) Functional role for c-*myc* in mitogenic response to platelet-derived growth factor. Nature 310:655–660

Bell JC, Jardine K, McBurney MW (1986) Lineage-specific transformation after differentiation of multipotential murine stem cells containing a human oncogene. Mol Cell Biol 6:617–625

Borrelli E, Hen R, Chambon P (1984) Adenovirus-2 E1A products repress enhancer-induced stimulation of transcription. Nature 312:608–612

Cooper G (1982) Cellular transforming genes. Science 217:801–806

Coughlin SR, Lee WMF, Williams PW, Giels GM, Williams LT (1985) c-*myc* gene expression is stimulated by agents that activate protein kinase C and does not account for the mitogenic effect of PDGF. Cell 43:243–251

Dolberg DS, Hollingsworth R, Hertle M, Bissell MJ (1985) Wounding and its role in RSV-mediated tumor formation. Science 230:676–678

Dotto GP, Parada LF, Weinberg RA (1985) Specific growth response of *ras*-transformed embryo fibroblasts to tumor promoters. Nature 318:472–475

Falcone G, Tato F, Alema S (1985) Distinctive effects of the viral oncogenes *myc, erb, fps,* and *src* on the differentiation program of quail myogenic cells. Proc Natl Acad Sci USA 82:426–430

Farber E, Cameron R (1980) The sequential analysis of cancer development. Cancer Res 31:125–226

Feramisco JR, Kamata T, Gross M, Rosenberg M, Sweet RW (1984) Microinjection of the oncogene form of the human Ha-*ras* (T24) protein results in rapid proliferation of quiescent cells. Cell 38:109–117

Hen R, Borrelli E, Chambon P (1985) Repression of the immunoglobulin heavy chain enhancer by the adenovirus-2 E1A products. Science 230:1391–1394

Kaczmarek L, Hyland JK, Watt R, Rosenberg M, Baserga R (1985) Microinjected c-*myc* as a competence factor. Science 228:1313–1315

Keath EJ, Caimi PG, Cole MD (1984) Fibroblast lines expressing activated c-*myc* oncogenes are tumorigenic in nude mice and syngeneic animals. Cell 39:339–348

Kelly K, Cochran BH, Stiles CD, Leder P (1983) Cell specific regulation of the *myc* gene by lymphocyte mitogens and platelet-derived growth factor. Cell 35:603–610

Land H, Parada LF, Weinberg RA (1983) Cellular oncogenes and multistep carcinogenesis. Science 222:771–778

Land H, Chen AC, Morgenstern JP, Parada LF, Weinberg RA (1986) Behavior of *myc* and *ras* oncogenes in transformation of rat embryo fibroblasts. Mol Cell Biol 6:1917–1925

Metha PP, Bertram JS, Loewenstein WR (1986) Growth inhibition of transformed cells correlates with their junctional communication with normal cells. Cell 44:187–196

Mougneau E, Lemieux L, Rassoulzadegan M, Cuzin F (1984) Biological activities of v-*myc* and rearranged c-*myc* oncogenes in rat fibroblast cells in culture. Proc Natl Acad Sci USA 81:5758–5762

Mulcahy LS, Smith MR, Stacey DW (1985) Requirement for *ras* proto-oncogene function during serum-stimulated growth of NIH 3T3 cells. Nature 313:241–243

Murray M, Cunningham J, Parada LF, Dautry F, Lebowitz P, Weinberg RA (1983) The HL-60 transforming sequence: A *ras* oncogene coexisting with altered *myc* genes in hematopoietic tumors. Cell 33:749–757

Rapp UR, Cleveland JL, Brightman K, Scott A, Ihle JN (1985) Abrogation of IL-3 and IL-2 dependence by recombinant murine retroviruses expressing v-*myc* oncogenes. Nature 317:434–438

Ruley HE, Moomaw JF, Maruyama K (1984) Avian myelocytomatosis virus *myc* and adenovirus early region 1A promote the in vitro establishment of primary cultured cells. Cold Spring Harbor Press Cancer Cells 2:481–486

Smith MR, DeGudicibus SJ, Stacey DW (1986) Requirement for c-*ras* proteins during viral oncogene transformation. Nature 320:540–543

Sporn MB, Todaro GJ (1980) Autocrine secretion and malignant transformation of cells. N Engl J Med 303:878 – 880

Stern DF, Roberts AB, Roche NS, Sporn MB, Weinberg RA (1986) Differential responsiveness of *myc*- and *ras*-transfected cells to growth factors: selective stimulation of *myc*-transfected cells by epidermal growth factor. Mol Cell Biol 6:870 – 877

Taya Y, Hosogai K, Hirohashi S, Shimosato Y, Tsuchiya R, Tsuchida N, Fushimi M, Sekiya T, Nishimura S (1984) A novel combination of K-*ras* and *myc* amplification accompanied by point mutational activation of K-*ras* in a human lung cancer. EMBO J 3:2943 – 2946

Velcich A, Ziff E (1985) Adenovirus E1a proteins repress transcription from the SV40 early promoter. Cell 40:705 – 716

Weinberg RA (1982) Oncogenes of spontaneous and chemically induced tumors. Adv Cancer Res 36:149 – 163

Weinberg RA (1985) The action of oncogenes in the cytoplasm and nucleus. Science 230:770 – 776

Yuspa SH, Kilkenny AE, Stanley J, Lichti U (1985) Keratinocytes blocked in phorbol ester-responsive early stage of terminal differentiation by sarcoma viruses. Nature 314:459 – 462

Zerler B, Moran B, Maruyama K, Moomaw J, Grodzicker T, Ruley HE (1986) Adenovirus E1A coding sequences that enable *ras* and *pmt* oncogenes to transform cultured primary cells. Mol Cell Biol 6:887 – 899

Individual and Combined Effects of Viral Oncogenes in Hematopoietic Cells

Patricia Kahn, Achim Leutz, and Thomas Graf

Activation of proto-oncogenes is believed to play an important role in the development of human leukemia. For example, chronic myelocytic leukemia is characterized by the presence of a specific translocation which leads to the activation of the c-*abl* proto-oncogene (see Ben-Neriah and Baltimore, this Vol.). This activation leads to the selective outgrowth of an apparently normal pluripotent stem cell clone and granulocytic cells derived from it. Progeny from this stem cell clone become leukemic in a later step(s) which usually affects myeloid and occasionally also lymphoid cells. Similarly, Burkitt's lymphoma is associated with a specific translocation which activates the c-*myc* gene, but this activation is not sufficient to induce the disease (see Klein, this Vol.). In both of these examples the nature of the secondary events required to generate the leukemic state is unclear.

This article describes one experimental approach to studying the multiple events required for leukemogenesis, which is to analyze the effects of oncogenes introduced singly or in pairs into cultured hematopoietic cells. Since hematopoietic cells cannot be readily transfected, this is generally done by using retroviral vectors. We emphasize results obtained in the chicken system, a species in which a large number of retroviruses containing either one or two oncogenes have been isolated. These viruses induce specific types of acute leukemia and transform hematopoietic cells of the same lineage(s) as in the animal (Graf and Beug 1978).

Assays for in Vitro Transformation of Hematopoietic Cells

Transformation assays with hematopoietic cells are generally performed by infecting cells from freshly isolated hematopoietic tissues with an oncogene-containing retrovirus and then seeding them in semi-solid medium. The cultures are screened 1–3 weeks later for the presence of colonies which differ in appearance from those seen in uninfected cultures; these are defined as "transformed". However, the outcome of such assays may depend on the culture conditions used, especially on whether or not appropriate growth factors are present. Furthermore, the transformed colonies show very different proliferative capacities when grown in liquid culture, with species appearing to be an important factor. The transformed colonies from mouse often develop into stable cell lines (thus becoming "immortalized"), while in other cases they do

Oncogenes and Growth Control
Edited by P. Kahn and T. Graf
© Springer-Verlag Berlin Heidelberg 1986

not grow at all. In contrast, transformed chick cells derived from a single colony grow for 10–40 generations and then undergo senescence. The fact that transformed chick cells do not become immortalized (as also observed for cultured human cells) has the advantage that the direct consequences of oncogene expression can be analyzed in the absence of secondary changes which frequently accompany the establishment of cell lines.

Features of Hematopoietic Cell Growth and Differentiation Which Can Be Altered by Oncogenes

Mature hematopoietic cells are generated from a small population of pluripotent stem cells which have the ability to either self-renew or to become committed to particular lineages. Once committed, the cells enter an irreversible maturation pathway during which their proliferative capacity gradually diminishes and is eventually lost. The survival, growth, and differentiation of hematopoietic cells depends on the presence of specific polypeptide growth factors (reviewed in Metcalf 1984, and Gough, this Vol.).

Although it is possible that oncogenes affect the self-renewal and/or commitment of pluripotent stem cells, there are as yet no clear examples of this. The better-understood effects of oncogenes occur at the level of the committed progenitors or, within certain lineages, in more mature cells. As outlined in Fig. 1, four basic effects can be distinguished, at least on an operational basis: (1) induction of self-renewal; (2) block of differentiation; (3) abolition

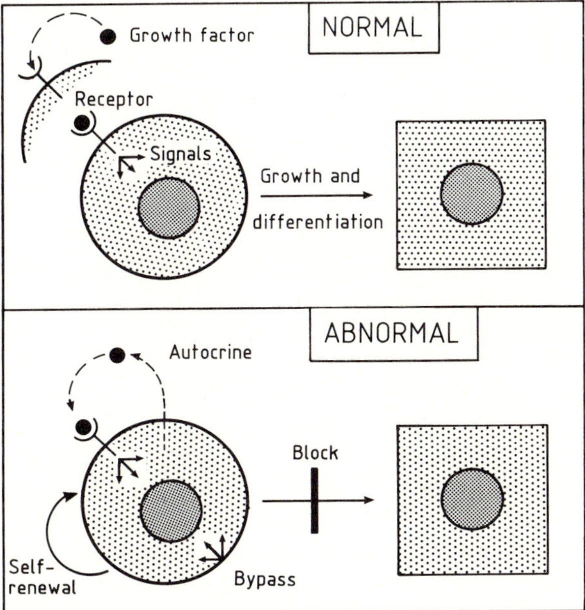

Fig. 1. Schematic representation of the regulation of growth and differentiation in normal hematopoietic cells, and possible alterations induced by oncogenes

of growth factor dependence by an autocrine mechanism and (4) abolition of growth factor dependence by a nonautocrine mechanism. Examples in which oncogenes affect avian erythroid or myelomonocytic cells in each of these ways are discussed in the following two sections and summarized in Fig. 2.

Induction of Self-Renewal and Block of Terminal Differentiation

Infection of chick bone marrow cells with viruses containing oncogenes that encode either tyrosine kinases (v-*erbB*, v-*src*, v-*fps*, v-*sea*) or serine kinases (v-*mil*) or the v-Ha-*ras* oncogene induces the formation of transformed erythroid colonies. These colonies exhibit extensive self-renewal capacity in liquid culture, although they contain a significant but variable proportion of cells that spontaneously differentiate into erythrocytes (Frykberg et al. 1983; Kahn et al. 1984, 1986; and Beug et al., this Vol.). In contrast, viruses containing the v-*myb* and v-*myc* oncogenes do not appear to affect erythroid cells but induce the (factor-dependent) self-renewal of myelomonocytic cells with an immature (myeloblast-like) or mature (macrophage-like) phenotype, respectively.

All oncogenes which induce self-renewal also induce an apparent (usually partial) block of terminal differentiation. At present it is impossible to determine whether or not these oncogenes directly interfere with differentiation. We consider it more likely that, in most cases, the block is simply the consequence of the induction of self-renewal. V-*myb* appears to be an exception in that, in addition to its capacity to induce self-renewal, it induces an extensive differentiation block and perhaps even leads to dedifferentiation of mature macrophages (reviewed by Moscovici and Gazzolo 1982; Graf 1985). An interesting question is whether any oncogenes can block differentiation *without* inducing proliferation. One candidate is v-*erbA*, the second oncogene encoded by the ES4 strain of the avian erythroblastosis virus (AEV), which carries v-*erbB* as its main transforming gene (see Beug et al., this Vol.). V-*erbA* completely suppresses the differentiation capacity of erythroblasts transformed with v-*erbB*, v-*src*, v-*sea*, or v-Ha-*ras* (Kahn et al. 1986), but by itself does not seem to induce self-renewal. It is not clear whether or not it can block the differentiation of normal erythroid cells. The recent demonstration that v-*erbA* exhibits partial sequence homology with steroid hormone receptors, which are known to regulate the expression of differentiation-specific genes (see Beato, this Vol.), raises the possibility that this oncogene has similar functions.

Induction of Growth Factor Independence

Oncogenes can abolish the requirement of their hematopoietic target cells for exogenous growth factors in one of two ways: by induction of an autocrine mechanism, in which cells acquire the ability to synthesize their own growth factor but strictly speaking remain factor-dependent, or by nonautocrine or "bypass"-type mechanisms, in which cells do not produce their own growth factors but the oncogene somehow replaces a growth factor signal. The simplest way for an oncogene to induce autocrine growth is for the oncogene itself to encode a growth factor, such as has been found for the *sis* oncogene (see Heldin and Westermark, this Vol.). Similarly, introduction of the gene for the murine granulocyte-macrophage colony-stimulating factor (GM-CSF) into a factor-dependent mouse myeloid cell line leads not only to autocrine growth, but also to tumor-forming capacity in nude mice (Lang et al. 1985; and Gough, this Vol.). These results raise the possibility that any growth factor gene, when inappropriately expressed, can function as an oncogene.

Another way in which oncogenes may trigger autocrine growth is by inducing cells to produce the growth factor(s) which they require. This type of mechanism has been demonstrated in v-*myb*- and v-*myc*-transformed chick myelomonocytic cells, which are dependent for growth on chicken myelomonocytic growth factor (cMGF; Leutz et al. 1984). Chick myelomonocytic cells transformed with v-*myb* or v-*myc* and then superinfected with viruses containing v-*erbB*, v-*src*, v-*fps*, v-*mil,* v-Ha-*ras*, v-*ros*, v-*yes* or v-*sea* (the latter two are tyrosinekinase-encoding oncogenes not listed in Fig. 2) become factor-independent. The doubly infected cells synthesize cMGF or a cMGF-like factor and require this factor to proliferate (Adkins et al. 1984; Graf et al. 1986). Furthermore, experiments using viral mutants that contain tempera-

v-onc	Self-renewal		Block		Autocrine		Bypass	
	E	M	E	M	E	M	E	M
erb B	+	–	–[a]	–	–	+	+	–
src,fps	+	–	–[a]	–	–	+	+	–
mil	+	–	–	–	–	+	+	–
Ha-ras	+	–	–[a]	–	–	+	+	–
myb	–	+	–	+	–	–	–	–
myc	–	+	–	–[a]	–	–	–	–
erb A	–	–	+	–	–	–	–	–

Fig. 2. Alterations in chick erythroid (*E*) and myelomonocytic (*M*) cells induced by viral oncogenes. Positive effects are indicated by a + and by the gray areas; absence of detectable effects is indicated by a – . [a] Partial differentiation block differentiation induced in these cells which probably reflects the induction of self-renewal

ture-sensitive v-*src* or v-*mil* oncogenes demonstrated that the continuous expression of the oncogene is necessary for the maintenance of autocrine growth (Adkins et al. 1984; v. Weizsäcker et al. 1986).

Can these oncogenes also induce cMGF production in normal cells? While this has not been directly demonstrated, two observations favor such a possibility. First, macrophage-like cells transformed by a mutant of MH2 virus which contains a wild-type v-*mil* gene and a temperature-sensitive v-*myc* gene continue to produce cMGF after shift to the nonpermissive temperature, when the cells are no longer transformed (v. Weizsäcker et al. 1986). Second, v-*fps* induces normal chick myelomonocytic progenitor cells to form colonies in the absence of exogenous growth factor; however, the mechanism in this instance is not known (Carmier and Samarut 1986).

One well-studied oncogene which induces factor independence by a bypass (nonautocrine) mechanism is the v-*abl* gene. This oncogene induces erythropoietin independence in mouse erythroid cells (Waneck et al. 1986) and independence of interleukin-3 (IL-3) or GM-CSF in mouse mast cells and myeloid cell lines. In the latter cases, it has been established that the cells do not express IL-3 or GM-CSF, nor do they synthesize other molecules with growth factor activity. Factor independence in these myeloid cells is also associated with the acquisition of tumorigenic capacity (Cook et al. 1985; Pierce et al. 1985; Oliff et al. 1985; and Gough, this Vol.). Recent experiments suggest that v-*myc*, but not v-Ha-*ras* or v-*raf* (the mammalian homolog of v-*mil*), induce factor independence in mouse myeloid and lymphoid cells, again by apparently nonautocrine mechanisms (Rapp et al. 1985; Ihle et al. 1985).

In the chicken system, v-*erbB*, v-*src*, v-*fps*, and v-Ha-*ras* abolish the erythropoietin dependence of erythroid cells for both proliferation and differentiation. This appears to be due to a bypass-type mechanism, since the cells do not synthesize detectable amounts of factor(s) with erythropoietin-like activity (H. Beug, A. Leutz and T. Graf, unpublished). However, since molecular probes for chick erythropoietin are not yet available, it cannot be excluded that these oncogenes induce expression of erythropoietin, which might intracellularly.

Naturally Occurring Retroviruses Containing Two Cooperating Oncogenes

Three acute avian leukemia virus strains have acquired two oncogenes during evolution. The fact that these viruses were originally selected for high oncogenicity by multiple in vivo passages (reviewed by Graf and Beug 1978) suggests that each of their oncogenes contributes to leukemia induction. As mentioned previously, the MH2 strain carries the *mil* and *myc* oncogenes and transforms macrophage-like cells that grow in a cMGF-independent fashion. In this virus, v-*myc* acts as a "primary" oncogene which efficiently induces the cells to proliferate but does not abolish their cMGF requirement, while v-*mil* acts as an "auxiliary" oncogene which induces the cells to grow in an autocrine fashion.

Furthermore, mutants carrying either v-*myc* or v-*mil* alone are only poorly oncogenic, while MH2 virus induces monocytic leukemias at high efficiencies (Graf et al. 1986).

AEV-ES4, which carries the v-*erbB* and v-*erbA* oncogenes, transforms erythroblasts and renders them erythropoietin-independent. As discussed earlier, v-*erbB* acts as the "primary" transforming gene in that it stimulates erythroid target cells to proliferate and to lose their factor requirement while v-*erbA* cooperates with v-*erbB* by inducing a complete differentiation block. The importance of this cooperativity in vivo was initially difficult to establish, since the v-*erbB* gene alone induces erythroleukemia (Frykberg et al. 1983). However, a cooperative effect was recently demonstrated with the v-*src* oncogene, which causes sarcomas but is inefficient in inducing erythroleukemia; in the presence of v-*erbA*, v-*src* also efficiently induced acute erythroblastosis similar to that obtained with AEV-ES4 (Kahn et al. 1986).

The E26 virus is the only known acute leukemia virus strain which transforms hematopoietic cells of two different lineages. It is not yet clear whether its two oncogenes, v-*myb* and v-*ets*, cooperate with one another to induce a more fully transformed phenotype within one lineage; rather, v-*myb* appears to be responsible for the myeloid transforming capacity and v-*ets* for the erythroid transforming potential of the virus (Beug et al. 1984).

Concluding Remarks

A major obstacle in establishing general models of oncogene action in hematopoietic cells is the variation among species. Results obtained with the v-Ha-*ras* oncogene illustrate this point. Infection of mouse bone marrow cells with retroviruses carrying the v-Ha-*ras* gene induces the formation of erythroid colonies which cannot be further propagated and which require erythropoietin (Waneck et al. 1986). In contrast, infection of chicken bone marrow with v-Ha-*ras*-carrying retroviruses induces erythroid colonies which can be propagated extensively in culture and which seem to be erythropoetin-independent (Kahn et al. 1986). This observation is paralleled by the recent finding that, unlike the RAS2 gene of baker's yeast (*S. cerevisiae*), the *RAS* gene of fission yeast (*S. pombe*) does not appear to regulate the activity of adenylate cyclase (Fukui et al. 1986; and Fasano, this Vol.), suggesting that *ras* has evolved to serve different functions in different organisms.

With this proviso, the following conclusions can be drawn from studies with hematopoietic cells. First, oncogenes may have different effects in different lineages, or they may have no effect at all in certain lineages. The molecular basis of lineage specificity is an important issue which remains to be resolved. Since oncogenes of avian leukemia viruses can be expressed in cells which they do not transform (Graf et al. 1980; Durban and Boettiger 1981; Gilmore and Temin 1986), it is unlikely that the viral regulatory sequences (LTRs) play a decisive role in target cell specificity. Instead, the failure of an oncogene to induce detectable effects in cells of a particular lineage may reflect the absence of appropriate molecular substrates in that cell type. Second, cooperativity in the full transformation of chicken hematopoietic cells by avian leukemia virus oncogenes generally involves the combination of genes from two functional classes: a primary oncogene, which functions as the main transforming gene by inducing self-renewal, and an auxiliary oncogene which, in one of several possible ways, enhances the transformed phenotype. Studies of oncogene cooperativity in other systems are needed to determine whether the requirement for these two functional classes is a more general phenomenon in multistep tumorigenesis.

Acknowledgments. We thank Hartmut Beug and Norman Iscove for helpful discussions and Birgit Blanasch for typing.

References

Adkins B, Leutz A, Graf T (1984) Autocrine growth induced by *src*-related oncogenes in transformed chicken myeloid cells. Cell 39:439–445

Beug H, Leutz A, Kahn P, Graf T (1984) E26 leukemia virus allow transformed myeloblasts, but not erythroblasts or fibroblasts, to differentiate at the nonpermissive temperature. Cell 39:579–588

Carmier JF, Samarut J (1986) Chicken myeloid stem cells infected by retroviruses carrying the v-*fps* oncogene do not require exogenous growth factors to differentiate in vitro. Cell 44:159–165

Cook W, Metcalf D, Nicola NA, Burgess AW, Walker F (1985) Malignant transformation of a growth factor-dependent myeloid cell line by Abelson virus without evidence for an autocrine mechanism. Cell 41:677–683

Durban E, Boettiger D (1981) Differential effects of transforming avian RNA tumor viruses on avian macrophages. PNAS USA 78:3600–3604

Fukui Y, Kozasa T, Kaziro Y, Takeda T, Yamamoto M (1986) Role of a *ras* homolog in the life cycle of Schizosaccharomyces pombe. Cell 44:329–336

Frykberg L, Palmieri S, Beug H, Graf T, Hayman MJ, Vennström B (1983) Transforming capacities of avian erythroblastosis virus mutants deleted in the *erbA* or *erbB* oncogenes. Cell 32:227–238

Gilmore T, Temin H (1986) Different localization of the product of the v-*rel* oncogene in chicken fibroblasts and spleen cells correlates with transformation by REV-T. Cell 44:791–800

Graf T (1985) Effects of acute leukemia viruses on the differentiation of hematopoietic cells. In: Weissman L (ed) Leukemia, Dahlem Konferenzen. Springer, Berlin Heidelberg New York, pp 131–145

Graf T, Beug H (1978) Avian leukemia viruses: interaction with their target cells in vivo and in vitro. BBA Revs Cancer 516:269–299

Graf T, Hayman MJ, Beug H (1980) Target cell specificity of defective avian leukemia viruses: hematopoietic target cells for a given virus type can be infected but not transformed by strains of a different type. PNAS USA 77:389–393

Graf T, v. Weizsäcker F, Grieser S, Coll J, Stehelin D, Patschinsky T, Bister K, Bechade C, Calothy G, Leutz A (1986) V-*mil* induce autocrine growth and enhanced tumorigenicity in v-*myc* transformed avian macrophages. Cell 45:357–364

Ihle JN, Keller J, Rein A, Cleveland J, Rapp U (1985) Interleukin-3 regulation of the growth of normal and transformed hematopoietic cells. Growth factors and transformation, Cancer cells, vol 3. Cold Spring Harbor Lab, Cold Spring Harbor, NY

Kahn P, Adkins B, Beug H, Graf T (1984) *Src-* and *fps*-containing avian sarcoma viruses transform chicken erythroid cells. PNAS USA 81:7122 – 7126

Kahn P, Frykberg L, Brady C, Stanley IJ, Beug H, Vennström B, Graf T (1986) V-*erbA* cooperates with sarcoma oncogenes in leukaemic cell transformation. Cell 45:349 – 356

Lang RA, Metcalf D, Gough NM, Dunn AR, Gonda TJ (1985) Expression of a hemopoietic growth factor cDNA in a factor-dependent cell line results in autonomous growth and tumorigenicity. Cell 43:531 – 542

Leutz A, Beug H, Graf T (1984) Purification and characterization of cMGF, a novel chicken myelomonocytic growth factor. EMBO J 3:3191 – 3197

Metcalf D (1984) The hemopoietic colony-stimulating factors. Elsevier, Amsterdam New York Oxford

Moscovici C, Gazzolo L (1982) Transformation of hemopoietic cells with avian leukemia viruses. In: Klein G (ed) Advances in viral oncology, vol 1. Raven, New York, pp 83 – 106

Oliff A, Agranovsky O, McKinney MD, Murty VVVS, Bauchwitz R (1985) Friend murine leukemia virus-immortalized myeloid cells are converted into tumorigenic cell lines by Abelson leukemia virus. PNAS USA 82:3306 – 3310

Pierce JH, DiFiore PP, Aaronson SA, Potter M, Pumphrey J, Scott A, Ihle JN (1985) Neoplastic transformation of mast cells by Abelson MuLV: abrogation of IL-3 dependence by a nonautocrine mechanism. Cell 41:677 – 683

Rapp UR, Cleveland JC, Brightman K, Scott A, Ihle J (1985) Abrogation of IL-3 and IL-2 dependence by recombinant retrovirus expressing v-*myc* oncogenes. Nature 317:434 – 438

Waneck G, Keyes L, Rosenberg N (1986) Abelson virus drives the differentiation of Harvey virus-infected erythroid cells. Cell 44:337 – 344

Weizsäcker F v, Beug H, Graf T (1986) Temperature-sensitive mutants of MH2 avian leukemia virus that map in the v-*mil* and v-*myc* genes of the virus. EMBO J 5:1521 – 1528

Multiple Factors Involved in B-Cell Tumorigenesis

GEORGE KLEIN

This chapter considers the hypothesis that activation of the c-*myc* locus by chromosomal translocation to a site adjacent to one of the immunoglobulin loci represents an essential, rate-limiting step in the development of Burkitt's lymphoma, mouse plasmacytoma, and rat immunocytoma. It also describes a possible mechanism by which an activated c-*myc* affects susceptible cells, as well as additional factors which may contribute to the acquisition of a malignant phenotype.

The existence of chromosomal rearrangements involving the long arm of chromosome 14 in human B-cell-derived tumors was first discovered in Burkitt's lymphoma (BL) by Manolov and Manolova (1972), and later also in a variety of other leukemias and lymphomas. They arise by the translocation of a terminal chromosome fragment to the distal part of the long arm of one chromosome 14. The breakpoint on chromosome 14 is at band q32, the site of the immunoglobulin heavy chain (IgH) locus (Kirsch et al. 1982). The transposed piece can originate from one of many possible donor chromosomes, depending on the histological type of the tumor, with chromosomes 1, 8, 11, and 18 being the most frequent donors. The BL-associated 14q + marker is always generated by the transposition of the distal piece of chromosome 8, which breaks at band q24 (Zech et al. 1976). It is present in approximately 80% of all BL cases, including both the high endemic, largely Epstein-Barr virus (EBV)-carrying form and the sporadic, usually EBV-negative form (Klein and Lenoir 1982) as well as in the L3 type of B-cell acute lymphoblastic leukemia (ALL) or "Burkitt's leukemia" that originates from the same type of cell as BL (Mitelman et al. 1979; Berger and Bernheim 1982). The 8;14 translocation is also occasionally found in histiocytic lymphoma, mixed cell type lymphoma, lymphosarcoma, B-cell chronic lymphocytic leukemia (CLL), and angioimmunoblastic lymphadenopathy (Mitelman 1981) but it is not a regular marker of any of these tumors. This translocation is unique among the lymphoma- or leukemia-associated chromosome markers, including the Philadelphia chromosome, in that it is exclusive to B-cell tumors (Mitelman 1981). Next to 8;14, the 11;14 and 14;18 translocations are most common among the B-cell lymphoma-associated 14q + markers. They occur in different types of malignant lymphoma, lymphocytic leukemia and occasionally in myeloma, but are not as regularly associated with any tumor as is the BL translocation.

Oncogenes and Growth Control
Edited by P. Kahn and T. Graf
© Springer-Verlag Berlin Heidelberg 1986

Regular Occurrence of c-*myc* Translocation and Activation in Burkitt's Lymphoma and Mouse Plasmacytoma

A broad analysis of chromosomal rearrangements in hematopoietic and other tumors led Rowley (1977) to propose that each type of chromosomal rearrangement provides a proliferative advantage predominantly to cells of a particular lineage and at a certain stage of maturation. From this hypothesis the idea evolved that the BL- and the closely analogous mouse plasmacytoma (MPC)-associated chromosomal translocations constitutively activate an oncogene by bringing it under the control of an immunoglobulin locus (Klein 1981). We also suggested that the same oncogene is activated in BL and in MPC and that the translocation is an essential step in the tumorigenic process. These ideas were based on cytogenetic studies which demonstrated that the translocations seen in MPC and BL are very similar, despite the fact that these tumors have been investigated in different species and affect B-cells at different stages of maturation. In MPC, the break occurs in chromosome 15, always at the D2/3 interface. The telomeric fragment moves to the IgH-carrying chromosome 12 in the typical translocations and to the kappa-bearing chromosome 6 in the less frequent variants (Ohno et al. 1979; Wiener et al. 1980). The BL-associated breaks occur at the q24 band of chromosome 8. Its telomeric fragment moves to one of the three chromosomes that carry immunoglobulin loci: chromosomes 14 (IgH), 2 (kappa) and 22 (lambda) (for review see Klein 1981; Mitelman 1981).

Molecular studies subsequently confirmed that both the MPC- and BL-associated translocations lead to the juxtaposition of c-*myc* and sequences derived from one of the three immunoglobulin loci (see Fahrlander and Marcu, this Vol.). Furthermore, extensive analysis of additional B-cell tumors has shown that the presence of either a typical or variant c-*myc*/Ig translocation is the only *regular* change detected in MPC (including those with highly complex rearrangements; Wiener et al. 1984a; Banerjee et al. 1985) in Burkitt's lymphoma and in the spontaneous immunocytoma of the Louvain rat (Pear et al. 1986; Sümegi et al. 1983) and therefore must be considered as the *key event*.

The molecular data indicate that the c-*myc*-carrying chromosome may break within or outside the gene, the former occurring more frequently in MPC than in BL. Breakage affects the noncoding first exon or the first intron, but never the coding second or third exons. This suggests that an intact c-*myc* protein is required for whatever selective advantage the translocation may confer upon a cell clone.

A small proportion of the MPCs carry no detectable translocations. Some of these exceptional tumors have an interstitial deletion in the c-*myc* region of chromosome 15, accompanied by a rearrangement of the c-*myc*-carrying DNA fragment and a high level of c-*myc* transcription (Wiener et al. 1984b). Cloning and sequencing of the rearranged fragment from one such tumor, ABPC45, showed that the c-*myc* gene was intact but that its 5′ end faced an S-alpha region head-to-head, as in the typical translocations, despite the fact that there was no cytogenetically detectable change on chromosome 12 (Fahr-lander et al. 1985). The S-alpha sequence was attached, tail-to-tail, to another IgH-derived sequence which contained S-mu and the IgH enhancer region further upstream. This complex rearrangement requires at least three cytogenetic events: two translocations and one inversion. The fact that it has nevertheless led to the juxtaposition of c-*myc* and IgH-derived sequences, including the enhancer, lends further support to the conclusion that c-*myc*/Ig juxtaposition plays an essential role in the genesis of MPC.

Role of Other Factors

It is likely that the breakage of the c-*myc*-carrying chromosomes is due to a random genetic accident. Since the risk of all genetic accidents increases in direct relation to the number of cell divisions, the precursor cells of BL and MPC (but not those of most other B-cell tumors) may be more likely to suffer such an accident because they are under chronic growth stimulation during the long pre-neoplastic period (Klein 1979; Klein and Klein 1984). In the African endemic form of BL, EBV and chronic malaria may act as the main predisposing combination. EBV extends the life of the short-lived B-cell toward potential immortality, increasing the probability that the relevant translocation will occur. The heavy parasite load associated with the hyperendemic form of tropical malaria may also increase the likelihood of the translocation event by a continuous process of B-cell activation and may facilitate the outgrowth of the highly immunogenic EBV-carrying cells by a relative T-cell suppression. The HTLV-III virus may play an analogous immunosuppressive role for the AIDS-associated, EBV-carrying form of BL (Ziegler and Levy 1985) that carries the same translocations (Chaganti et al. 1983; Ernberg et al. 1986), while the numerous other infections responsible for the chronic lymphadenopathy in AIDS may act as B-cell activators.

Various factors may also influence the induction of mouse plasmacytomas, a process which is dependent on the establishment of a chronic foreign body granuloma (Potter et al. 1984). Infection with the Abelson leukemia virus (A-MuLV) shortens the latency period and increases the tumor frequency. Since A-MuLV is an immortalizing agent for pre-B cells, it may play a role analogous to that of EBV in high endemic BL by raising the number of the precursor cells and by prolonging their lifespan, thereby increasing the risk of genetic error. There is also some evidence that secondary transpositions may contribute to an increasingly autonomous phenotype in MPCs (for review see Klein and Klein 1985).

The possible involvement of oncogenes other than c-*myc* in the genesis of non-BL human B-cell-derived tumors is suggested by the recent identification of non-BL lymphomas or leukemias with 14q+ markers where other sequences have translocated to the IgH locus. Croce's group has detected two putative new oncogenes, *bcl-1* and *bcl-2*, by analyzing leukemias with 11;14 and 14;18

translocations, respectively (Erikson et al. 1984; Pegoraro et al. 1984; Tsujimoto et al. 1984, 1985a, b). *Bcl-1* is located on chromosome 11q13 and *bcl-2* on 18q21. Both sequences were found to recombine with a JH region on chromosome 14. There was considerable sequence homology between the donor and the recipient regions and it was therefore suggested that these translocations may utilize the VDJ recombination mechanism (Tsujimoto et al. 1985b). These two translocations have been found in a certain proportion of cases with diffuse large cell lymphoma, diffuse small cell lymphoma, follicular lymphoma, B-CLL and multiple myeloma (for refs. see Pegoraro et al. 1984). They do not occur as regularly in any of these tumors as the c-*myc*/Ig translocations in BL, but the potential significance of the *bcl-2* locus has been reinforced by the recent finding (Tsujimoto et al. 1985a) that 60% of the follicular lymphomas have DNA rearrangements within a 2.1-kb-wide region that hybridized with a B-ALL translocation-derived probe containing chromosome 18 breakpoint-flanking regions.

Taken together, the information presented so far suggests that the precursor cell of both BL and MPC (and probably of the less extensively investigated rat immunocytoma as well) can only escape normal growth control mechanisms if c-*myc* but not any other oncogene is activated by juxtaposition to an Ig locus. One must assume that similar translocations involving other oncogenes occur but do not generate tumors in the same precursor cell populations and/or with the same phenotypic pattern. This could be related to the fact that c-*myc*, unlike most of the other known proto-oncogenes, can become activated by changes which affect the levels of its protein product without altering its coding sequence (reviewed in Weinberg 1985).

Precursor Cells of Burkitt's Lymphoma and Mouse Plasmacytoma and Possible Mechanism of Transformation by c-*myc*

There is much evidence which suggests that c-*myc* plays an important role in cell growth and differentiation (see Moelling, this Vol.; andBravo and Müller, this Vol.). In spite of the variations between the experimental systems in which this evidence has been accumulated, one consistent finding is that c-*myc* expression appears to be regularly associated with continuous proliferation and is turned off in terminally differentiated and other cells that are indefinitely arrested in G_0/G_1. In contrast, an activated c-*myc* gene appears to prevent the normal transition of cells from the proliferative to the resting stage. B-lymphocytes containing a c-*myc* gene activated by translocation might therefore become locked into proliferation at a point where such a shift to a resting state would normally occur, such as during terminal differentiation, senescence, or the transition of the clonally expanding immunoblast to the resting memory cell which occurs following the waning of the antigenic stimulus. The memory cell is therefore a good candidate as the B-cell that is most susceptible to transformation by an activated c-*myc*.

The idea that the transformation-sensitive cell has already followed a program involving downregulation of c-*myc* is supported by the finding that the normal, nontranslocated c-*myc* allele is regularly switched off in both BL and MPC. The possibility that BL and MPC may arise from candidate memory cells rather than from pre-B cells is also suggested by the fact that all BL-tumors and derived lines make a heavy chain. Since the c-*myc*/Ig juxtaposition is a special case of a functional rearrangement, this would be expected only if the precursor cell was pre-selected for a functional Ig product. Alternatively, the regular heavy chain production of all BLs could be reconciled with the pre-B cell translocation hypothesis by assuming that the translocation-carrying cell can grow into a tumor only if exposed to an appropriate antigenic stimulus. A third possibility is that B-cells without a functional heavy chain are eliminated by a physiological scavenging process, whether they are translocation carriers or not.

References

Banerjee M, Wiener F, Spira J, Babonits M, Nilsson M-G, Sumegi J, Klein G (1985) Mapping of the c-*myc*, *ptv-1* and immunoglobulin kappa genes in relation to the mouse plasmacytoma-associated variant (6;15) translocation breakpoint. EMBO J 4:3183–3188

Berger R, Bernheim A (1982) Cytogenetic studies on Burkitt's lymphoma-leukemia. Cancer Gen Cytogen 7:231–244

Chaganti RSK, Jhanwar SC, Koziner B, Arlin Z, Mertelsman R, Clarkson B (1983) Specific translocations characterize Burkitt's like lymphoma of homosexual men with the acquired immunodeficiency syndrome. Blood 61:1265–1268

Erikson J, Finan J, Tsujimoto Y, Nowell PC, Croce CM (1984) The chromosome 14 breakpoint in neoplastic B cells with the t(11;14) translocation involves the immunoglobulin heavy chain locus. Proc Natl Acad Sci USA 81:4144–4148

Ernberg I, Björkholm M, Zech L, Sandstedt B, Szigeti R, Andersson J, Henle W, Klein G (1986) An EBV genome currying pre B cell leukemia in a homosexual man with characteristic Karyotype and impaired EBV-specific immunity. J Clin Oncol (in press)

Fahrlander PD, Sümegi J, Yang JQ, Wiener F, Marcu KB, Klein G (1985) Activation of the c-*myc* oncogene by the immunoglobulin heavy-chain gene enhancer after multiple switch region-mediated chromosome rearrangements in a murine plasmacytoma. Proc Natl Acad Sci USA 82:3746–3750

Kirsch IR, Morton CC, Nakahara K, Leder P (1982) Human immunoglobulin heavy chain genes map to a region of translocations in malignant B lymphocytes. Science 216:301–303

Klein G (1979) Lymphoma development in mice and humans: diversity of initiation is followed by convergent cytogenetic evolution. Proc Natl Acad Sci USA 76:2442–2446

Klein G (1981) The role of gene dosage and genetic transpositions in carcinogenesis. Nature 294:313–318

Klein G, Klein E (1984) The changing faces of EBV research. In: Melnick JL, Hummeler K (eds) Prog Med Virol. Karger, Basel, pp 87–106

Klein G, Klein E (1985) Myc/Ig juxtaposition by chromosomal translocations: Some new insights, puzzles and paradoxes. Immunology Today 6:208–215

Klein G, Lenoir G (1982) Translocation involving Ig-locus carrying chromosomes: A model for genetic transposition in carcinogenesis. Adv Cancer Res 37:381–387

Manolov G, Manolova Y (1972) Marker band in one chromosome 14 from Burkitt lymphomas. Nature 237:33–34

Mitelman F (1981) Marker chromosome 14q + in human cancer and leukemia. Adv Cancer Res 34:141–170

Mitelman M, Anderson-Anvret M, Brandt L, Catovsky D, Klein G, Manolov G, Manolova Y, Mark-Vendel E, Nilsson PG (1979) Reciprocal 8;14 translocation in EBV-negative B-cell acute lymphocytic leukemia with Burkitt-type cells. Int J Cancer 24:27–33

Ohno S, Babonits M, Wiener F, Spira K, Klein G (1979) Non-random chromosome changes involving the Ig gene-carrying chromosomes 12 and 6 in pristane-induced mouse plasmacytomas. Cell 18:1001–1007

Pear W, Ingvarsson S, Steffen D, Münke M, Francke U, Bazin H, Klein G, Sümegi J (1986) Proc Natl Acad Sci USA (in press)

Pegoraro L, Palumbo A, Erikson J, Falda M, Giovanazzo B, Emanual BS, Rovera G, Novell PC, Croce CM (1984) A 14;18 and an 8;14 chromosome translocation in a cell line derived from an acute B-cell leukemia. Proc Natl Acad Sci USA 81:7166–7170

Potter M, Wiener F, Mushinski JF (1984) Recent developments in plasmacytomagenesis in mice. Adv Viral Oncol 4:139–162

Rowley JD (1977) Mapping of human chromosomal regions related to neoplasia: Evidence from chromosomes 1 and 17. Proc Natl Acad Sci USA 74:5729–5733

Sümegi J, Spira J, Bazin H, Szpirer J, Levan G, Klein G (1983) Rat c-*myc* oncogene is located on chromosome 7 and rearranges in immunocytomas with t(6;7) chromosomal translocation. Nature 306:497–498

Tsujimoto Y, Finger LR, Yunis J, Nowell PC, Croce CM (1984) Cloning of the chromosome breakpoint of neoplastic B cells with the t(14;18) chromosome translocation. Nature 226:1097–1099

Tsujimoto Y, Cossman J, Croce CM (1985a) Involvement of the *bcl-2* gene in human follicular lymphoma. Science 228:1440–1443

Tsujimoto Y, Jaffe E, Cossman J, Gorham J, Nowell PC, Croce CM (1985b) Clustering of breakpoints on chromosome 11 in human B-cell neoplasmas with the t(11;14) chromosome translocation. Nature 315:340–342

Weinberg RA (1985) The action of oncogenes in the cytoplasm and nucleus. Science 230:770–776

Wiener F, Babonits M, Spira J, Klein G, Potter M (1980) Cytogenetic studies on IgA/lambda-producing murine plasmacytomas: regular occurrence of a T(12;15) translocation. Somatic Cell Gen 6:731–738

Wiener F, Babonits M, Bregula U, Klein G, Leonard A, Wax JS, Potter M (1984a) High resolution banding analysis of the involvement of strain Balb/c-and AKR-derived chromosomes no. 15 in plasmacytoma specific translocations. J Exp Med 159:276–291

Wiener F, Ohno S, Babonits M, Sümegi J, Wirschubsky Z, Klein G, Mushinski JF, Potter M (1984b) Hemizygous interstitial deletion of chromosome 15 (band D) in three translocation-negative murine plasmacytomas. Proc Natl Acad Sci USA 83:1159–1163

Zech L, Haglund U, Nilsson K, Klein G (1976) Characteristic chromosomal abnormalities in biopsies and lymphoid-cell lines from patients with Burkitt and non-Burkitt lymphomas. Int J Cancer 17:47–56

Ziegler JL, Levy JA (1985) Acquired immunodeficiency syndrome and cancer. Adv Virol Oncol 5:239–255

Molecular Events Associated with Tumor Initiation, Promotion, and Progression in Mouse Skin

ALLAN BALMAIN

The terms initiation, promotion and progression originated largely from studies on chemical carcinogenesis in mouse skin (Boutwell 1974), but the general applicability of these concepts to tumorigenesis in other animal systems and to humans is now recognized (Hecker et al. 1982). This review outlines the recent advances made in our understanding of the molecular genetic events associated with each step of tumor development. The conclusions reached are by necessity over-simplistic and in some cases highly speculative, but may serve to focus attention on hypotheses which have become amenable to experimental test.

Genetic Events Associated with Initiation

Initiation of tumorigenesis in mouse skin is accomplished by a single treatment with a chemical carcinogen. The capacity of many carcinogens to cause point mutations in DNA, together with the irreversible nature of the initiation event (Van Duuren et al. 1975), led to the not unreasonable proposal that initiation involves the induction of point mutations in a gene or genes which can confer some selective growth advantage on the target cell(s). Molecular techniques have led to the identification of a number of genes which are frequently altered by translocation or mutation in human or animal tumors (for review, see Klein and Klein 1984; Balmain 1985). One of the most informative techniques has been DNA transfection, which has led to the isolation of genes that have acquired the ability to confer the transformed phenotype on NIH/3T3 cells. Most genes in this category are members of the *ras* family, comprising Harvey- (Ha-), Kirsten- (Ki-) and N-*ras*, although several other genes have also been identified (Weinberg 1985).

Experiments in this laboratory focussed on the use of the transfection assay to determine whether any proto-oncogene becomes reproducibly activated at a defined stage of chemical carcinogenesis. The testing of a series of DNA samples from chemically induced tumors showed that this was indeed the case, since a high percentage of carcinomas and papillomas had an activated Ha-*ras* oncogene (Balmain et al. 1984a). Moreover, the fact that the gene was activated in pre-malignant papillomas led to the conclusion that the mutation must have occurred at a relatively early stage.

Oncogenes and Growth Control
Edited by P. Kahn and T. Graf
© Springer-Verlag Berlin Heidelberg 1986

The mutations in tumors induced by treatment with chemical initiators (dimethylbenzanthracene; DMBA) and promoters (12-O-tetradecanoyl-phorbol-13-acetate; TPA) have been analyzed by a combination of DNA sequencing, *ras* p21 analysis and a search for tumor-specific restriction fragment length polymorphisms (RFLPs). The mouse cellular Ha-*ras* gene was cloned and the sequences of coding exons 1 and 2 (M. Ramsden, K. Brown, F. Fee, R. Krumlauf and A. Balmain, manuscript in preparation) were used to predict RFLPs diagnostic of single base changes at codons 12 and 61 (Zarbl et al. 1985). These experiments gave the surprising result that over 90% of the tumors tested which were initiated with DMBA and promoted with TPA had the same specific A-T mutation at codon 61 (Quintanilla et al. 1986). The proportion of tumors with this mutation was not changed when a different tumor promoter, chryserobin, was used, which does not bind to the same cellular receptor as TPA (Delclos et al. 1980), but was altered when another initiating agent, N-methyl-N'-nitro-N-nitrosoguanidine (MNNG), was used instead of DMBA. Of 12 tumors tested, none exhibited the tumor-specific polymorphism seen after DMBA initiation. This correlation between the mutation observed and the type of initiator used supports the proposal by Barbacid and co-workers (Zarbl et al. 1985) that the carcinogen can interact directly with the DNA to induce the observed mutations. Rat mammary carcinomas which developed after a single treatment with nitrosomethylurea (NMU), but not those resulting from DMBA treatment, all had the same G-A transition at the second base of codon 12 in the Ha-*ras* gene.

DMBA is known to form adducts with both guanosine and adenosine residues in DNA (Dipple et al. 1983). Interestingly, Dipple and co-workers have previously predicted that the critical carcinogenic adduct involves adenosine rather than guanosine (Dipple et al. 1983). This was based on the observation that benzo(a)pyrene (BP) and DMBA form similar amounts of G-adducts when applied to mouse skin, but DMBA forms 30 times more A-adducts. Correspondingly, DMBA is approximately 30 times more carcinogenic, on a molar basis, than BP.

One possible, but highly speculative, explanation for the apparently different specificities of DMBA and NMU for codons 61 and 12, respectively, might be found in the nucleotide sequences of these codons. Codon 61 has the sequence CAA and encodes the amino acid glutamine. Although there is a G on the non-coding strand which could constitute a potential "target" for an NMU-induced mutation, induction of a G-A transition would result in the introduction of the stop codon TAA at this position. Similarly, codon 12 (GGA) might not be frequently altered by DMBA (if A-adducts are critical for carcinogenesis) because the only A residue at this position is at the third base, mutation of which would not lead to any change in coding capacity.

Many questions relating to specific mechanisms involved in generating these mutations or to the general applicability of the conclusions to other systems remain unanswered. For example, why should predominantly the second base positions of codons 12 and 61 be mutated? Mutations at the first G of codon 12 or the third position of codon 61 could in theory also activate the transforming potential of Ha-*ras*, but these appear to be very rarely observed. It is possible that some physical aspects of chromatin structure might restrict mutations to certain positions, or, alternatively, in vivo selection might operate to favour particularly these two sites.

A Test for *ras* Gene Involvement in Initiation

The hypothesis that *ras* gene mutation is an early event in carcinogenesis and that it is causally related to tumor development predicts that introduction of the mutated gene into the appropriate target cells in vivo could substitute for the chemical carcinogen. We have recently completed a series of experiments which demonstrate that viral *ras* genes can act as initiators of two-stage mouse skin carcinogenesis (K. Brown, M. Quintanilla, M. Ramsden, I. B. Kerr, S. Young and A. Balmain, submitted). When Harvey- or Balb murine sarcoma viruses are applied directly to mouse skin, followed by treatment with TPA, papillomas developed after 4 – 5 weeks, some of which subsequently progressed to carcinomas. No skin tumors were seen in the absence of promoter treatment, and the virus-initiated cells were stable within the epidermis for several months. The fact that the viral *ras* genes can reproduce the effects of a chemical initiator, when taken together with the above evidence for carcinogen-specific mutations in the c-Ha-*ras* gene, lends strong support to the idea that the mutation is a critical event in initiation.

Alternative Mechanisms of Initiation

It is unlikely that an all-embracing theory of initiation will be found which reconciles all of the published results on this topic. Although the above arguments favour a direct involvement of the carcinogen in initiation, other results can be cited which apparently do not agree with this interpretation. Among these are the experiments of Guerrero et al. (1985) on the mutations in the N-*ras* genes of thymic lymphomas induced by treatment with NMU. These authors found that the N-*ras* gene was activated by a C-A transversion, in contrast to the G-A change consistently observed by Zarbl et al. (1985) in mammary carcinomas induced by the same carcinogen. However, it is difficult to make a direct comparison between these systems. Induction of thymic lymphomas apparently requires repeated carcinogen treatment, whereas only a single treatment is required for initiation in both the skin and mammary gland systems. In the latter cases, a promoting stimulus, which may be endogenous hormones or exogenous TPA, then amplifies the initial change to produce visible tumors. In view of these differences it is possible that thymic lymphomas develop by a different route and that some gene other than N-*ras* constitutes the initial "target". The occurrence of a *ras* mutation during a later phase of tumor growth might not be unexpected, since in vitro experiments have shown that *ras* gene mutations can take place in the absence of carcinogen treatment (Vousden and Marshall 1984).

The nature of the alternative "target" for activation by carcinogens remains unknown. It may be another proto-oncogene, for example, the *myc* gene, which can "initiate" tumorigenesis when introduced into cells in a retrovirus or in transgenic mice (Stewart et al. 1984). Another possible explanation is that initiation in some systems does not involve a rare mutation of a single base in a specific gene, but is a more general event which takes place in a high proportion of treated cells. Convincing evidence has been presented that such mechanisms may be operative in the transformation of C3H 10T1/2 cells in vitro by chemicals or radiation (Kennedy et al. 1984) although the nature of this initial event and its relevance to tumor induction in vivo remain unclear.

In conclusion, it is almost certain that no universal mechanism of initiation exists. The particular route will depend upon a host of different factors, including the ability of different cells or tissues to metabolize carcinogens, the tissue-specific susceptibility of different genes to mutation (Ramsden et al. 1985), and the selection pressures applied. In some cases, *ras* gene mutations are heavily implicated in initiation, but it would not be expected that all tumors conform to the same pattern.

Molecular Mechanisms of Tumor Promotion

One of the most striking advances in this area has been the recognition that a major cellular binding site for the tumor promoter TPA is protein kinase C (Castagna et al. 1982; and Parker and Ullrich, this Vol.). However, an understanding of how promoters function is still lacking, and in particular the roles of so-called first-stage and second-stage promoters remain obscure (for review see Parkinson 1985). It has been proposed that in addition to providing a mitogenic stimulus, TPA fulfills some other function, possibly related to chromosomal changes, which constitutes a necessary "first stage". However, others have argued that the existence of discrete stages in promotion has not been proven, since the results obtained, which depend upon the combination of mouse strain and chemical agent used, could be explained in part by selective toxicity towards the initiated cell (Hennings and Yuspa 1985; Parkinson 1985).

If the activation of a *ras* gene is in fact the initiating event, the equivalent of the initiated cell could be obtained by inserting an activated *ras* gene into primary cells. These should provide the ideal test system for the effects of first- and second-stage promoters. Some experiments along these lines have already been carried out. The results demonstrate synergism between *ras* gene activation and TPA in transformation of primary fibroblasts (Dotto et al. 1985) and a TPA-responsive block in terminal differentiation induced by a viral *ras* gene in primary epidermal cells (Yuspa et al. 1985). These co-operative effects of TPA and *ras* are also seen in vivo when viruses containing activated *ras* are used as initiators (Brown et al., submitted). The promoter might act by stimulating mitogenesis along a pathway in which *ras* p21 acts as the transducing signal (Berridge and Irvine 1984; and Masters and Bourne, this Vol.), or alternatively by supplying the function normally provided by the *myc* gene in cotransfections with *myc* and *ras* (Land et al. 1983; and Land, this Vol.). In any case, these recent results provide an array of testable hypotheses which should ultimately lead to the elucidation of the mechanism of promoter action.

Molecular Changes During Tumor Progression

In clinical terms, progression from a benign to a more malignant tumor is one of the most important stages, but relatively little is known about the molecular changes involved. Chromosomal aberrations involving rearrangements, deletions or amplification of specific genes may be important at this stage, since there is a general correlation between tumor behavior and the degree of aneuploidy (Barlogie 1984). In the mouse skin system, the pre-malignant papillomas are diploid, although they may have as-yet undetected chromosomal changes, whereas carcinomas are aneuploid (Balmain et al. 1984b). Some chromosomal changes in malignant tumors appear to involve additional events at the Ha-*ras* gene locus, since a series of carcinomas have been found which show either homozygosity or amplification of the mutated Ha-*ras* allele (Quintanilla et al. 1986). This is not observed in every case, however, since some carcinomas have heterozygous mutations. Presumably in these tumors, changes at other loci can lead to the same result. It is possible that the dosage of the *ras* gene product is critical for progression and that this can be altered by amplification of the mutated allele, loss of the normal allele, or increasing the relative expression level of the mutated allele. An exciting possibility is that repressor genes (or "anti-oncogenes") may also be involved. The existence of such genes has been inferred from studies on fusion products between normal and transformed cells (Stanbridge 1985), and on certain forms of hereditary cancer (Knudson 1985; and article by Wyke and Green, this Vol.). The critical factor may therefore not be the absolute level of mutated *ras* p21, or even the relative levels of normal and mutated product, but the balance between the amounts of the activated oncogene product and its associated repressor protein. Animal model systems and the development of pre-malignant cell lines where defined molecular changes have already taken place should be very useful in the ultimate resolution of these questions.

Acknowledgements. The Beatson Institute is supported by the Cancer Research Campaign. I am grateful to a number of colleagues including Ken Brown, Miguel Quintanilla, Frances Fee and Sheena Young for discussions, and to Dr. J. Paul for critical reading of the manuscript.

References

Balmain A (1985) Transforming *ras* oncogenes and multistage carcinogenesis. Br J Cancer 51:1 – 7

Balmain A, Ramsden M, Bowden GT, Smith J (1984a) Activation of the mouse cellular Harvey-*ras* gene in chemically induced benign skin papillomas. Nature 307:658 – 660

Balmain A, Sauerborn R, Ramsden M, Pragnell IB, Bowden GT, Smith J, Cole G (1984b) Oncogene activation at different stages of chemical carcinogenesis in mouse skin. In: Omenn G, Harris C, Gelboin H (eds) Banbury report 16: Genetic variability in responses to chemical exposure. Cold Spring Harbor Lab, Cold Spring Harbor, NY, pp 243–255

Barlogie B (1984) Abnormal cellular DNA content as a marker of neoplasia. Eur J Cancer Clin Oncol 20:1123–1125

Berridge MJ, Irvine RF (1984) Inositol triphosphate, a novel second messenger in cellular signal transduction. Nature 312:23–38

Boutwell R (1974) Function and mechanism of promoters of carcinogenesis. Crit Rev Toxicol 2:419–431

Castagna M, Takai Y, Kaibuchi K, Sano K, Kikkawa U, Nishizuka Y (1982) Direct activation of calcium-activated phospholipid-dependent protein kinase by tumor-promoting phorbol esters. J Biol Chem 257:7847–7851

Delclos KB, Nagle DS, Blumberg PM (1980) Specific binding of phorbol ester tumor promoters to mouse skin. Cell 19:1025–1032

Dipple A, Sawicki JT, Moschel RC, Bigger CAH (1983) 7,12-dimethylbenzanthracene-DNA interactions in mouse embryo cultures and mouse skin. In: Rydstrom J, Montelius J, Bengtsson M (eds) Extrahepatic drug metabolism and chemical carcinogenesis. Elsevier, Amsterdam, pp 439–448

Dotto GP, Parada LF, Weinberg RA (1985) Specific growth response of *ras*-transformed fibroblasts to tumor promoters. Nature 318:472–475

Guerrero I, Villasante A, Corces V, Pellicer A (1985) Loss of the normal N-*ras* allele in a mouse thymic lymphoma induced by a chemical carcinogen. Proc Natl Acad Sci USA 82:7810–7814

Hecker E, Fusenig NE, Kunz W, Marks F, Thielmann HW (eds) (1982) Carcinogenesis – a comprehensive survey, vol 7. Raven, New York

Hennings H, Yuspa SH (1985) Two-stage tumor promotion in mouse skin. An alternative interpretation. J Natl Canc Inst 74:735–739

Kennedy AR, Cairns J, Little JB (1984) Timing of the steps in transformation of C3H10T1/2 cells by X-irradiation. Nature 307:85–86

Klein G, Klein E (1984) Oncogene activation and tumor progression. Carcinog Compr Surv 5:429

Knudson A (1985) Hereditary cancer, oncogenes and antioncogenes. Cancer Res 45:1437–1443

Land H, Parada LF, Weinberg RA (1983) Tumorigenic conversion of primary embryo fibroblasts requires at least two cooperating oncogenes. Nature 304:596–602

Parkinson EK (1985) Defective responses of transformed keratinocytes to terminal differentiation stimuli. Their role in epidermal tumor promotion by phorbol esters and by deep skin wounding. Br J Cancer 52:479–493

Quintanilla M, Brown K, Ramden M, Balmain A (1986) Carcinogen-specific mutations in mouse skin tumors; *ras* gene involvement in both initiation and progression. Nature 322:78–80

Ramsden M, Cole G, Smith J, Balmain A (1985) Differential methylation of the c-*ras*^{H} gene in normal mouse cells and during skin tumor progression. EMBO J 4:1449–1454

Stanbridge E (1985) A case for tumor suppressor genes. Bioessays 3:252–255

Stewart TA, Pattengale PK, Leder P (1984) Spontaneous mammary adenocarcinomas in transgenic mice that carry and express MTV/*myc* fusion genes. Cell 38:627–637

Van Duuren BL, Sivak A, Katz C, Seidman I, Melchionine S (1975) The effect of ageing and interval between primary and secondary treatment in two-stage carcinogenesis in mouse skin. Cancer Res 35:502–505

Vousden KH, Marshall CJ (1984) Three different activated *ras* genes in mouse tumors: evidence for oncogene activation during progression of a mouse lymphoma. EMBO J 3:913–917

Weinberg RA (1985) The action of oncogenes in the cytoplasm and nucleus. Science 230:770–776

Yuspa SH, Kilkenny AE, Stanley J, Lichti U (1985) Keratinocytes blocked in a phorbol ester-responsive early stage of terminal differentiation by sarcoma viruses. Nature 314:459–462

Zarbl H, Sukumar S, Arthur AV, Martin-Zanca D, Barbacid M (1985) Direct mutagenesis of Ha-*ras*-1 oncogene by N-nitroso-N-methylurea during initiation of mammary carcinogenesis in rats. Nature 315:382–386

Amplification of Proto-Oncogenes and Tumor Progression

MANFRED SCHWAB

The term gene amplification refers to the selective increase of the gene copy number and is better designated as DNA amplification. It should not be confused with elevated gene expression, although amplification generally does result in enhanced levels of the products encoded by the amplified gene. Amplification is one of the mechanisms by which cells can meet the demand for synthesis of specific gene products in amounts exceeding the transcriptional capacity of a single copy gene.

Cytogenetic studies of human and animal tumor cells have provided evidence for mysterious chromosomal abnormalities including double minutes (DMs), C-bandless chromosomes (CMs) or homogeneously staining chromosomal regions (HSRs) diagnostic of amplified DNA (for details and photographs see Schwab 1985). First indications for a causal relationship between the presence of DMs and CMs in tumor cells and the malignant phenotype emerged from work with murine SEWA cells (reviewed by Levan et al. 1981). The reproducibility with which SEWA cells generated CMs and DMs following transfer from tissue culture conditions into an animal host led to the suggestion that these chromosomal abnormalities contain genes which contribute to the malignant phenotype.

Amplified Proto-Oncogenes

It is probably more than coincidence that proto-oncogenes are amplified consistently in animal and human tumor cells containing DMs, CMs or HSRs. The list of amplified proto-oncogenes which have been identified so far includes c-*myc*, c-*abl*, c-Ha-*ras*, c-Ki-*ras*, N-*myc*, L-*myc*, c-*erbB*, c-*myb* and N-*ras*. However, they are not always present in DMs or HSRs (reviewed by Alitalo and Schwab 1986). The copy number of amplified proto-oncogenes ranges from 5 to 700. (Data indicating amplification of less than fivefold should be viewed with caution.) Interestingly, *myc* genes account for over 90% of the amplifications which have been detected so far.

Cells derived from at least two different types of tumor have been found to carry two co-existing amplified proto-oncogenes. Both c-*myc* and c-Ki-*ras* were amplified in a human giant lung cell carcinoma (Taya et al. 1984), while amplified c-*myc* and c-Ha-*ras* genes have been detected in eight squamous cell carcinomas of the uterine cervix (Riou et al. 1984; see also Table 1). At least

Oncogenes and Growth Control
Edited by P. Kahn and T. Graf
© Springer-Verlag Berlin Heidelberg 1986

Table 1. Amplification of *myc* genes in different human cancers

		Incidence of cells with amplification of		
		N-*myc*	L-*myc*	c-*myc*
Neuroblastoma	Tumor stage I	0/13	0	0
	Tumor stage II	2/24	0	0
	Tumor stage III	14/43	0	0
	Tumor stage IV	32/45	0	0
	Tumor stages IV-S	0/5	0	0
	Cell lines	22/27	0	0[a]
Retinoblastoma	Tumors	3/27	0	0
	Cell lines	2/16	0	0
Small cell	Tumors	3	1	0
lung	Cell lines (classical)	4	4	0
carcinoma	Cell lines (variant)	1	0	5
Cervical carcinoma	Tumors	0	0	8/12

The table summarizes data as of February 1986 from several authors cited in the text. Only amplifications frequently associated with specific tumors are indicated. For a list including sporadic cases see Alitalo and Schwab (1986). Those values where the total number of tumors or cell lines tested is not indicated represent a sample size of at least 20. [a] Cell line MC-IX reported to have a c-*myc* amplification (Kohl et al. 1983) is not a neuroblastoma line (F. Alt, personal communication).

many copies of amplified c-Ki-*ras* were mutated at codon 12, which so far appears to be one of the most common sites at which this gene becomes activated (see Marshall, this Vol.).

The topography of the amplified proto-oncogene is usually indistinguishable from that of the single-copy counterpart in the normal cell, in spite of the fact that the amplification is often associated with relocation of the DNA encompassing the proto-oncogene from its original chromosomal position to diverse locations on other chromosomes (Schwab et al. 1984b). Only a few instances are known in which amplified proto-oncogenes have undergone structural rearrangements. These exceptional cases include c-*abl* in the myelogenous leukemia line K-562, c-*myc* in the APUDoma line COLO 320, and c-*erbB* in the human epidermoid carcinoma cell line A431 and in several human brain tumors (reviewed by Alitalo and Schwab 1986). The structural rearrangements are associated with the presence of an abnormal mRNA.

myc-Box Genes: N-*myc* and L-*myc*

The discovery of *myc*-box genes is the outcome of the search for a proto-oncogene that was assumed to be present in DMs or HSRs, which are a frequent feature of human neuroblastoma cells (Schwab 1985). N-*myc* was found using

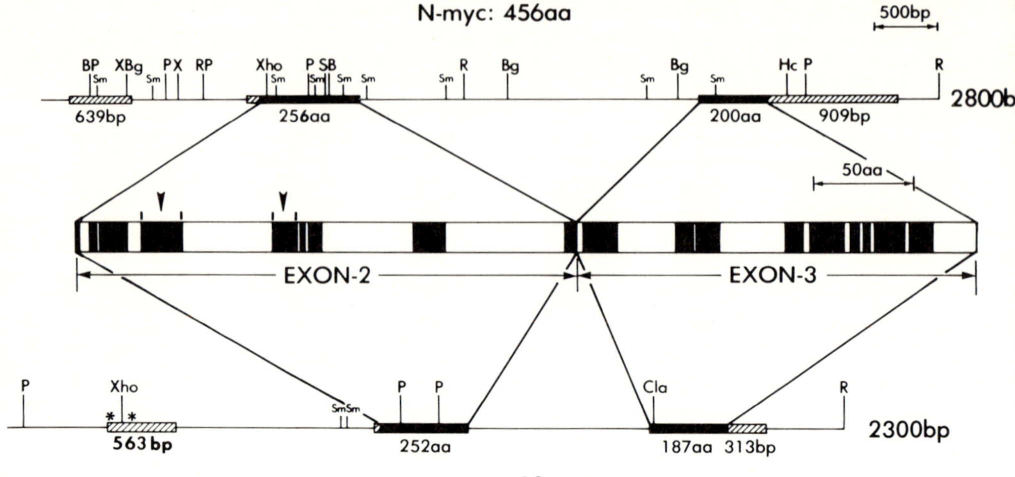

Fig. 1. Structural comparison between human N-*myc* and c-*myc*. The top and bottom drawings display the topography of the N-*myc* and c-*myc* genes, respectively. Solid areas present coding regions, hatched areas non-coding portions, and lines intervening sequences and flanking DNA (for more detailed information and for restriction endonuclease sites in N-*myc* and c-*myc* see Stanton et al. 1986; Battey et al. 1983). Asterisks over non-coding exon 1 in c-*myc* indicate the two promoter sites. N-*myc* contains multiple initiation sites (at least 14), causing a variation in the size of the exon (between roughly 450 to 600 base pairs). The numbers at the extreme right refer to the average size of the N-*myc* and c-*myc* mRNA without considering the poly(A)-tail.
The center drawing presents a comparison of the deduced amino acid sequences of N-*myc* and c-*myc*. Black areas indicate regions where the amino acid homology exceeds 70%. Arrowheads over top of 2 black areas in exon 2 indicate the positions of the *myc*-boxes in which amino acids homology is greater than 90%

a viral *myc* probe and employing reduced hybridization stringency conditions (Schwab et al. 1983b), and L-*myc* was subsequently detected in small cell lung cancer cells by employing principally the same approach (Nau et al. 1985). Further analysis of N-*myc* revealed the presence of two highly conserved nucleotide boxes (designated as *myc* boxes) within the 5′-end of v-*myc*, or c-*myc*, and N-*myc* (Schwab et al. 1983b; see Fig. 1). These *myc*-boxes are also conserved in the L-*myc* gene (Nau et al. 1985).

At the present time the complete structure is known only for N-*myc*. Like c-*myc*, N-*myc* consists of three exons separated by two intervening sequences (Kohl et al. 1986; Stanton et al. 1986), and the frist exon is noncoding (Fig. 1). N-*myc* and L-*myc* may be members of a family of *myc*-related genes that are evolutionary conserved among vertebrates (Schwab 1985). The human *myc* genes identified to date are dispersed on different chromosomes: c-*myc* is localized at 8q24-ter (Dalla-Favera et al. 1982), N-*myc* at 2p23-24 (Schwab et al. 1984b), and L-*myc* at 1p32 (Nau et al. 1985).

As summarized in Table 1, amplification of N-*myc* has been found in neuroblastoma cell lines and in tumors derived directly from patients not subjected to chemo- or radiotherapy (reviewed by Schwab 1985). DMs and HSRs, the presence of which originally stimulated the studies of neuroblastoma, are the sites of the amplified N-*myc* (Kohl et al. 1983; Schwab et al. 1984b). Amplification of N-*myc* has also been found in cell lines derived from small cell lung cancers (Nau et al. 1986), from a few retinoblastoma tumors and cell lines (Lee et al. 1984; Squire et al. 1985; and Schwab, unpublished) and from one astrocyte tumor (Garson et al. 1985).

To determine the stage of tumor progression at which amplifications are first detectable, several groups have examined neuroblastomas which have been classified on the basis of various pathological and clinical criteria into Evans stages I – IV, with stage IV being the most advanced. In studies of more than 130 tumors, nearly all of the amplifications detected were present in the more advanced tumors (stages III and IV; Brodeur et al. 1984; Seeger et al. 1985). The estimated progression-free survival over a period of 18 months of standardized therapy was 70% for patients whose tumors lacked amplification and dropped to 30% in cases with 3 – 10 copies and to 5% in cases with more than 10 copies (Seeger et al. 1985). It is conceivable that N-*myc* copy number may eventually turn out to be a useful marker for assessing prognosis of neuroblastoma and for devising specific therapeutical regimens.

Expression of N-*myc* in different cell types appears to be strictly regulated (reviewed by Schwab 1986). Deregulation, for example as the result of amplification, could disrupt this intrinsic control, thereby contributing to the neoplastic phenotype. It is possible that cellular factors involved in regulating expression of single-copy N-*myc* genes are not available in sufficient amounts to control expression of multiple N-*myc* copies. However, strong extrinsic stimuli such as retinoic acid are capable of inducing a dramatic decrease in the steady-state level of N-*myc* mRNA even in cells with a high degree of amplification (Thiele et al. 1985). The decrease of N-*myc* mRNA levels is followed by morphological differentiation of the neuroblastoma cells, indicating that amplified N-*myc* responds to regulation by differentiation-inducing signals. In situ hybridization studies of neuroblastoma tumors in which N-*myc* was amplified have provided evidence for high N-*myc* expression in undifferentiated neuroblasts and for low expression in differentiated ganglion cells (Schwab et al. 1984a). It is tempting to speculate that high expression of N-*myc* results in a block of differentiation of neuroblasts into ganglion cells.

Experimentally it could be shown that constitutively high N-*myc* expression, achieved using expression vectors carrying N-*myc* from normal tissue or from neuroblastoma cells, can cooperate with mutationally activated c-Ha-*ras* 1 in neoplastic transformation of normal cells in culture (Schwab 1985).

Much less data are available concerning L-*myc* amplification, which has so far been encountered only in small cell lung cancer (SCLC) (see Table 1). SCLC represents an interesting example of a type of tumor in which all three members of the *myc* gene family have been found amplified, although never in the same tumor (reviewed by Brooks et al. 1986). The c-*myc* gene appears to be amplified exclusively in the more malignant variants, although the number of lines tested may be too small to definitively prove such a correlation. N-*myc* and L-*myc* have been detected in both the classical and variant forms of SCLC cell lines.

Amplification of Proto-Oncogenes As a Late Event During Evolution of Cancer Cells

Experimental carcinogenesis studies, as well as clinical evidence, suggest that the development of a malignant tumor is a gradual evolutionary process during which tumor cells progressively acquire permanent, qualitatively different characteristics (Foulds 1958). Genetic instability of tumor cells has been found to be greatly enhanced over that of normal cells (reviewed by Nowell 1976), and amplification of DNA may be one of the mechanisms which leads to the emergence of clonal populations with increasingly malignant properties. There is now good experimental evidence that gene amplification occurs in tumorigenic cells at a higher rate than in nontumorigenic ones (Sager et al. 1985).

The basal rate of spontaneous amplification of a given genetic locus may be as high as 10^{-3} events per cell per generation. The rate at which amplification takes place can be increased severalfold by various extrinsic agents, including metabolic inhibitors, tumor promoters, and carcinogens. The persistence of amplified DNA suggests that there is a selective pressure for their retention. The nature of the selective mechanisms which maintain amplification of proto-oncogenes is at present unknown. A simple but not unlikely explanation could be that the enhanced expression of the amplified proto-oncogene confers a growth advantage upon the host cells within a specific tissue architecture. Amplification of genes could allow the host cells to escape growth control, to become mobile and invasive, or to escape immune surveillance.

The role of amplification of *myc* sequences can also be examined in light of the multistage nature of tumorigenesis, which often involves the successive steps of initiation and progression. The tight correlation of amplified N-*myc* with advanced disease stages is best explained by assuming that it contributes to malignant progression; however, it is unclear whether this amplification correlates with early or with late phases or progression. Tumor progression, in the sense of Foulds (1958), is independent of tumor growth. Staging of neuroblastomas is merely based upon the degree to which the tumor extends beyond the organ or tissue of origin and does not consider the qualitative differences in the tumor cells themselves at early and late phases of progression. It is therefore possible that a particular stage II tumor, for instance, is in a late phase of progression, while a stage III tumor is still in an earlier phase.

Similar correlations between amplification of c-*myc* and degree of malignancy have been made for carcinoma of the uterine cervix (Riou et al. 1984). Whether additional correlations will emerge, e.g., for small cell lung cancer in which c-*myc* or one of the *myc* box genes is frequently amplified, remains to be seen when more primary tumors have been analyzed. In SEWA cells amplification of c-*myc* does not seem to be involved in the early stages of initiation, although it is not clear whether amplification occurring at later stages contributes to establishment of the tumor cells in the animal or to progression of the developing tumor.

Amplification appears to show some specificity in that certain tissues tend to contain a specific amplified proto-oncogene. Analyses of neuroblastomas have revealed amplification only of N-*myc* (Table 1), although N-*myc* and c-*myc* do not differ detectably in their capacity to contribute to the neoplastic transformation of normal rodent cells in tissue culture. L-*myc* has so far been encountered exclusively in small cell lung carcinomas. It remains to be seen whether this specificity is due to preferential amplification at the N-*myc* or L-*myc* locus or whether the particular environment surrounding the host cells selects for amplification of these genes.

Taken together, neuroblastomas and cervical carcinomas provide well-studied systems in which proto-oncogene amplification is correlated with an advanced stage of disease and in which the role of DNA amplification in tumor progression can be further investigated.

Future Perspectives

Cytogenetic studies of human cancer cells have uncovered the presence of chromosomal abnormalities diagnostic of amplified DNA and have provided an entrance point for molecular biological studies of various human cancers. Amplification of proto-oncogenes appears to be a frequent correlate of malignant progression in some naturally occurring tumors and appears to be a late event unrelated to initiation. These studies have brought to attention a new class of cellular genes related to c-*myc* in very distinct, conserved regions. These *myc*-box genes appear to play an important role during development and cellular differentiation, and their amplification may contribute to progression of tumors derived from embryonal cell lineages in which the *myc* box is expressed during development. The molecular study of tumor cells which show cytogenetic evidence of amplified DNA but where amplification of a known proto-oncogene has not been detected could represent a novel strategy for identifiying as-yet unknown genes involved in growth control and able to contribute to carcinogenesis.

Acknowledgments. I thank the many colleagues, too numerous to be mentioned individually (but adequately referenced, I trust), for collaborating on the projects discussed here. Major parts of my original work was done while a Heisenberg-Fellow of the Deutsche Forschungsgemeinschaft.

References

Alitalo K, Schwab M (1986) Oncogene amplification in tumor cells. Adv Cancer Res (in press)

Battey J, Moulding C, Taub R, Murphy W, Stewart T, Potter H, Lenoir G, Leder P (1983) The humane c-*myc* oncogene: Structural consequences of translocation into the IgH locus in Burkitts lymphoma. Cell 34:779 – 787

Brodeur G, Seeger RC, Schwab M, Varmus HE, Bishop JM (1984) Amplification of N-*myc* in untreated human neuroblastomas correlates with advanced disease stage. Science 224:1121 – 1124

Brooks B, Battey J, Nau M, Gazdar A, Minna J (1986) Amplification and expression of the *myc* gene in small cell lung cancer. Adv Viral Oncol (in press)

Dalla-Favera R, Bregni M, Erikson J, Patterson D, Gallo RC, Croce C (1982) Human c-*myc onc* gene is located on the region of chromosome 8 that is translocated in Burkitt lymphoma cells. Proc Natl Acad Sci USA 79:7824 – 7827

Foulds L (1958) The natural history of cancer. J Chronic Dis 8:2 – 37

Garson JA, McIntyre PG, Kemshead JT (1985) N-*myc* amplification in malignant astrocytoma. Lancet 28:718 – 719

Kohl N, Kanda N, Schreck RR, Bruns G, Latt SA, Gilbert F, Alt F (1983) Transposition and amplification of oncogene-related sequence in human neuroblastomas. Cell 35:359 – 367

Kohl NE, Legouy E, De Pinho RA, Nisen PD, Smith RK, Gee CE, Alt FW (1986) Human N-*myc* is closely related in organization and nucleotide sequence to c-*myc*. Nature 319:73 – 77

Lee WH, Murphee AC, Benedict WF (1984) Expression and amplification of the N-*myc* gene in primary retinoblastoma. Nature 309:458 – 460

Levan A, Levan G, Mandahl N (1981) Double minutes and C-bandless chromosomes in a mouse tumor. In: Arrighi FE, Rao RN, Stubblefield E (eds) Genes, chromosomes, and neoplasia. Raven, New York, pp 223 – 251

Nau MM, Brooks BJ, Battey J, Sausville E, Gazdar AF, Kirsch I, McBride OW, Bertness V, Hollis GF, Minna JD (1985) L-*myc*, a new *myc*-related gene amplified and expressed in human small cell lung cancer. Nature 318:69 – 73

Nau MM, Brooks B, Carney D, Gazdar A, Battey J, Sausville E, Minna J (1986) Human small cell lung cancers show amplification and expression of the N-*myc* gene. Proc Natl Acad Sci USA 83:1092 – 1096

Nowell P (1976) The clonal evolution of tumor cell populations. Science 194:23 – 28

Riou G, Barrois M, Tordjman I, Dutronquay V, Orth G (1984) Detection of papillomavirus genomes and evidence for amplification of the c-*myc* and c-Ha-*ras* in invasive squamous cell carcinoma of the uterine cervix. C R Acad Sci Paris ser III, nuod 14:575 – 580

Sager R, Gadi IK, Stephens L, Grabwy CT (1985) Gene amplification: An example of accelerated evolution in tumorigenic cells. Proc Natl Acad Sci USA 82:7015 – 7019

Schwab M (1985) Amplification of N-*myc* in human neuroblastomas. Trends Genet 1:271 – 275

Schwab M (1986) *Myc*-box genes: N-*myc* and L-*myc*. In: Reddy P, Skalka A, Curran T (eds) The oncogene handbook. Elsevier, Amsterdam (in press)

Schwab M, Alitalo K, Klempnauer KH, Varmus HE, Bishop JM, Gilbert F, Brodeur G, Goldstein M, Trent J (1983b) Amplified DNA with limited homology to *myc* cellular oncogene is shared by human neuroblastoma cell lines and a neuroblastoma tumor. Nature 305:245 – 248

Schwab M, Ellison J, Busch M, Rosenau W, Varmus HE, Bishop JM (1984a) Enhanced expression of the human gene N-*myc* consequent to amplification of DNA may contribute to malignant progression of neuroblastoma. Proc Natl Acad Sci USA 81:4940 – 4944

Schwab M, Varmus HE, Bishop JM, Grzeschik KH, Naylor S, Sakaguchi A, Brodeur G, Trent J (1984b) Chromosome localization in normal human cells and neuroblastomas of a gene related to c-*myc*. Nature 308:288 – 291

Seeger RC, Brodeur GM, Sather H, Dalton A, Siegel SE, Wong KY, Hammond D (1985) Association of multiple copies of the N-*myc* oncogene with rapid progression of neuroblastomas. N Engl J Med 313:1111 – 1116

Squire J, Gallie BL, Phillips RA (1985) A detailed analysis of chromosomal changes in heritable and non-heritable retinoblastoma. Hum Genet 70:291 – 301

Stanton LW, Schwab M, Bishop JM (1986) Nucleotide sequence of the human N-*myc*. Proc Natl Acad Sci USA 83:1772–1776

Taya Y, Hosogai K, Hirohashi S, Shimosato Y, Tsuchiya R, Tsuchida N, Fushimi M, Sekiya T, Nishimura S (1984) A novel combination of K-*ras* and *myc* amplification accompanied by point mutational activation of K-*ras* in a human lung cancer. EMBO J 3:2943–2946

Thiele CJ, Reynolds CP, Israel M (1985) Decreased expression of N-*myc* precedes retinoic acid-induced morphological differentiation of human neuroblastoma. Nature 313:404–406

Suppression of the Neoplastic Phenotype

John A. Wyke and A. Richard Green

The Alternatives to Positively Acting Oncogenes

The concept that neoplasia results from an accumulation of somatic mutations in the tumour lineage is heuristically extremely useful. Strong support for this hypothesis has been provided by the demonstration that suitably altered proto-oncogenes can induce a neoplastic phenotype when introduced into normal recipient cells. However, activated oncogenes that behave in this phenotypically "dominant" fashion have been detected so far in only a minority of tumours and therefore the genetic basis of neoplasia in most cancers remains obscure. There are at least four explanations for the frequent failure to implicate positively acting oncogenes.

1. There is no a priori reason to assume that oncogenesis results solely from mutations in the tumour lineage. Inherited defects in molecules with an exocrine effect on tumours, such as tumour-inhibitory factors, may predispose the host animal to neoplasia, although similar defects arising from somatic mutations are less likely to be involved. Furthermore, neoplasia may sometimes be caused not by mutations but by stochastic aberrations in epigenetic mechanisms, leading to changes that have the phenotypic stability characteristic of cell differentiation, that do not involve a change in the nucleotide sequence of the genome, and that may not be transferable to normal recipients. The difficulty in studying such hypothetical events explains their neglect by experimental biologists.

2. Dominant oncogenes may be present but current assays are unable to detect them.

3. Pertinent to the previous point is the considerable evidence that the full neoplastic phenotype usually requires alterations in several genes (see other parts of Chap. V). These may be hard to identify because they would rarely be co-transferred into normal cells.

4. There is mounting evidence, in species from *Drosophila* to man, that some mutations which lead to a tumorigenic phenotype may be recessive at the cellular level, but nonetheless inherited in a Mendelian-dominant fashion (Gateff 1978; Anders 1983; Knudson 1985). Most mutations impair or ablate gene functions but in somatic cells the presence of a normal allele on the homologous chromosome would usually compensate for deleterious effects unless further genetic or epigenetic changes lead to structural or functional homozygosity for the recessive mutant allele. These considerations, together

Oncogenes and Growth Control
Edited by P. Kahn and T. Graf
© Springer-Verlag Berlin Heidelberg 1986

with the non-random chromosomal abnormalities that typify many tumours (see Klein, this Vol.) and the frequent loss of heterozygosity in tumour cell lines (Dracopoli et al. 1985), underly the growing realisation that gene loss may play a vital and hitherto underestimated role in neoplasia.

The genes in which recessive mutations are associated with a predisposition to cancer have been termed anti-oncogenes (Knudson 1985) but their identities and functions are unknown. However, in vitro studies of somatic cell hybrids between normal and tumour cells have repeatedly demonstrated extinction of the neoplastic phenotype, suggesting the existence of suppressor genes capable of modulating the activity of the genes responsible for this phenotype. It is tempting to speculate that anti-oncogenes and suppressor genes may be related (Green and Wyke 1985; Sager 1985). Their characterisation is an important challenge that is currently being tackled in two ways. One approach is based on the use of tumours in which there is evidence for recessive mutations; it attempts the technically daunting task of direct identification of the relevant gene (van Heyningen et al. 1985). An alternative strategy, on which this chapter will concentrate, utilises the techniques of somatic cell genetics in vitro.

Spontaneous and Induced Reversibility of the Neoplastic Phenotype

Cells grown in culture are freed from many selective pressures that afflict them in the whole animal. This fact precludes the study of certain important features of tumour formation, but it permits examination of some fundamental aspects of neoplastic growth that are difficult to study in vivo. In particular, cultured cells may revert towards phenotypic normality, either infrequently but spontaneously, or at high frequency after fusion with certain normal cells. Pioneering earlier work as well as some current investigations have focussed on testing how in vitro manipulation of cells originally derived from tumours affects their tumorigenicity (reviewed in Stanbridge et al. 1982). Spontaneous revertants of tumour cells may reflect rare mutations in positively acting oncogenes rather than the restitution of a suppressive effect. In view of the paucity of knowledge about the molecular basis of transformation, let alone tumour formation, a study of reversion in these cells is unlikely to be informative.

In contrast, cell fusion experiments have been very useful in showing that tumorigenicity and other phenotypes which correlate with neoplasia may be suppressed at a high frequency in hybrids with normal cells (Sager 1985). The subsequent segregation of chromosomes derived from the normal parent leads to re-expression of tumorigenicity and in some cases the putative suppressor genes act in a dose-dependent manner (Benedict et al. 1984). Results obtained by fusing tumour cells to other tumour cells suggest that tumours share common lesions in some cases but in other instances can complement one another and mutually suppress tumorigenicity (Stanbridge et al. 1982). In view of the complex phenotypes under study, it is not surprising that suppression of tumorigenicity in some instances seems to require multiple genes located on several chromosomes (Klinger and Shows 1983). It has been pointed out that the human chromosomes implicated in suppression of tumorigenicity in cell hybrids are frequently also the sites of characteristic abnormalities in human tumours (Klinger 1982) and it is but a short step to equate the suppressor genes whose effect is apparent in cell hybrids with the anti-oncogenes that are the targets for recessive mutations in naturally occurring cancers.

Further definition of suppressor genes and their mechanism of action is limited by the unknown nature of the genetic lesions that contribute to the neoplastic phenotype in most tumour lines. As a result, many studies have used cells in which the transforming principles are partly or wholly defined, being either positively acting oncogenes identified in tumour lines, such as the human HT 1080 fibrosarcoma containing an activated N-*ras* gene, or exogenous oncogenes introduced into established cell lines. Cell bearing such dominantly acting oncogenes are still subject to phenotypic reversion, both spontaneous (Wyke et al. 1980; Noda et al. 1983; H. Paterson and C. Marshall, personal communication) and hybrid-induced (Marshall and Dave 1978; Dyson et al. 1982; Benedict et al. 1984; Craig and Sager 1985; Griegel et al., submitted). The major advantage of this approach is that it should be possible to ascertain directly the molecular mechanism of oncogene modulation.

Such studies have revealed suppression at two levels. In our work on Rous sarcoma virus transformed rat cells, reversion and hybrid suppression are accompanied by greatly reduced levels of both the oncogene protein product and its transcripts whilst the provirus, although remaining intact, is converted to an inactive chromatin configuration (Chiswell et al. 1982a, b; Dyson et al. 1982, 1985). This suggests that suppression involves *trans*-acting negative regulation of transcription. With other revertants (Noda et al. 1983) and phenotypically normal hybrids (Craig and Sager 1985; Griegel et al., submitted), expression of the oncogene product, in these cases Ha-*ras* p21, remains high and suppression presumably operates at a post-translation level. Although a diversity of suppressor mechanisms is to be expected in view of the functional heterogeneity of oncogenes, it immediately raises interesting questions about the nature of the suppressor genes. These cannot yet be answered, but let us follow some of the speculations invited by these findings.

What Are the Relationships Between Oncogenes, Anti-Oncogenes and Suppressor Genes?

Three types of mutation, singly or in concert, might convert a proto-oncogene to a dominant activated oncogene that is detectable by gene transfer into normal recipients:

1. Regulatory mutations at the level of gene transcription. These might result in proto-oncogene activity either in cells that do not usually express the normal allele or at deleterious levels.

2. Coding region mutations which drastically alter function so that the altered product participates in metabolic pathways distinct from those of its normal counterpart.

3. Regulatory mutations at post-translational levels conferring, for instance, increased affinity for a target molecule, exemption from feedback control or enhanced stability. These mutant proteins may compete with the normal product. Activated *ras* genes and v-*src* are likely examples.

Other proto-oncogene mutations might lead to alterations that are only partially dominant or are even recessive in the presence of an active normal allele because of regulatory interactions or competition between the two alleles. Several observations are consistent with this possibility. Some human tumours seem to have lost one allele of certain proto-oncogenes (Yokota et al. 1986 and references therein), although this loss may represent a random consequence of genetic instability in tumours (Dracopoli et al. 1985). However, the normal N-*ras* allele is lost in a mouse thymic lymphoma containing an activated N-*ras* gene (Guerrero et al. 1985), and Benedict et al. (1984) find that segregation of tumorigenic variants from non-malignant hybrids between the sarcoma line HT 1080 and diploid fibroblasts is associated with loss of chromosome 1, the site of N-*ras*. Moreover, phenotypic reversion in HT 1080 is associated with an increase in the dose of the normal N-*ras* allele relative to the mutant form (H. Paterson and C. Marshall, personal communication).

It thus seems that the phenotypic consequences of activated *ras* may be affected by the normal gene product. Indeed, the finding that hybrid suppression of Ha-*ras* transformation occurs in the continued presence of p21 could also reflect a dosage effect of the normal over the transforming gene product (Craig and Sager 1985; Griegel et al., submitted). Even the unusual revertants that are resistant to retransformation by v-Ha-*ras* and whose phenotype is dominant in cell hybrids (Noda et al. 1983) may be partially explicable as a further mutation in a Ha-*ras* allele. Clearly, determining whether post-translational suppression is due to an excess of the normal product (or one with a compensatory mutation) or to an effect on some molecule that interacts directly or indirectly with p21 must await identification of the suppressor gene. In the meantime we should consider that oncogenes and anti-oncogenes may not always be distinct classes, as suggested by Knudson (1985), but may sometimes reflect different mutations in the same target.

Thus, neither the anti-oncogenes defined in vivo nor the suppressor genes that affect *ras* activity in vitro can be clearly distinguished at present from oncogenes. However, the genes that seem to regulate transcription of integrated Rous sarcoma proviruses in a site-specific manner (Dyson et al. 1982, 1985) appear to be distinct from dominantly acting oncogenes. The suppression they mediate is not specific to the provirus or to the v-*src* oncogene, but probably to the integration locus. These *trans*-acting negative regulators of cell gene transcription represent a class in which recessive mutations in some members might be expected to lead to neoplasia (Comings 1973; Green and Wyke 1985). The potential mechanism of this effect can be envisaged by considering the control of c-*mos*. Activity of this proto-oncogene is tightly regulated by a region 5′ to its coding sequences that apparently downregulates its transcription (see Blair, this Vol.). Such regulation may be mediated by DNA: protein interactions and ablation of both alleles of the gene encoding the relevant *trans*-acting protein (an anti-oncogene?) might lead to uncontrolled c-*mos* activity.

We now need to identify suppressor genes acting at both the transcriptional and post-translational levels. Their activity in hybrids between normal and transformed cells provides the basis for attempts at genetic complementation, which in turn should lead to the isolation of the genes involved. In particular the identification and characterisation of negative regulatory genes may shed much needed light on fundamental mechanisms of both eukaryotic gene control and carcinogenesis.

Acknowledgements. We are grateful to Drs. H. Paterson and R. Schafer for telling us of their unpublished data.

References

Anders F (1983) The biology of an oncogene, based upon studies in neoplasia in *Xiphophorus*. In: Neth R, Gallo R (eds) Modern trends in human leukemia 5. Springer, Berlin Heidelberg New York, p 186

Benedict WF, Weissman BE, Mark C, Stanbridge EG (1984) Tumorigenicity of human HT 1080 fibrosarcoma × normal fibroblast hybrids: chromosome dosage dependency. Cancer Res 44:3471 – 3479

Chiswell DJ, Enrietto PJ, Evans S, Quade K, Wyke JA (1982a) Molecular mechanisms involved in morphological variation of avian sarcoma virus-infected rat cells. Virology 116:428 – 440

Chiswell DJ, Gillespie DA, Wyke JA (1982b) The changes in proviral chromatin that accompany morphological variation in avian sarcoma virus-infected rat cells. Nucl Acids Res 10: 3967 – 3979

Comings DE (1973) A general theory of carcinogenesis. Proc Natl Acad Sci USA 70:3324 – 3328

Craig RW, Sager R (1985) Suppression of tumorigenicity in hybrids of normal and oncogene-transformed CHEF cells. Proc Natl Acad Sci USA 82:2062 – 2066

Dracopoli NC, Houghton AN, Old LJ (1985) Loss of polymorphic restriction fragments in malignant melanoma: implications for tumor heterogenicity. Proc Natl Acad Sci USA 82:1470 – 1474

Dyson PJ, Quade K, Wyke JA (1982) Expression of the ASV *src* gene in hybrids between normal and virally transformed cells: specific suppression occurs in some hybrids but not others. Cell 30:491 – 498

Dyson PJ, Cook PR, Searle S, Wyke JA (1985) The chromatin structure of Rous sarcoma proviruses is changed by factors that act in *trans* in cell hybrids. EMBO J 4:413 – 420

Gateff E (1978) Malignant neoplasms of genetic origin in *Drosophila melanogaster*. Science 200:1448 – 1459

Green AR, Wyke JA (1985) Anti-oncogenes − a subset of regulatory genes involved in carcinogenesis? Lancet ii:475 – 477

Griegel S, Traub O, Willecke K, Schafer R Suppression and reexpression of transformed phenotype in hybrids of Ha-*ras* 1 transformed Rat-1 cells and early passage rat embryo fibroblasts. Submitted

Guerrero I, Villasante A, Corces V, Pellicer A (1985) Loss of the normal N-*ras* allele in a mouse thymic lymphoma induced by a chemical carcinogen. Proc Natl Acad Sci USA 82:7810 – 7814

van Heyningen V, Boyd PA, Seawright A, Fletcher JM, Fantes JA, Buckton KE, Spowart G, Porteous DJ, Hill RE, Newton MS, Hastie ND (1985) Molecular analysis of chromosome 11 deletions in aniridia-Wilms tumor syndrome. Proc Natl Acad Sci USA 82:8592–8596

Klinger HP (1982) Suppression of tumorigenicity. Cytogenet Cell Genet 32:68–84

Klinger HP, Shows TB (1983) Suppression of tumorigenicity in somatic cell hybrids. Human chromosomes implicated as suppressors of tumorigenicity in hybrids with Chinese hamster ovary cells. J Natl Cancer Inst 71:559–569

Knudson AG (1985) Hereditary cancer, oncogenes, and anti-oncogenes. Cancer Res 45: 1437–1443

Marshall CJ, Dave H (1978) Suppression of the transformed phenotype in somatic cell hybrids. J Cell Sci 33:171–190

Noda M, Selinger Z, Scolnick EM, Bassin RH (1983) Flat revertants isolated from Kirsten sarcoma virus-transformed cells are resistant to the action of specific oncogenes. Proc Natl Acad Sci USA 80:5602–5606

Sager R (1985) Genetic suppression of tumor formation. Adv Cancer Res 44:43–68

Stanbridge EJ, Der CJ, Doersen C-J, Nishimi RY, Peehl DM, Weissman BE, Wilkinson JE (1982) Human cell hybrids: analysis of transformation and tumorigenicity. Science 215:252–259

Wyke JA, Beamand JA, Varmus HE (1980) Factors affecting phenotypic reversion of rat cells transformed by avian sarcoma virus. Cold Spring Harbor Symp Quant Biol XLIV:1065–1075

Yokota J, Tsunetsugu-Yokota Y, Battifora H, LeFevre C, Cline MJ (1986) Alterations of *myc, myb,* and *ras*[Ha] proto-oncogenes in cancers are frequent and show clinical correlation. Science 231:261–265

VI Oncogenesis in Transgenic Mice

The concluding article of the book describes a new approach to the problem of oncogenesis. Transgenic mice are defined as mice in which the genome of each cell contains specific DNA sequences that were introduced experimentally during early embryogenesis. They provide a model system for analysis of the expression and phenotypic effects of precisely engineered genes whose expression can, by inclusion of appropriate *cis*-acting regulatory sequences, be directed to specific tissues. These mice therefore represent tools which are likely to be invaluable in resolving many of the questions which have been raised in other sections of the book, such as the properties of the DNA sequences that determine tissue-specific gene expression, the molecular basis of oncogene activation, and elucidation of the multiple events which are required for full malignant transformation.

Oncogenesis in Transgenic Mice

DOUGLAS HANAHAN

The study of the diseases collectively labeled cancer has involved a diverse set of approaches, ranging from clinical evaluation and treatment to tumor induction in experimental organisms, and, perhaps most significantly, into the molecular genetics of cell proliferation and growth control. This book is organized around the remarkable body of knowledge which has developed in recent years from studies of the genes involved in cancer and cell regulation. There is now a diverse collection of genes known which, when inappropriately expressed, produce dramatic effects on cells and in animals. A considerable number are directly implicated in various naturally occurring cancers. Many of these oncogenes have been initially identified through their presence in tumor viruses, employing assays which have demonstrated their capabilities to transform cells and thereby subvert processes which regulate normal cellular activities.

In spite of this exciting plethora of oncogenes and growth factor/regulatory genes, the detailed process of oncogenesis remains obscure. The classical epidemiology of cancer has strongly supported multistep models of oncogenesis which postulate that a series of events is required to convert a normal cell into an abnormally proliferating one, and finally into a malignant tumor cell (Knudson 1977; Peto 1977). The statistics suggest that the number of steps may be on the order of 3–4. A considerable collection of experiments on induction of tumors in experimental animals is consistent with this view of multistep carcinogenesis (Foulds 1969; Hecker et al. 1982). However, it has been difficult to separate the individual steps and thereby identify each one and its role in the progession of events. In particular, it is unclear where different kinds of oncogenes fit into the overall picture. The purpose of this chapter is to explore the prospects that gene transfer into the germ line of mice may contribute to the analysis of the function and dysfunction of oncogenes and related genes. The production of stable lineages of transgenic mice may provide a new approach to studying the initiation and progression of tumors, one which will directly address the role of oncogenes in induced cancers and perhaps allow the identification of secondary (or additional) events which are necessary for the development of natural cancers.

Oncogenes and Growth Control
Edited by P. Kahn and T. Graf
© Springer-Verlag Berlin Heidelberg 1986

Oncogenes in Transgenic Mice

The stable transfer of genetic information into the mouse germ line has become a well-established experimental technique (Gordon and Ruddle 1983; Palmiter and Brinster 1985; Hogan et al. 1986; Brinster and Palmiter 1986). Gene transfer is accomplished by microinjecting a solution of DNA into one of the pronuclei of a fertilized egg (one cell embryo), after which the injected embryo is inserted into an oviduct of a pseudopregnant female mouse and allowed to develop. About 20% of the mice born have acquired the injected DNA, which is generally integrated as a head-to-tail tandem array in a single (random) location. The result is a line of mice which inherit and transmit to their progeny the aquired "transgene" and usually express the phenotype it endows.

Two types of genes have been used to study the properties and consequences of oncogene expression in transgenic mice. These could be called "natural" oncogenes and "hybrid" or "recombinant" oncogenes. In this context a natural oncogene refers to one in which the protein coding information is expressed under the control of the regulatory sequences with which it is normally associated, as in cancer cells in vivo or in oncogene detection assays which employ transfection or infection to produce either transformed foci or tumorigenic potential in cultured cells.

In contrast, hybrid oncogenes recombine (or fuse) the protein coding information for oncogenes or related genes which affect growth with regulatory information derived from unrelated genes. A primary purpose of such hybrids is to target expression to specific tissues, so as to examine in detail the consequences of oncogene expression on a specific cell type (or types). Table 1 summarizes the published results obtained to date using these two types of oncogenes in transgenic mouse experiments. There are merits to each approach, and the sections below present pertinent examples of each.

Natural Oncogenes

Simian Virus 40. The first attempt to study oncogenes in transgenic mice involved injection of the genome of simian virus 40 (SV40) into the blastocyl cavity of mouse embryos, thereby producing mosaic mice which harbored SV40 DNA but showed no phenotypic consequences (Jaenisch and Mintz 1974). Ten years later the early region of SV40 was shown to be a potent oncogene in transgenic mice (Brinster et al. 1984). The SV40 early region encodes two proteins, large T and small t, which are translated from two messages derived by differential splicing of a primary transcript (Tooze 1981). SV40 large T is a 96 kDa protein which has a diverse set of functions. It is capable of transforming cultured cells in that it relieves them from contact inhibition and dependence on serum growth factors, and confers on them the ability to form tumors in syngeneic animals. Large T also can transform primary cells to

Table 1. Oncogenes in transgenic mice

Oncogene (protein coding information)	Regulatory region (promoter/enhancer)	Phenotype	Reference
SV40 early region (large T antigen)	SV40 early region	Choroid plexus tumors; kidney abnormalities	Brinster et al. 1984; Palmiter et al. 1985; Small et al. 1985
	Rat insulin II	β-cell tumors (insulinomas)	Hanahan 1985
	Rat elastase	Pancreatic acinar cell tumors	Ornitz et al. 1985
	SV40 promoter (w/o enhancer) + mouse metallothionein I 5′ flanking	Peripheral neuropathy, multi-focal tumors, esp. hepatomas, islet cell adenomas (occasional choroid plexus tumors)	Messing et al. 1985
	Adenovirus early region 1	Glioblastomas	Kelly et al. 1986
c-*myc*	c-*myc*	No effects	Adams et al. 1985
	Mouse mammary tumor virus LTR	Mammary tumors	Stewart et al. 1984
	Mouse immunoglobulin gene enhancers (heavy or light chain)	B-cell lymphomas	Adams et al. 1985
	SV40 early region	Occasional tumors	Adams et al. 1985
	Mouse metallothionein I	No effects	Adams et al. 1985
Bovine papilloma virus 1	Bovine papilloma virus 1	Skin tumors (fibropapillomas)	Lacey et al. 1986

Notes: (1) Several other oncogenes have been established in the mouse germline (including H-*ras* and v-*src*), and both c-*myc* and the SV40 early region have been placed under the control of a variety of other regulatory elements. The analyses of transgenic mice harboring these genes are, in general, still in progress. Some of the preliminary observations are described in the text. (2) The lack of phenotype associated with the two c-*myc* genes listed above has not been accompanied by analyses of gene expression (as RNA or protein).

growth in culture, can stimulate cellular and viral DNA replication, shows ATPase activity, and is a DNA-binding protein. The large T protein is localized primarily in the nucleus, but a small fraction is found associated with the cell surface. SV40 small t is an 18 kDa cytoplasmic protein whose function is unknown. Both of these proteins are expressed under the control of the early region transcriptional promoter/enhancer, which has been shown to be active in a wide variety of cultured cell types and is generally viewed as a promiscuous or broad-specificity regulatory element.

Brinster and Palmiter and their colleagues have established and studied a number of lines of transgenic mice harboring the SV40 early region (Brinster et al. 1984; Palmiter et al. 1985; van Dyke et al. 1985). Most of these lineages show a characteristic phenotype: the mice develop tumors of the choroid plexus, a layer of epithelial cells lining the ventricles of the brain. The mice die at 4−6 months of age bearing large tumors which distort the cranial cavity and presumably disrupt a variety of brain functions. The tumor cells express appreciable levels of SV40 large T antigen. A common secondary effect is the appearance of kidney abnormalities. These characteristics have been observed in lineages of mice arising from independent insertions of the SV40 early region, and the phenotype segregates with the transgene, indicating that the early region elicits a phenotype of tissue-specific formation of choroid plexus tumors. Additional experiments have demonstrated that small t protein is not required for the phenotype of choroid plexus tumors, while the presence of the SV40 transcriptional enhancer element is important but not strictly obligatory (Palmiter et al. 1985).

An unusual quality associated with these transgenic mice is the lack of demonstrable expression of large T prior to the onset of abnormalities. In one family which has been studied in detail, T antigen is not detected in normal choroid plexus tissue of newborn mice, and tissue adjacent to the tumor remains inactive, in contrast to the high level of expression observed in the tumor itself (R. Palmiter, personal communication). Young mice in a second family develop focal clusters of abnormal choroid plexus cells which show moderate levels of T antigen expression, and one or more of these foci appears to progress into overt tumors which express T at high levels (A. Levine, personal communication). It thus appears that the SV40 early region is inactive in normal tissues of a transgenic mouse, but can undergo a rare activation event which then allows continuous expression of this oncogene and its consequent manifestations (see also Small et al. 1985; Van Dyke et al. 1985). This idea is further supported by the observation that normal tissues from these mice, when placed in culture, produce transformed colonies which express large T antigen, whereas the primary cells do not show detectable levels of this protein. Interconvertability between nonexpressing and expressing states of the SV40 early region has previously been observed by Hanahan et al. (1980) in experiments where the early region was transferred into and maintained in cultured mouse cells under conditions that did not select for transformation, which is also the case in gene transfer into mice. Both observations are consistent with an epigenetic mechanism that stably represses the expression of certain regulatory elements (which seem to include SV40 and several murine retroviruses), but in a manner which does not preclude their subsequent expression following a rare event of gene activation.

From the perspective of oncogenesis, these observations indicate that specific tissues are particularly susceptible to undergo rare de-repression of this dormant oncogene. The activation of the SV40 early region in only a few cell types of transgenic mice is particularly provocative in view of the wide spectrum of cells in which the early region can be expressed in vitro. The stable repression of an oncogene and its rare and tissue-specific escape from that repression may represent a model for studying the control of (inappropriate) gene expression in an organism and the mechanisms underlying the occasional failures to maintain that regulation.

Bovine Papilloma Virus 1. When the bovine papilloma virus type I genome was established in a line of transgenic mice, a phenotype of tissue-specific tumor formation was again observed (Lacey et al. 1986). Transgenic mice harboring integrated BPV-1 DNA develop fibropapillomas of the skin. BPV-1 normally infects cutaneous tissue in cattle, thereby eliciting the development of fibropapillomas composed of dermal fibroblasts and epidermal keratinocytes. The BPV-1 genome encodes at least two oncogenic proteins, which are derived from the E5 and E6 open reading frames (Yang et al. 1985; Schiller et al. 1986). These two oncogenes have been identified in cultured murine cells using transfection assays and screening for transformed foci.

BPV DNA stably maintained in the mouse germ line induces a phenotype which is quite similar to that which it naturally produces following infection of cattle. This observation suggests that the BPV-1 regulatory elements are acting in a correct tissue-specific manner in transgenic mice, directing expression of the BPV oncogenic proteins to skin tissue. Alternatively, these proteins may be expressed uniformly in all tissues but have oncogenic capacity only in skin.

The characteristics of skin tumor formation make this lineage of transgenic mice a useful model for studying certain aspects of oncogenesis, particularly since skin is quite accessible to experimental manipulation. The tumors arise relatively late in life, beginning at about 8–9 months of age. Multiple protuberant tumors can appear on an individual animal, although there is no indication of metastasis to internal organs. Tumors arise most frequently in areas prone to wounding: the face, neck, mouth, and the tip of the tail, which is clipped at 3 weeks of age for DNA analysis. Rearrangements in the BPV DNA are observed in the tumors. In normal tissue the BPV genome is integrated in a head-to-tail tandem array, which is characteristic of DNA insertions produced in transgenic mice. Tumor tissues show amplification of the BPV-1 genome, both in situ and by the generation of extrachromosomal copies of the unit length genome. In bovine fibropapillomas, BPV-1 replicates as a free extrachromosomal element; similarly, BPV-1 DNA is capable of plasmid replication in cultured murine cells. It is possible that the excision and extrachromosomal replication observed in these skin tumors is somehow causal to their genesis.

The multifocal development of skin tumors at sites prone to wounding and with consequent excision and amplification of the BPV genome, properties characteristic of a heritably induced skin cancer, suggest that this line of transgenic mice may provide an assay system for the initiation and progression of oncogenesis in this tissue and for the roles played by expression of the BPV oncogenes in that process.

myc Hybrid Oncogenes

The *myc* oncogene was originally identified in a series of avian retroviruses (MC29, OK10, MH2) and was subsequently implicated in a variety of B and T cell leukemias because of its presence in chromosomal translocations which placed it under control of B and T cell specific regulatory elements (see Fahrlander and Marcu, this Vol., and Klein, this Vol.). In this sense the *myc* translocations represent natural hybrid oncogenes, and therefore activation of potentially oncogenic protein coding information by recombination with cell specific regulatory elements is not only a useful experimental approach, but is also part of the mechanism in certain types of naturally occurring cancers. The normal c-*myc* protein is localized in the nucleus and is cell-cycle-regulated in normal cells, being expressed shortly after the G_0/G_1 transition (see Mölling, this Vol., and Bravo and Müller, this Vol.). The *myc* proteins are able to immortalize primary cells to continuous growth in culture, a quality shared with a number of other oncogenes (see Land, this Vol.).

MMTV-myc. A variety of hybrid *myc* oncogenes have been established in lines of transgenic mice. Several of these have proven to be oncogenic, and the characteristics of tumor formation provides some insight into the activity of the *myc* protein. Stewart et al. (1984) placed c-*myc* under control of the long terminal repeat (LTR) of mouse mammary tumor virus (MMTV) and observed the heritable development of mammary tumors in female mice harboring the MMTV-*myc* gene. The fusion genes employed the coding sequences from the mouse c-*myc* gene, which precluded analysis of protein expression. RNA analysis demonstrated expression of the transgene in several tissues. The pattern of expression varied among all the lines tested but a few tissues, such as salivary gland and mammary gland, always showed MMTV-*myc* RNA.

Two characteristics of breast tumor formation in these mice are particularly notable. The first is that tumorigenesis is slow. Characteristically, only one of the ten mammary glands develops a tumor, which generally appears during the second or third pregnancy or lactation period. This long latency suggests that a secondary event must occur before the tumor can form. The second noteworthy characteristic is that expression of the transgene (as mRNA) in a tissue does not necessarily lead to tumors. For example, high expression levels are detected in the salivary glands of both male and female mice in several lineages, yet salivary gland tumors have never been observed. One interpretation of this result is that expression of c-*myc* has no effect on certain tissue types (such as salivary glands) due to some unkown feature of their cellular physiology. Another is that the frequency of somatic events required to complement the actions of c-*myc* is significantly lower in such tissues.

Ig-myc. Recombinant oncogenes which mimic the natural *myc* translocations by placing the c-*myc* transcription unit in juxtaposition with transcriptional enhancer elements from immunoglobulin (Ig) genes produce lymphoid tumors in transgenic mice (Adams et al. 1985). The heavy chain enhancer is particularly effective at helping to induce multicentric lymphomas which can metastasize to other organs and be transplanted into syngeneic mice. The lymphomas belong to the B cell lineage, as evidenced by the characteristic immunoglobulin gene rearrangements observed. Both pre-B and mature B cells can be represented and the tumors are often monoclonal, using the criteria of unique Ig rearrangements.

The Ig-*myc* hybrid oncogenes induce lymphomas which cause death by approximately 11 weeks and 44 weeks of age, in the cases employing the heavy chain enhancer or light chain enhancer, respectively. The onset of the tumors, first detectable as enlarged and abnormal lymph nodes, ranges from 3 – 18 weeks of age. This is well past the proliferative period of B cell development, which occurs during late embryogenesis and the first few weeks after birth, thus suggesting the requirement for other changes before tumor formation begins.

Transgenic mice harboring *myc* coding sequences under the control of the normal c-*myc* regulatory elements (with or without the noncoding first exon, which is implicated in attenuation of *myc* expression; see Fahrlander and Marcu, this Vol.) show no change in phenotype, suggesting that multiple copies of the normal cellular gene are insufficient to disrupt its proper regulation and function. The SV40 promoter/enhancer region driving c-*myc* elicits occasional tumors in diverse tissues. Significantly, no choroid plexus tumors have been observed, which is somewhat puzzling in view of the fact that the intact SV40 early region is silent in almost all tissues but preferentially activated in the choroid plexus. Another surprise is that the c-*myc* gene fused to a mouse metallothionein (MT) promoter produces no effects whatsoever, despite the fact that the MT promoter, like MMTV, mediates expression in a wide variety of tissues. This suggests that the levels of *myc* synthesis could be insufficient, or simply unable, to disrupt normal function in a considerable number of cell types. An important qualification is that the expression of these hybrid genes has not yet been studied in detail. Thus, lack of expression cannot be excluded as the explanation for the absence of effects. Taken together, these observations nevertheless imply that both the levels of *myc* expression and the sensitivity of particular cell types to *myc* expression are likely to be significant factors in elucidating the functions and dysfunctions of this protein.

Targeted Expression of SV40 Large T via Hybrid Oncogenes

An increasing number of hybrid oncogenes have used cell type-specific regulatory information to control the expression of the SV40 early region, which has generally been assessed by expression of large T. These genes, when established in the mouse germ line, are typically potent oncogenes. The analyses of large T expression and of its consequences are providing insight into mechanisms which control of gene expression and into various aspects of oncogenesis. Several pertinent examples are described below.

Insulin – Large T. Recombinant genes containing ~700 bp of 5′ flanking DNA derived from the rat insulin II gene linked to the SV40 early region elicit the heritable formation of insulinomas (Hanahan 1985). These tumors are encapsulated, highly vascularized, transplantable, and are composed almost exclusively of insulin-producing pancreatic β-cells. The sole known source of insulin synthesis is the β-cells, which are distributed in islands (called the islets of Langerhans) along with several other endocrine cell types which synthesize polypeptide hormones involved in carbohydrate metabolism. Biochemical and immunohistochemical analyses indicate that only the β-cells synthesize detectable levels of large T. The "bioassay" of tumor development or other abnormalities supports the conclusion that this small portion of the insulin gene is sufficient to direct expression to the correct cell type. A fragment which is truncated to -455 bp from the point of initiation of transcription behaves similarly, indicating that a major determinant of tissue specificity of insulin gene expression lies in the small region which includes the transcriptional promoter and enhancer elements.

The formation of β-cell tumors is the inevitable fate for mice which inherit an insulin-T antigen fusion gene. Among the several established lineages, the rate of tumor formation (and the rate of death due to consequent hypoglycemia) can be classified as either slow or fast. Mice in the fast lineage are invariably dead by 16 weeks, while mice in the slow lineages die from $20-60$ weeks. Preliminary results suggest that there are reproducible differences in the levels of large T expression in fast and slow lineages, which may, in part, account for the heritably different rates of oncogenesis.

There is a pattern to the formation of β-cell tumors in these families of transgenic mice, one which may be pertinent to the general process of oncogenesis. There are approximately 500 islets of Langerhans scattered throughout the exocrine pancreas, varying in size and containing from about 10 to 1000 β-cells. The islets of the transgenic mice are all full of β-cells expressing large T antigen, and all islets develop hyperplasia as a result of β-cell proliferation. Yet only a few hyperplastic isles (ranging from $1-8$, which is only a few percent of the number of islets) progress into encapsulated, vascularized tumors. Furthermore, large T expression begins in late embryogenesis, concomitant with the initiation of insulin synthesis (S. Alpert and D. Hanahan, unpublished observations), but tumors do not arise until at least 9 weeks after birth, that is, 10 weeks after high expression of this potent oncogene is detected. These observations are again consistent with a multistep process of oncogenesis in which large T provides only some of the steps.

Elastase – Large T. A gene composed of the elastase promoter-enhancer region linked to the SV40 early region elicits the formation of pancreatic acinar cell tumors in a number of lineages of transgenic mice (Ornitz et al. 1985). There are several parallels to the results obtained with insulin-large T hybrid genes, which suggests that certain aspects of these observations have some generality. The elastase-large T lineages develop acinar tumors at characteristic rates and can perhaps be classified as fast or slow. Again large T is expressed well before the appearance of solid tumors, and general hyperplasia precedes tumor formation.

The regulation conferred by the 5′ flanking region is remarkably specific: expression is detected only in the acinar cells of the exocrine pancreas. The finding that two hundred bp of DNA extending upstream of the cap site for elastase gene transcription is sufficient for correct cell type specific expression further strengthens the hypothesis that tissue specificity is a quality of the transcriptional promoter/enhancer elements, and as a consequence greatly facilitates targeted expression in transgenic mice.

Other T Antigen Fusion Genes. An increasing number of promoter/enhancer regions are being linked to the SV40 early region and transferred into the mouse germ line. Many induce various abnormalities, including solid tumors, in specific cell types, and it can be concluded that SV40 large T is a very potent oncogene when expressed in most mouse tissues. This statement is remarkable in view of the fact that exogenous infection of both mice and monkeys with SV40 generally fails to induce tumors.

Experiments in progress using large T expression in transgenic mice employ the regulatory regions from mouse metallothionein I, human adenovirus early region I (E1A), and rat prolactin. Metallothionein is expressed in a wide variety of tissues, and the metallothionein/SV40 early region hybrid gene produces a considerable number of abnormalities in transgenic mice, including hepatomas, insulinomas, and peripheral neuropathy, a nerve disorder associated with the lack of peripheral nerve sheath myelination (Messing et al. 1985).

The SV40 early region under the control of the adenovirus serotype 5 E1A promoter/enhancer produced glioblastomas in some of the transgenic mice which harbored this hybrid gene (Kelly et al. 1986). Expression studies have not yet been performed, but the results suggest tissue-specific expression (or activation) in the glial cells of the brain. Prolactin is normally produced in the lactotroph cells of the anterior pituitary, and a prolactin/T antigen hybrid gene elicits fatal pituitary tumors (G. Rosenfeld and R. Evans, personal communication). The tumors are complex but appear to arise from the prolactin-producing endocrine cells in the pituitary. Similar results have been obtained with a T antigen gene fused to the 5′ flanking region of the α glycoprotein gene for the common subunit of the pituitary hormones TSH, FSH, and LH, where a pituitary tumor arose in one transgenic mouse (P. Mellon and D. Hanahan, unpublished observations). Finally, the lens alpha crystallin promoter targets expression of large T to the eye, inducing tumors of the crystallin-producing cells (H. Westphal, personal communication). Taken together, these observations suggest that large T expression can be targeted to a variety of tissues by cell type-specific regulatory elements, and thereby elicit heritable tumors in mice carrying the hybrid oncogene.

Additional Hybrid Oncogenes

The results described above clearly demonstrate that c-*myc* and SV40 large T are capable of inducing tumors in specific cell types, and one can easily predict that this approach will be extended to a wide variety of other oncogenic and growth regulatory proteins. Some initial studies have been performed using v-*src* and activated Ha-*ras* isolated from the EJ bladder carcinoma cell line (EJ-*ras*). Use of either the elastase or metallothionein promoters to express v-*src* produces no obvious effects in transgenic mice (R. Brinster and R. Palmiter, personal communication). The activated Ha-*ras* gene is apparently ineffective when expressed either under its own promoter (D. Solter, personal communication), or under control of the metallothionein promoter (R. Brinster and Palmiter, personal communication). However, expression of the hybrid oncogenes has not yet been examined in detail and therefore one cannot exclude the trivial explanation that there is no expression and therefore no consequences.

In contrast, a mammary gland specific promoter (derived from the whey acidic protein gene) linked to EJ-*ras* coding information occasionally elicits mammary tumors in transgenic mice (M. LeMeur and P. Gerlinger, personal communication). Furthermore, an elastase-promoted EJ-*ras* gene is a very potent oncogene, as demonstrated by the finding that the transgenic mice are born with dramatic hyperplasia of the exocrine pancreas and die within a few days (R. Brinster and R. Palmiter, personal communication). All of these results are preliminary, with, in many cases, detailed expression studies still in progress. Yet the emerging picture suggests that different oncogenes may well be effective (or penetrant) only in specific cell types, and thus, that cellular sensitivity to those oncogenes, as well as the relative or absolute levels of their expression, are likely be significant factors in their tumorigenic capacity.

Transgenic Predisposition: Oncogenes Are Not Enough

From the studies summarized above it is clear that cloned oncogenes can predispose transgenic animals to tumor formation following their establishment as new heritable genetic elements present in the mouse genome in every cell of the body. However, the consistent result is that the presence of these oncogenes is necessary to produce the noted effects but is generally not sufficient; in virtually every case, there are clear indications that secondary events are also necessary for oncogenesis. The character of these complementing processes varies with the transgenic oncogene.

In the case of targeted expression of SV40 T antigen to the β-cells in the islets of Langerhans, all islets show high level expression of this oncogene, yet only a few are transformed into solid, encapsulated, vascularized tumors. Similarly, only one of the ten mammary glands in mice harboring a MMTV-*myc* gene becomes the source of a mammary tumor, and then only during the second (or later) pregnancy or lactation cycle. The B cell lymphomas produced by the Ig-*myc* fusion genes are generally clonal, as judged by unique immunoglobulin gene rearrangements, again suggesting that a secondary event has produced a clonal, tumorigenic cell which populates the lymphoma. With the bovine papilloma virus I genome, slow oncogensis occurs in discrete locations of the large area of skin tissue and at sites prone to wounding, which strongly implicates co-factors in the development of skin tumors. SV40 undergoes an apparent tissue-specific gene activation event which produces brain tumors expressing large T from a cell type that was apparently not expressing it prior to this event.

These findings support the idea that oncogenesis is a multistage process and further strengthen the notion that oncogenes play important roles in this process. Normal mice rarely, if ever, develop any of the tumors produced by these hybrid oncogenes, while mice inheriting these transgenes invariably develop characteristic types of tumor. Thus the transgenic mice can be considered to be predisposed to specific cancers, which then reduces the number of variables in the multistep process of oncogenesis (Fig. 1). This predisposition should allow the secondary events to be studied in new ways which distinguish them from those produced by oncogenes themselves. The fact that transgenic mice heritably recapitulate the events of specific tumor formation will allow detailed studies on the onset and progression of cancers, and may provide an assay system in which to identify and examine the nature of these secondary events. The study of tumor progression is likely to become a major focus in elucidating the mechanism of this complex set of diseases.

Future Prospects: Genetic Complementation

The results described in this review clearly indicate that gene transfer into the mouse germ line provides a new approach to studying oncogenes and the pro-

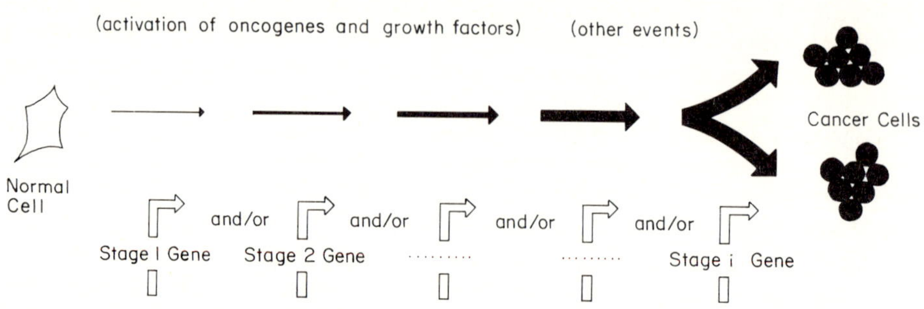

Fig. 1. Transgenic predisposition to cancer. The multiple steps believed to be required for tumorigenesis can be grouped into ones of *initiation*, promotion or *maintenance* of an initiated state (hyperplasia, proliferative capacity), *progression* to malignancy, and *metastasis* to other organs. Specific genes can be postulated to mediate these changes, as a stage 1 gene might be an oncogene, secondary event or stage 2 gene could be an angiogenesis factor, and finally a stage i gene could confer the ability to metastasize. Their involvement can be tested by establishing each in the mouse germ line, thus genetically predisposing the transgenic mouse to cancers which employ that gene

gression of cancer. It seems certain that targeted expression using cell-specific regulatory elements will be a very powerful tool with which to study the initiation and progression of tumors, given the reproducible recapitulation of those events in mice which inherit such genes. Surely an increasing number of cell types will be targeted for oncogenesis, and with an ever-larger collection of oncogenic proteins.

Another aspect of oncogenesis studies in transgenic mice which has thus far received little attention but is probably quite significant is the role of the immune system. There are already indications of differences when the same transgene is backcrossed to different inbred strains. Tumor immunity is probably involved in suppressing certain naturally occurring cancers, and the use of appropriate transgenic mice is likely to provide additional insight in this important area.

An additional characteristic of targeted oncogene expression in transgenic mice which is likely to become a valuable technique in the future involves employing the proliferation-inducing capacity of certain oncogenes to facilitate the establishment of rare cell types as stable cell lines in culture. Only a small fraction of the cell types found in mammalian organisms are currently represented as cultured cell lines which can be propagated in vitro. The directed expression of appropriate oncogenes to effect selective proliferation could become a powerful approach to studying rare cells, especially those of the brain, where techniques for expanding cell number are probably necessary for detailed analysis of these cells. Cell lines which express low levels of their char-

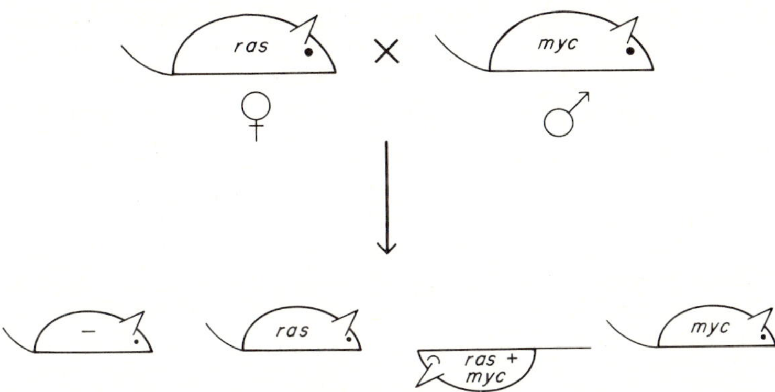

Fig. 2. A genetic complementation assay for oncogenesis. Given genes which are suspected of mediating separate events in the initiation and progression of specific cancers, transgenic mice can be produced to test their involvement

acteristic gene products have been established from several of the tumor types induced in transgenic mice, including pituitary, pancreatic acinar, and choroid plexus, while pancreatic β-cell tumors have not yet proved amenable to culture. It is possible that the proper combination of targeted oncogene expression and appropriate cell culture conditions will allow the expansion of a many cell types in a manner that will maintain qualities (and quantity) of their differentitative state. However, this may require considerable experimentation with defined growth media and substrates and with different combinations of oncogenes.

The genetic approach will also be essential for experiments which address the specific roles played by different oncogenes/growth factors in tumor progression. Given the same targeting sequences, putative mediators of specific events can be separately established in transgenic mice, and then mice of each trans-genotype can be mated to each other to test the effects when both gene products are co-expressed in the same cell type. For example, *ras* and *myc* have been shown to cooperate in the transformation of certain primary cells (see Land, this Vol.) and the generality of this effect can be analyzed by genetic complementation experiments in transgenic mice. If one was using the insulin promoter, separate lines of transgenic mice harboring either the insulin-*ras* or insulin-*myc* genes would be produced and then mice from each family mated. This would allow the effects of each individual hybrid oncogene to be separately examined in each family, and then the consequences of their combination in the progeny which inherit both genes from their parents can be studied in detail. This concept is schematically presented in Fig. 2. Perhaps more significantly, it will be possible to prove genetically the role played by putative "secondary event" genes (Sporn and Roberts 1985). When maintained in a separate line of mice, no dramatic effects should be observed from a secondary event gene, since the primary event(s) would be lacking. However, in F1 mice which carry both a targeted oncogene and a similarly targeted secondary event gene, tumor induction should occur without the slow latency observed in many of the current studies of induced oncogenesis in transgenic mice. Eventually one may be able to maintain genes determining all of the events in the induction and progression of specific tumor types as transgenes in separate families. These could then be appropriately combined to demonstrate their essential and cooperative roles in the development of cancer.

Acknowledgments. I wish to thank Erwin Wagner and Richard Palmiter for comments on the manuscript, Mike Ockler and Dave Greene for art work and photography, and Marilyn Goodwin for preparation of the manuscript.

References

Adams JM, Harris AW, Pinkert CA, Corcoran LM, Alexander WS, Cory S, Palmiter RD, Brinster RL (1985) The c-*myc* oncogene driven by immunoglobulin enhancers induces lymphoid malignancy in transgenic mice. Nature 318:533 – 538

Brinster RL, Palmiter RD (1986) Introduction of genes into the germ line of animals. Harvey Lect Alan R Liss, Inc NY, pp 1 – 38

Brinster RL, Chen HY, Messing A, van Dyke T, Levine AJ, Palmiter RD (1984) Transgenic mice harboring SV40 T-antigen genes develop characteristic brain tumors. Cell 37:367 – 379

Foulds L (1969) Neoplastic development, vol I. Academic Press, London New York

Gordon JW, Ruddle FH (1983) Gene transfer into mouse embryos: production of transgenic mice by pronuclear injection. Methods Enzymol 101C:411 – 433

Hanahan D (1985) Heritable formation of pancreatic β-cell tumors in transgenic mice expressing recombinant insulin/simian virus 40 oncogenes. Nature 315:115 – 122

Hanahan D, Lane D, Lipsich L, Wigler M, Botchan M (1980) Characteristics of an SV40-plasmid recombinant and its movement into and out of the genome of a murine cell. Cell 21:127 – 139

Hecker E, Fusenig NE, Kunz W, Marks F, Thielmann HW (1982) Cocarcinogenesis and biological effects of tumor promoters. Raven Press, New York

Hogan B, Costantini F, Lacy L (1986) Manipulating the mouse embryo – a laboratory manual. Cold Spring Harbor Lab, Cold Spring Harbor, NY (in press)

Jaenisch R, Mintz B (1974) Simian virus 40 DNA sequences in DNA of healthy adult mice derived from preimplantation blastocysts injected with viral DNA. Proc Natl Acad Sci USA 71:1250 – 1254

Kelly F, Kellermann O, Mechali F, Gaillard J, Babinet C (1986) Expression of simian virus 40 oncogenes in F9 embryonal carcinoma cells, in transgenic mice and in transgenic embryos. Cancer cells, vol IV. Cold Spring Harbor Lab Cold Spring Harbor, NY (in press)

Knudson AG Jr (1977) Genetic predisposition to cancer in origins of human cancer. Cold Spring Harbor Lab, Cold Spring Harbor, NY, pp 45 – 52

Lacey ML, Alpert S, Hanahan D (1986) The bovine papilloma virus genome elicits skin tumors in transgenic mice (in press)

Messing A, Chen HY, Palmiter RD, Brinster RL (1985) Peripheral neuropathies, hepatocellular carcinomas, and islet cell adenomas in transgenic mice. Nature 316:461 – 463

Ornitz DM, Palmiter RD, Messing A, Hammer RE, Pinkert CA, Brinster RL (1985) Elastase-I promoter directs expression of human growth hormone and SV40 T-antigen genes to pancreatic acinar cells in transgenic mice. Cold Spring Harbor Symp Quant Biol 50:411 – 416

Palmiter RD, Brinster RL (1985) Transgenic mice. Cell 41:343 – 345

Palmiter RD, Chen HY, Messing A, Brinster RL (1985) SV40 enhancer and large-T antigen are instrumental in development of choroid plexus tumors in transgenic mice. Nature 316:457 – 460

Peto R (1977) Epidemiology, multistage models, and short term mutagenicity tests in origins of human cancer. Cold Spring Harbor Lab, Cold Spring Harbor, NY, pp 1403–1428

Schiller JT, Vass WC, Vousden KH, Lowy DR (1986) E5 open reading frame of bovine papilloma virus type 1 encodes a transforming gene. J Virol 57:1–6

Small JA, Blair DG, Showalter SD, Scangos GA (1985) Analysis of a transgenic mouse containing simian virus 40 and v-*myc* sequences. Mol Cell Biol 5:642–648

Sporn MB, Roberts AB (1985) Autocrine growth factors and cancer. Nature 313:745–747

Stewart TA, Pattengale PK, Leder P (1984) Spontaneous mammary adenocarcinomas in transgenic mice that carry and express MTV/*myc* fusion genes. Cell 38:627–637

Tooze J (1981) DNA tumor viruses. Cold Spring Harbor Lab, Cold Spring Harbor, NY

Van Dyke T, Finlay C, Levine AJ (1985) A comparison of several lines of transgenic mice containing the SV40 early genes. Cold Spring Harbor Symp Quant Biol 50:671–677

Yang YC, Okayama H, Howley PM (1985) Bovine papilloma virus contains multiple transforming genes. Proc Natl Acad Sci USA 82:1030–1034

Subject Index